International Meeting on High-Dimensional Data-Driven Science (HD3-2015)

Journal of Physics: Conference Series Volume 699

Kyoto, Japan
14 – 17 December 2015

Editors:

Tomoyuki Obuchi
Takuma Kasai
Masamichi J. Miyama
Masayuki Ohzeki
Makoto Uemura

ISBN: 978-1-5108-2282-5
ISSN: 1742-6588

Printed from e-media with permission by:

Curran Associates, Inc.
57 Morehouse Lane
Red Hook, NY 12571

Some format issues inherent in the e-media version may also appear in this print version.

Printed by Curran Associates, Inc. (2016)

For permission requests, please contact the Institute of Physics at the address below.

Institute of Physics
Dirac House, Temple Back
Bristol BS1 6BE UK

Phone: 44 1 17 929 7481
Fax: 44 1 17 920 0979

techtracking@iop.org

Additional copies of this publication are available from:

Curran Associates, Inc.
57 Morehouse Lane
Red Hook, NY 12571 USA
Phone: 845-758-0400
Fax: 845-758-2633
Email: curran@proceedings.com
Web: www.proceedings.com

Table of contents

Volume 699

International Meeting on High-Dimensional Data-Driven Science (HD^3-2015)
14–17 December 2015, Kyoto, Japan

Accepted papers received: 29 February 2016
Published online: 6 April 2016

Preface

Papers

Perspective

Bioscience

Astronomy

Brain science

Algorithms and applications

Methodologies

Journal of Physics: Conference Series **699** (2016) 011001 doi:10.1088/1742-6596/699/1/011001

Preface

A never-ending series of innovations in measurement technology and evolutions in information and communication technologies have led to the ongoing generation and accumulation of large quantities of high-dimensional data every day. While detailed data-centric approaches have been pursued in respective research fields, situations have been encountered where the same mathematical framework of high-dimensional data analysis can be found in a wide variety of seemingly unrelated research fields, such as estimation on the basis of undersampled Fourier transform in nuclear magnetic resonance spectroscopy in chemistry, in magnetic resonance imaging in medicine, and in astronomical interferometry in astronomy. In such situations, bringing diverse viewpoints together therefore becomes a driving force for the creation of innovative developments in various different research fields.

This meeting focuses on "Sparse Modeling" (SpM) as a methodology for creation of innovative developments through the incorporation of a wide variety of viewpoints in various research fields. The objective of this meeting is to offer a forum where researchers with interest in SpM can assemble and exchange information on the latest results and newly established methodologies, and discuss future directions of the interdisciplinary studies for High-Dimensional Data-Driven (HD^3) science.

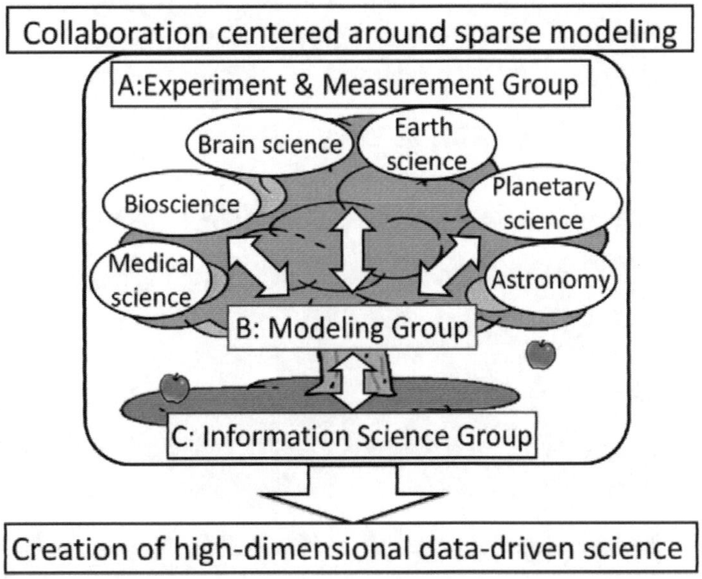

The meeting was held in Kyoto from 14-17 December 2015. We are pleased to publish 22 papers contributed by invited speakers in this volume of *Journal of Physics: Conference Series*. We hope that this volume will promote further development of HD^3 sciences.

Journal of Physics: Conference Series **699** (2016) 011001 doi:10.1088/1742-6596/699/1/011001

Editors:

Tomoyuki Obuchi, (Tokyo Institute of Technology, Editor in Chief)

Takuma Kasai (RIKEN)

Masamichi J. Miyama (The University of Tokyo)

Masayuki Ohzeki (Kyoto University)

Makoto Uemura (Hiroshima University)

Organizing Committee:

Yoshiyuki Kabashima (Tokyo Institute of Technology, General Chair)

Kenji Nagata (The University of Tokyo)

Tomoyuki Obuchi (Tokyo Institute of Technology)

Masayuki Ohzeki (Kyoto University)

Masato Okada (The University of Tokyo, Head Investigator of SpM project)

Toshiyuki Tanaka (Kyoto University, Local Chair)

On Dec. 16, at Mielparque Kyoto

Journal of Physics: Conference Series **699** (2016) 011002 doi:10.1088/1742-6596/699/1/011002

Peer review statement

All papers published in this volume of *Journal of Physics: Conference Series* have been peer reviewed through processes administered by the proceedings Editors. Reviews were conducted by expert referees to the professional and scientific standards expected of a proceedings journal published by IOP Publishing.

Three levels of data-driven science

Yasuhiko Igarashi[1], Kenji Nagata[1], Tatsu Kuwatani[2], Toshiaki Omori[3], Yoshinori Nakanishi-Ohno[1] and Masato Okada[1,4]

[1]Graduate School of Frontier Sciences, the University of Tokyo, 5-1-5, Kashiwanoha, Kashiwa, Chiba, 277-8561, Japan
[2]Department of Solid Earth Geochemistry, Japan Agency for Marine-Earth Science and Technology, 2-15 Natsushima-cho, Yokosuka 237-0061, Japan
[3]Department of Electrical and Electronic Engineering, Graduate School of Engineering, Kobe University, 1-1 Rokkodai-cho, Nada-ku, Kobe, Hyogo 657-8501, Japan
[4]ERATO, Okanoya Emotional Information Project, Japan Science Technology Agency, 2-1 Hirosawa, Wako, Saitama, 351-0198, Japan

E-mail: okada@k.u-tokyo.ac.jp

Abstract. A research project, called "the Initiative for High-dimensional Data-Driven Science through Deepening of Sparse Modeling" is introduced. A concept, called the *three levels of data-driven science*, is proposed to untie the complicated relation between many fields and many methods. This concept claims that any problem of data analysis should be discussed at different three levels: computational theory, modeling, and representation/algorithm. Based on the concept, how to choose a suitable method among several candidates is discussed through our study on spectral deconvolution. In addition, how to find a universal problem across the disciplines is presented by explaining our proposed ES-SVM method. Moreover, it is illustrated that the hierarchical structure of data analysis should be visualized and shared. From these discussions, we believe that data-driven science is *mother of science*, namely, a scientific framework that drives many fields of science.

1. Introduction

Recent progress in the technology of experimentation and measurement makes it possible to obtain a huge amount of high-dimensional data. Needless to say, this trend has forced scientists to change their attitudes toward data. Effective use of high-dimensional data involves a strong framework and a flexible methodology to make the tight connection of information science to the original purpose of data analysis derived from various scientific disciplines. The aim of this project is to establish a novel framework for natural science, namely, *data-driven science.*

We noticed that *sparse modeling* (SpM) is a key technology of data-driven science. In the field of information science, SpM has recently received much attention. The basic notions of SpM are as follows. First, high-dimensional data are sparse and can be essentially expressed by a small number of variables. Second, the number of explanatory variables should be reduced without loss of accuracy. Finally, explanatory variables are selected objectively, and models of target phenomena are constructed automatically.

In this study, extracting latent structures from high-dimensional data by SpM is focused on. An effective framework of data-driven science in various fields of natural science is proposed, and some activities by the organizers of the conference, HD^3-2015, Grant-in-Aid for Scientific Research on Innovative Areas, also known as Sparse Modeling [1], are described.

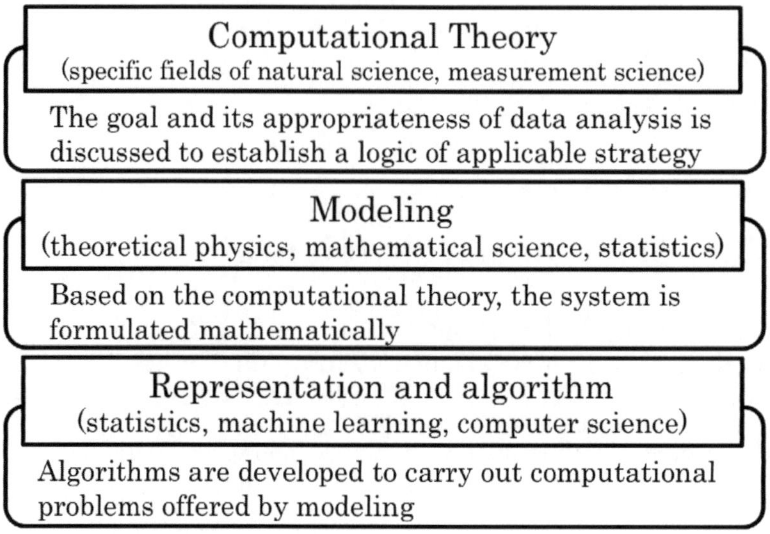

Figure 1. Three levels of data-driven science.

This paper consists of five sections. First, the concept of *three levels of data-driven science*, namely, a guiding principle of data-driven science, is proposed in section 2. This concept claims that any problem of data analysis should be discussed at different three levels: computational theory, modeling, and representation/algorithm. Section 3 explains how to choose a suitable representation/algorithm in the case where there are several algorithms for a certain computational theory of data analysis, giving the example of spectral deconvolution shared by various disciplines [2]. In section 4, we demonstrate that it is important to find an essential problem of data-driven approach that is latent in a variety of fields and develop a universal modeling to the problem, while the applications of our proposed "exhaustive-search method for support vector machines (ES-SVM)" to geoscience, psychology, and so forth are introduced. In sections 3 and 4, the initiative for data-driven science and the deepening of SpM, respectively, are described [1]. In section 5, a three-step approach to materials science is presented, and an example of applying the three levels of data-driven science to specific fields of natural science is described. Finally, the topics covered in the preceding sections are summarized, and the future prospects of data-driven science are discussed.

2. The three levels of data-driven science

Recently, data science was featured in one of the leading journals *Science.* In its special issue, it was reported that a common method for analyzing image data performs well in the fields of both astronomy and life science, though these fields completely differ in terms of target objects and spatial and temporal scales [5]. The versatility and broad scope of data-analysis methods inspire us to believe that a general principle exists behind a wide variety of data, and that universally applicable algorithms can be developed on the basis of this principle. It follows that one of the most important purposes of data-driven science is the pursuit of a fundamental principle. As described in section 3, on the other hand, the same set of spectroscopic data can be analyzed by several different methods. It is thus necessary to choose the most suitable method. That choice hinges on the question, gAccording to what criteria should the appropriateness of data-analysis methods to specific purposes be evaluated?h Moreover, the relation between target data and analysis methods is not one-to-one but many-to-many. Accordingly, the promotion of data-driven science will untie this complex correspondence systematically.

To develop a guiding principle of data-driven science, it is useful to cast a spotlight on the three levels proposed by a famous neuroscientist David Marr. In his book *Vision*, he proposed the three levels at which any machine carrying out an information-processing task must be understood, and since the publication of that book, the concept of the three levels has considerably influenced neuroscience [4]. Also in computer science, there was a paradigm shift called "object orientation" and, since then, it has been argued that the hierarchy of problems should be recognized and utilized. Through the scientific activities of Sparse Modeling, we have found an importance of hierarchy for data-driven science as the object orientation in computer science, and have become convinced that the three levels shown in figure 1 give a novel insight into data-driven science.

The top (first) level, gcomputational theory,h in figure 1 comprises the goal and strategy for information processing as well as the appropriateness of data analysis. This level is the linchpin connecting the specific fields of natural science and data-driven science. For example, materials informatics aims to predict physical properties (such as electric conductivity, permittivity and magnetic permeability) from data obtained by large-scale computation regarding electronic states of materials. It is impossible to set a proper objective of data analysis without a background in materials science, on the basis of which the meaning of these physical properties and the mechanism of material functions are discussed in terms of physics and chemistry. This specific example shows that constructing computational theory requires a deep understanding of a target field of natural science in addition to knowledge about techniques for data analysis.

The significance of strategy setting at the top level (computational theory) is explained as follows. In general, data do not always include a sufficient amount of information to achieve the objective of data analysis. Let us consider the computational theory of stereopsis by binocular disparity. The relevant data are two images formed on retinas by visual inputs into both eyes. Actually, it is impossible to calculate the depth of what is seen from these two images only, because the correspondence between the two retinal images is still ambiguous. Marr and Poggio developed a computational theory for depth perception by assuming that the correspondence between retinal images is spatially smooth [6]. The assumption of smoothness is a strategy at the top level (computational theory). In addition, the appropriateness of the strategy is ensured by the fact that the spatial distributions of physical quantities usually vary smoothly. The above example shows that the objective, strategy, and appropriateness of data analysis are based on the knowledge of the target field of natural science.

The bottom level in figure 1 consists of two concepts: one is the choice of mathematical representation for the goal and strategy of computational theory, and the other is the algorithm to be used to execute the representation. This layer corresponds to machine-learning techniques and data-analysis methods. Between the computational-theory level and the grepresentation and algorithmh level, a gmodelingh level, which enables smooth translation of computational theory into representation and algorithm, is necessary to construct an information-processing system called gdata-driven scienceh (figure 1). The modeling level to formulate what is common to various purposes of data analysis plays a leading role in connecting various fields of natural science and information science. Thus, the modeling level forms the basis of data-driven science. In this study, the concept of three novel levels, which consist of computation theory, modeling, and representation/algorithm, namely, *the three levels of data-driven science* was set as a guiding principle for data-driven science.

As stated above, the discovery of interdisciplinary structural similarity regarding data analysis is necessary to untie the complex relation between scientific data and analysis methods, and it is beyond conventional academic areas dealing with only a specific field. When researchers from different disciplines discuss the issue of data analysis, they often miss which level it oriented from, only to lose sight of the route to the solution. The breakthrough regarding this situation

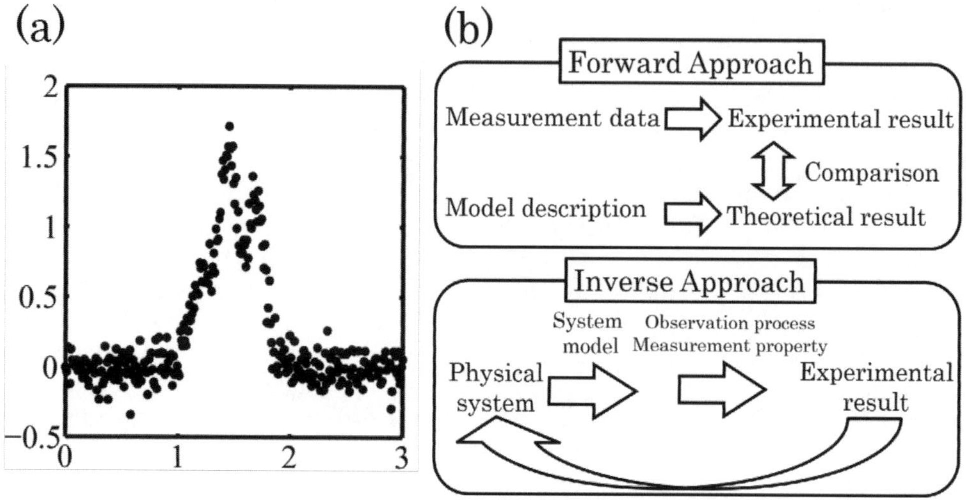

Figure 2. (a) Spectral deconvolution and (b) Bayesian sensing: Inverse approach for sensing.

involves removing the barriers between different disciplines, conveying common knowledge to different fields, and revealing interdisciplinary structural similarity. Hence, we insist that it is necessary to keep the three levels of data-driven science in mind constantly.

3. Spectral deconvolution

Spectral deconvolution is a framework for deconvoluting multimodal spectral data, such as that shown in figure 2(a), into the linear sum of unimodal basis functions such as Gaussian functions[2]. In this section, we consider spectral deconvolution, and describe how to apply the three levels of data-driven science (described in section 2) to a specific data-analysis problem.

Firstly, the computational theory of spectral deconvolution is described. In spectroscopy, each unimodal basis function is contained in a complex spectrum, such as that shown in figure 2(a), and corresponds to the electron energy levels of the target material. Therefore, the computational theory of spectral deconvolution is to determine the unknown electronic state of a measuring material. As the strategy of spectral deconvolution, the multimodal spectral data, such as that shown in figure 2(a), is deconvoluted into the sum of unimodal basis functions, and the position of each unimodal basis function, which corresponds to the energy state, is estimated. In the following, two representations, Bayesian estimation in figure 2(b) and L1VM (which stands for l_1 regularization vector machine and is one kind of SpM [7]), and their algorithms for this computational theory and its strategy, are described. Moreover, the appropriateness of each representation is evaluated by mathematically investigating the advantages and disadvantages of the representations and algorithms.

3.1. Bayesian spectral deconvolution

Bayesian spectral deconvolution [2] is explained as follows. The following equation is the mathematical representation for the strategy of the spectral deconvolution,

$$G(x;\theta,K) = \sum_{k=1}^{K} a_k \exp\left(-\frac{(x-\mu_k)^2}{2\sigma_k^2}\right), \tag{1}$$

Journal of Physics: Conference Series **699** (2016) 012001 doi:10.1088/1742-6596/699/1/012001

where the number of Gaussian functions (K) shows the number of energy levels for the measured material. Parameter set θ is defined as $\theta = \{a_k, \mu_k, \sigma_k\}_{k=1}^K$. The mathematical representation for the computational theory of the spectral deconvolution is to appropriately determine K and θ from data set $D = \{x_i, y_i\}_{i=1}^N$ with N data points, which compose the multimodal spectral data shown in figure 2(a).

Parameter set θ is often optimized by minimizing the following mean-square error function:

$$E(\theta, K) = \frac{1}{2N}\sum_{i=1}^{N}(y_i - G(x_i; \theta, K))^2. \tag{2}$$

This process is called the method of least squares. It can be used to determine not only θ but also K as follows. Although $E(\theta, K)$ can be reduced to almost zero by increasing K, the obtained curve $G(x; \theta, K)$ is an excellent fit to not only signal but also noise, so it gives a very poor representation of the measured material. Bayesian estimation can optimize not only θ but also K by stochastically formulating an observation model and by tracking back causal laws with Bayes' theorem.

To be more precise, the causal laws, which produce observation data y, are stochastically formulated. First, it is assumed that the number of energy levels for the measured material (K) is generated according to probability $p(K)$, and θ is obtained from conditional probability $p(\theta|K)$. For simplicity , $p(K)$ and $p(\theta|K)$ are set as constants. The observed data (y) are assumed to be the sum of a signal following equation (1) and a Gaussian noise (ϵ) with zero mean and variance 1; that is, $y = G(x; \theta, K) + \epsilon$. Given K and θ, y is generated from the conditional probability of input x as

$$p(y|x, \theta, K) = \frac{1}{\sqrt{2\pi}}\exp\left(-\frac{(y - G(x; \theta, K))^2}{2}\right). \tag{3}$$

Under the assumption that each data, (x_i, y_i), is independently obtained, the conditional probability of obtaining set D of N data given θ can be written as $p(D|\theta, K) = \prod_{i=1}^{n} p(y_i|x_i, \theta, K) \propto \exp(-NE(\theta, K))$. Then, applying Bayes' theorem and marginalization, it is possible to obtain the posterior probability of the parameter set, $p(\theta|D, K)$, given data set D and K as

$$p(\theta|D, K) = \frac{p(D, \theta, K)}{p(D, K)} = \frac{p(D|\theta, K)p(\theta|K)p(K)}{p(D, K)} \propto \exp(-NE(\theta, K)), \tag{4}$$

where it is assumed that $p(K)$ and $p(\theta|K)$ are constant. ¿From equation (4), θ that maximizes $p(\theta|D, K)$ is found to be equal to θ obtained by the least-square method.

Posterior probability $p(K|D)$ of K given the data D is given by

$$p(K|D) = \frac{p(D, K)}{p(D)} = \frac{\int d\theta p(D, \theta, K)}{p(D)} \propto \int d\theta \exp(-NE(\theta, K)). \tag{5}$$

Bayesian estimation uses the value of K that maximizes $p(K|D)$ on the basis of the *maximum a posteriori estimation.* The right term of equation (5) corresponds to the partition function in statistical mechanics. Hence, free energy $F(K)$ is defined by using the partition function as follows:

$$F(K) = -\log\int d\theta \exp(-NE(\theta, K)). \tag{6}$$

According to the monotonicity of logarithms, the value of K that maximizes $p(K|D)$ is equal to the one that minimizes $F(K)$.

Using Bayesian estimation for spectral deconvolution has two difficulties. One is obtaining the correct θ for minimizing error function $E(\theta, K)$ (because of the local minima of the error function). The other is how to calculate the integral for θ in equations (5) and (6).

To overcome these difficulties, the exchange Monte Carlo (EMC) method [2], an algorithm of the Markov-chain Monte Carlo method, is used. It can efficiently sample from a probability distribution with complex local minima such as the spin glass system [8, 9]. Moreover, EMC has another advantage that it can calculate free energy $F(K)$ in equation (6) by using the sampling results[2].

3.2. Spectral deconvolution by L1VM

Spectral deconvolution by L1VM is explained as follows. The following function is a mathematical representation for the strategy of the spectral deconvolution using Gaussian functions:

$$G(x; \{a_i\}_{i=1}^N, \sigma) = \sum_{i=1}^{N} a_i \exp\left(-\frac{(x - x_i)^2}{2\sigma^2}\right), \tag{7}$$

where a_i is the magnitude of the ith Gaussian function, and it is assumed that the variance of each Gaussian function is constant (σ) to reduce the number of estimated parameters.

It is assumed that a data set is much larger than the number of energy levels of the measured material (K), as shown in figure 2(a). A subset of the training samples is then efficiently selected by using sparsity-promoting priors, and the multimodal spectral data are deconvoluted into the sum of a few unimodal basis functions. It is considered that the position of each unimodal basis function corresponds to the energy state.

As denoted above, l_1 regularization is used, and parameters $\{a_i\}_{i=1}^N$ are optimized by minimizing the following l_1 regularized function:

$$E(\{a_i\}_{i=1}^N, \sigma, \lambda) = \frac{1}{2N}\sum_{j=1}^{N}(y_j - G(x_j; \theta_2))^2 + \lambda \sum_{i=1}^{N} |a_i|. \tag{8}$$

Here, to avoid over-fitting and optimize the parameters of L1VM, namely, λ and σ, 10-fold cross-validation is used, and the one-standard-error rule, namely, the simplest (most regularized) model is taken, and parameters whose error is within one standard error of the minimal cross validation error are chosen[10], is used.

In the case of spectral deconvolution by L1VM, K is obtained as a by-product. Although the results can lead to inappropriate spectral deconvolution, the evaluation function in equation (8) gives no local solution, and a global minimum solution can be obtained by using a polynomial time algorithm of number of data points N.

3.3. Comparison between results by Bayesian estimation and L1VM

An appropriate representation of the spectral deconvolution is determined by comparing the results obtained by the two representations, Bayesian estimation and L1VM, as explained above. Applying Bayesian estimation and L1VM to the data set (figure 2(a)), which are generated data and composed of the sum of three Gaussian functions, obtained the results as shown in figure 3.

Figure 3(a) shows that the Bayesian estimation selects $K = 3$ as plausible result by minimizing free energy $F(K)$. This result is considered a suitable outcome since the generated data are composed of three Gaussian functions. On the other hand, as shown in figure 3(b), the number of peaks abstracted from L1VM is 19, which is far from the true number of peaks,

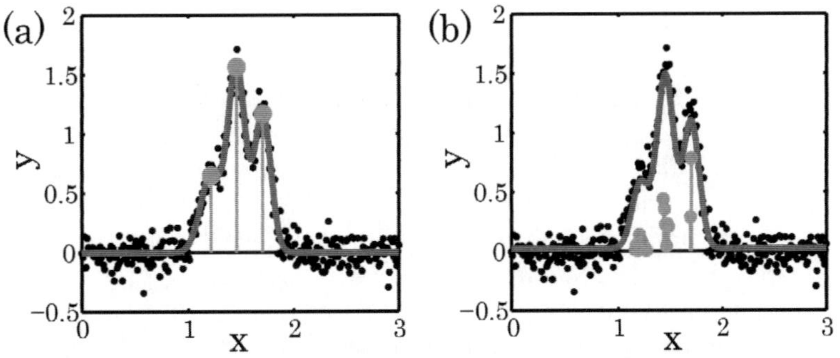

Figure 3. Results of spectral deconvolution by Bayesian estimation (a) and L1VM (b), respectively. The parameters of L1VM are bandwidth $\sigma = 0.11$ and $\lambda = 0.026$. The black dots represent the multimodal data shown in figure 2(a). The solid line indicates the estimated functions, and the impulses show the estimated height of spectrum a_k.

Table 1. Comparison between computational costs, advantage and disadvantage by Bayesian estimation and L1VM

Algorithm	Cores	Time (mins)	Advantage	Disadvantage
Bayes est.	10	157.4	Effective representation	High computational cost
L1VM	10	21.3	$O(N^3)$ computational cost	Biased estimation result

$K = 3$. This is because L1VM uses a restricted model in which the variance of each Gaussian function is constant σ and the position of the ith peak of the spectrum is fixed data x_i. The peak positions obtained by L1VM are seemingly divided into three regions, indicated by three peaks. Moreover, peak-fitting results obtained by L1VM are almost equal to those obtained by Bayesian estimation, as shown in figures 3(a) and (b).

The results of Bayesian estimation and L1VM are compared in table 1. Using only the data set and stochastic formulation, Bayesian estimation can estimate not only θ but also K, and the method can be considered an effective representation for the spectral deconvolution. However, Bayesian estimation suffers from the disadvantages that error function $E(\theta, K)$ in equation (2) has local minima and high computational cost is incurred in deriving free energy $F(K)$. Actually, as shown in table 1, Bayesian estimation takes about 7 times longer to obtain the results than L1VM. Thus, it cannot easily handle a large-scale data set such as 2-dimensional spectral data and multimodal data with more than a hundred peaks. On the other hand, L1VM has the advantage of a polynomial computational cost, namely, $O(N^3)$, where the number of data sets is N, and is thus applicable to such large-scale data. At the same time, since L1VM uses the restricted model, the obtained parameters, such as the number of peak functions, could be different from those given by a true function. Due to this disadvantage, post-processing, for example, clustering neighboring peak functions, must be performed as shown in figure 3(b).

Multimodal spectral data can be analyzed by several different representations. It is necessary to appropriately recognize the advantages and disadvantages of each representation and keep the three levels of data-driven science in mind constantly, as shown in figure 1(b). Computational theory and strategy must therefore be taken into account when choosing the most suitable representation and algorithm.

4. ES-SVM:Exhaustive Search with Support Vector Machine

The deepening of SpM encouraged by data-driven science is illustrated from the viewpoint of the problem of feature selection solved by the framework of supervised learning [3]. The purpose of feature selection is to select a significant combination of features among all given features and variables. In light of the three levels of data-driven science, the problem of feature selection is derived from the level of computational theory where it is requested to extract essential variables to describe data, and should be discussed at the level of representation and algorithm. Cover and Van Campenhout demonstrated that the only algorithm that can find the optimal subset of feature variables is the exhaustive-search method, which explores all of the solution space at the expense of a computational cost of $O(2^N)$ with dimension of data N [11].

To avoid computational complexity, fast approximate algorithms, such as least absolute shrinkage and selection operator (LASSO), whose computational cost is $O(N^3)$ [12], have been proposed. The recent spread of SpM in various fields was triggered by the mathematical equivalence between the results of l_1-norm minimization and those of l_0-norm minimization under a certain condition. However, we cannot be too careful of the equivalence condition in applying SpM to raw data. For example, the case where SpM is utilized for extreme measurement in natural science is considered as follows. The amount of data obtained by MRI and black-hole imaging does not always satisfy the condition for success of approximate algorithms.

The task of latent-structure extraction is a representative application of feature selection and is very complicated. For example, in electrophysiology, a field of brain science, selecting neurons performing pattern recognition is one task done by SpM. In such a case, several sets of neurons can have a function of pattern recognition in the first place. In short, at the level of computational theory, it is assumed that the problem of feature selection has multiple solutions. LASSO is designed on the assumption that the problem has a unique solution. Consequently, when LASSO is applied to such a system with multiple solutions, it is unlikely to find the proper solutions. In a more serious case where no neuron is involved in the task of pattern recognition, LASSO is urged into a phantom set of neurons. Nagata *et al.* proposed ES-SVM (exhaustive search with support vector machine) to address these kind of problems with multiple or no solutions properly.

When a monkey was performing a pattern-recognition task requiring the identification of facial images, Eifuku *et al.* observed the neural activities of 23 neurons in the anterior inferior temporal (AIT) cortex of the monkey[13]. The neural activities are electrophysiologically measured by conducting a single-unit recording. In this study, the face images of 4 different identities viewed from 7 different angles were presented. Then, view-invariant neurons working on the pattern-recognition task with SpM were selected [3]. The neural activities were used to solve the binary-classification problem for face identification, and tried any combination of face images of 4 different identities, that is, ${}_4C_2 = 6$ combinations. Two approximate algorithms of SpM, LASSO and SLR (sparse logistic regression) [14], were used for solving the feature-selection problem. To select the parameters of LASSO and SLR, LOOCV (leave-one-out cross validation) [7] was used. In LOOCV, 14 samples of 23 neurons were divided into 13 training data and 1 test data for calculating the generalization performance of unknown data. This training and validation operation was iterated with different partitioning 14 times to reduce variability. The parameters whose generalization error is minimal were then optimized. It was found that using LASSO and SLR makes it possible to select almost the same neurons working on 5 of the 6 binary classification problems irrespective of training data. On the other hand, LASSO can select no neuron working on the last problem of the 6 binary classification problem and SLR selected neurons for the last problem, which unsteadily vary as training data.

To investigate the difficulties concerning selecting neurons working on the pattern-recognition

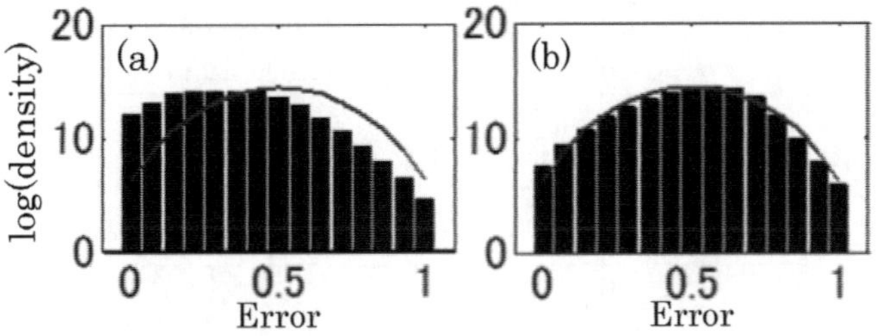

Figure 4. The histograms of $CVE(\boldsymbol{c})$.

task, Nagata *et al.* proposed ES-SVM [3]. Let us formulate ES-SVM concretely. An indicator variable is defined as $\boldsymbol{c}$, which represents a combination of the selected neurons working on the pattern-recognition task and has the same dimensions as the number of neurons, that is, 23-dimension vectors. The ith variable c_i takes 0 or 1 and indicates that $c_i = 1$ if the i-th neuron is contained in the combination, and $c_i = 0$ if it is not. By using the combination of features represented by the indicator variable $\boldsymbol{c}$, the average of LOOCV error, which is defined as prediction error $CVE(\boldsymbol{c})$, is calculated. The prediction performance is evaluated by calculating $CVE(\boldsymbol{c})$ for all combinations of neurons except $\boldsymbol{c} = (0, 0, 0, \cdots, 0)$, whose number is $2^{23} - 1 = 8388607$. Figure 4 shows histograms of $CVE(\boldsymbol{c})$. Figure 4(a) shows the result of one of the 6 binary classification problems on which almost the same neurons working are selected by the LASSO and SLR, irrespective of training data. Figure 4(b) represents the histogram of $CVE(\boldsymbol{c})$, where no neuron should be selected by LASSO, and the selected neurons of SLR unsteadily vary as training data in LOOCV. Comparing figures 4(a) and (b) reveals that there is a certain connection between the shape of the histograms and the solution stability of LASSO and SLR. Accordingly, a method for comparing the obtained histograms with the histogram given by random guessing [3] is proposed in the following. More specifically, it is assumed that given data are randomly labeled and the probability that each of the data is correctly labeled was 0.5. We take a prediction of binary classification 14 times in LOOCV for all combinations, the averaged prediction error of all combinations are given by the following binomial distribution shown by the solid line in figures 4(a) and (b).

$$p(k) = \frac{14!}{k!(14-k)!}(0.5)^k(1-0.5)^{14-k}, \tag{9}$$

where k is the number of misclassifications. Figure 4(b) shows that the histogram and the binomial distribution do not differ by much. This means that the neural activities data do not include information about view-invariant binary classification.

The above results are summarized as follows. As pointed out by Cover and Van Campenhout, no non-exhaustive sequential feature-selection procedure can be guaranteed to produce the optimal subset in the feature-selection problem[11]. Based on this knowledge, ES-SVM[3], in which a support vector machine (SVM) is used as the linear discriminant for the binary classification, was proposed. ES-SVM calculates the prediction error $CVE(\boldsymbol{c})$ for all $2^N - 1$ combinations of N dimensional features and makes a histogram of $CVE(\boldsymbol{c})$. The histogram shows whether the given data contain the desired information or not. ES-SVM does not cause the problem of multiple solutions since it checks all candidates of feature combination without approximation. Moreover, the histogram of $CVE(\boldsymbol{c})$ makes it possible to evaluate both existing and future approximate algorithms. Thus, ES-SVM can be considered a gmeta-algorithmh for judging the performance of approximate algorithms. Although the main disadvantage of ES-SVM is an exponential explosion of computational cost, recently, ES-SVM has been able to

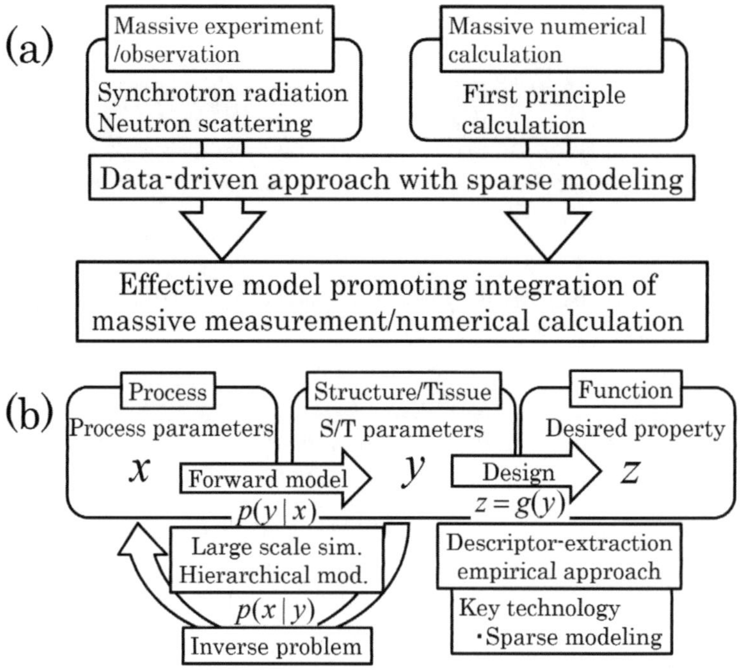

Figure 5. (a) Integration of numerical calculation and instrumental data by effective model and (b) three levels of data-driven approach for materials science

calculate low-dimensional data, such as 23-dimension data of neural activities, in a realistic calculation time. ES-SVM is thus considered better than approximate algorithms, which do not assure accuracy. Since ES-SVM is an algorithm for solving the feature-selection problem, which has broad applicability, it has already been applied to geochemical discrimination of tsunami deposits and selection of NIRS channels [15, 16]. It is thus paving the way for a new trend in various academic disciplines. Moreover, the ES (exhaustive search) framework can easily be extended to regression problems. Actually, we have recently noticed a potential application to regression problems in astronomy [17].

To overcome the above-mentioned problem of computational cost, a statistical mechanical method, which consists of the exchange Monte Carlo (EMC) method and the method of multiple histograms [3], is proposed here. The basic idea behind this method is that $CVE(\boldsymbol{c})$ can be considered an energy when $\boldsymbol{c}$ is seen as a binary spin. In this system, unlike a general spin system, $CVE(\boldsymbol{c})$ is not given the analytical function of $\boldsymbol{c}$. This structure is similar to approximate Bayesian computation (ABC), which is of current interest. This statistical mechanical method is thus called gAES-SVM (approximate exhaustive search with support vector machine),h and its effectiveness was demonstrated by estimating a histogram of $CVE(\boldsymbol{c})$ while incurring a computational cost of about 1/100 of that of ES-SVM [3]. The histogram makes it possible to estimate the number of states which have, for example, $CVE(\boldsymbol{c}) = 0$ and to verify whether the given data include information for binary classification even when the data size is large and computational cost exponentially explodes. Moreover, similar to ES-SVM, AES-SVM is a meta-algorithm for judging the performance of approximate algorithms.

5. Discussion and Perspective

One of the future developments in data-driven science is the data-driven approach to numerical data obtained by large-scale simulations such as first-principles calculations in condensed matter physics. SpM was utilized to study the dynamics of anharmonic phonons and to extract an effective model of an alloy [18]. In these studies, numerical data play the role of measurement data shown in figure 2(b). It is important to project measurement data and numerical data onto the same effective model. Accordingly, we propose a framework based on effective models obtained by the data-driven approach shown in figure 5(a), to unify numerical data from large-scale simulations and experimental data from the most-advanced measuring equipment seamlessly. This framework was applied to the study of $NiGa_2S_4$, which is a triangular-lattice antiferromagnet, and is considered to be a typical system described by Anderson's theory of resonating valence bonds (RVB). Recently, Takenaka *et al.* proposed a Bayesian framework of model selection in which an effective model described by a classical spin system is automatically extracted from Hartree-Fock calculation of electronic states [19]. In experimental studies of neutron scattering, on the other hand, the dispersion relation of a spin wave is related to the magnitude of exchange interactions between classical spins in effective models. In short, in the study of $NiGa_2S_4$, it is impossible to directly compare the numerical results of electronic states and the experimental results of neutron scattering; instead, the concept of effective models makes it possible to relate both experimental and numerical results, as shown in figure 5(a). In the framework of figure 5(a), the key to seamless integration of large-scale simulations and advanced experiments is Bayesian sensing and Bayesian estimation of effective models.

In terms of the characteristics of the specific science and its inherent universality which goes beyond the border of scientific fields, we will describe a concrete way to incorporate the data-driven approach into the specific fields of science. The following hierarchical structures exist in many fields of natural sciences: hierarchies from a genome via a protein, a cell to an individual in life science, hierarchies from behavior of a neuron to psychology and behavior science in brain science, and hierarchies of various scales of time and space in geosciences. These scientific fields have similarities: For all scientific fields, explanation ability has already reached a high level for many phenomena within each layer due to the significant technological progress of measurement instruments and computational equipment. On the other hand, systematic ways to connect among different layers for these scientific fields have not yet been developed. Data-driven science and its fundamental concept, namely, a three-level structure, can be a promising start point which connects among different layers. The procedure is as follows: First, each scientific field is described hierarchically according to the characteristics of the field. The characteristics of the field are represented in the obtained description. Next, data-driven science is used as the universal framework that connects adjacent layers.

The interpretation of the three levels of data-driven science is well illustrated in the field of materials science. It is useful to divide the development process of structural materials such as iron into three steps, as shown in the hierarchical expression in figure 5(b). The structural materials engineering aims to find the optimal values of process parameters in production for making and designing good materials in terms of durability such as strength and heat resistance. The process parameters and the desired properties are put at the first and third steps, respectively, in the hierarchical expression, as shown in figure 5(b). In the development of structural materials, it is important to parameterize the structure of material tissues [20]. Concretely, we enumerate tissue parameters such as grain size, grain shape, lath width, and grain boundary from images of metallographic structure obtained by electron microscopy. Then, we search for the optimal set of these structure parameters to predict the desired properties of materials, such as strength and heat resistance. The structure and tissue parameters are not always causally related to the material properties, but just a correlation between them can perform well in improving the function predictability. This framework has the same mathematical structure that appears in a brain machine interface, where the intention of users is read out from electroencephalography [21], and in disaster management, where the height of a tsunami is estimated from the data obtained by sensors of DONET (Dense Oceanfloor Network system for Earthquakes and Tsunamis) [22]. We emphasize that the tissue and structure parameters enumerated by intuition do not always form a necessary and sufficient condition for explaining material functions. Therefore, we discuss a framework of supervised learning in which all the structure and tissue parameters that have been suggested so far are taken as inputs, and the desired functions are taken as outputs. In such a case, the technology of SpM will lead to the appropriate set of structure parameters.

Next, we focus on the connection between the first and second steps in figure 5(b). The objective of computational theory is to optimize the process parameters in order to generate the above-obtained structure parameters. In this case, it is not important to estimate process parameters that completely restore the observed electron microscopic images, but it is necessary to estimate the process parameters that restore the appropriate structure parameters, that is, the statistical parameters of the observed electron microscopic images. This task requires the framework of approximate Bayesian computation (ABC) that has recently been developed. ES-SVM proposed in section 4 can be regarded as one example of ABC. Because forward models, which are incorporated into the Bayesian estimation, are generally phenomenological equations such as the phase-field model, it is necessary to systematically extract an effective model from a first-principles model.

This paper reviews the present situation of the project called gInitiative for High-dimensional Data-Driven Science through Deepening of Sparse Modeling.h We proposed three levels, which can be regarded as the fundamental concept of data-driven science, and introduced a spectral deconvolution as a specific example. Then, we showed that the creation of data-driven science made it possible to propose the ES-SVM method, which results in deepening of sparse modeling. Finally, we proposed a framework that solves scientific problems in various fields in order to practice data-driven science. In this framework, hierarchical structures of a scientific field are visualized as shown in figure 5(b), and adjacent hierarchical structures are linked by data-driven approaches. ¿From the above discussion, we believe that data-driven science is *mother science*, namely, a scientific framework that will drive many fields of science.

Acknowledgments

We thank S. Tokuda, H. Takenaka and H. Sakamoto for useful comments and discussion to this paper. This work was supported by a Grant-in-Aid for Scientific Research on Innovative Areas (No. 25120005,25120009,25120010) from the Ministry of Education, Culture, Sports, Science and Technology of Japan, and funding Exploratory Research for Advanced Technology, Okanoya Emotional Information Project from the Japan Science and Technology Agency.

References

[1] Grant-in-Aid for Scientific Research on Innovative Areas Initiative for High-dimensional Data-Driven Science through Deepening of Sparse Modeling, http://sparse-modeling.jp/
[2] Nagata K, Sugita S, and Okada M 2012 *Neural Netw.* **28** 82–89
[3] Nagata K, Kitazono J, Nakajima S, Eifuku S, Tamura R and Okada M 2015 *IPSJ Transactions on Mathematical Modeling and Its Applications* **8** 23–30
[4] Marr D 1982 *Vision* (Cambridge MA: MIT press)
[5] Reed S 2011 *Science* **331** 696–697
[6] Marr D and Poggio T 1976 *Science* **194** 283–287
[7] Murphy K P 2012 *Machine learning: a probabilistic perspective* (Cambridge MA: MIT press)
[8] Geyer C J 1991 *in Comput. Sci. Stat. Proc. 23rd Symp. Inter.* ed E M Keramidas (Fairfax Station) 156–163
[9] Hukushima K and Nemoto K 1996 *J. Phys. Soc. Jpn* **65** 980–988
[10] Hastie T, Tibshirani R and Friedman J 2001 *The Elements of Statistical Learning* (Berlin: Springer) 216
[11] Cover T M and Canpenhout J M 1977 *IEEE Trans. Syst., Man, Cybern., Syst.* **7** 657–661

Journal of Physics: Conference Series **699** (2016) 012001 doi:10.1088/1742-6596/699/1/012001

[12] Tibshirani R 1996 *J. R. Statist. Soc.* B **58** 267–288
[13] Eifuku S, Souza, W C, Tamura R, Nishijo H and Ono T 2004 *J. Nuueorphysiol.* **91** 358–371
[14] Yamashita O, Sato M A, Yoshioka T, Tong F and Kamitani Y 2008 *Neuroimage* **42** 1414–1429
[15] Kuwatani T, Nagata K, Okada M, Watanabe T, Ogawa Y, Komai T and Tsuchiya N 2014 *Scientific Reports* **4** 7077
[16] Ichikawa H, Kitazono J, Nagata K, Manda A, Shimamura K, Sakuta R, Okada M, Yamaguchi M K and Kakigi R 2014 *Front. Hum. Neurosci.* **8** 480
[17] Uemura M, Kawabata K S, Ikeda S, Maeda K, Wu H-Y, Watanabe K, Takahashi S and Fujishiro I 2015 *Proc. Int. Meeting of HD3-2015* 22
[18] Nelson L J, Hart G L W, Zhou F, and Ozolins V 2013 *Phys. Rev.* B **87** 035125
[19] Takenaka H, Nagata K, Mizokawa T and Okada M 2014 *J. Phys. Soc. Jpn.* **83** 124706
[20] Simmons J P, Chuang P, Comer M, Spowart J E, Uchic M D and Greaf M De 2009 *Model. Simul. Mater. Sci. Eng.* **17** 025002
[21] Yokota Y, Igarashi Y, Okada M and Naruse Y 2015 *Proc. IEEE Eng. Med. Biol. Soc.* in press
[22] Igarashi Y, Baba T, Sato K, Hori T and Okada M 2015 to be submitted

Causal analysis, Correlation-Response, and Dynamic cavity

Erik Aurell[1] and Gino Del Ferraro[2]

[1]Dept Computational Biology and Center for Quantum Materials, KTH-Royal Institute of Technology, SE-100 44 Stockholm, Sweden *and* Depts of Information and Computer Science and Applied Physics, Aalto University, P.O. Box 15400, FI-00076 Aalto, Finland
[2]Dept Computational Biology, KTH-Royal Institute of Technology, SE-100 44 Stockholm, Sweden

E-mail: eaurell@kth.se[1], gino@kth.se[2]

Abstract. The purpose of this note is to point out analogies between causal analysis in statistics and the correlation-response theory in statistical physics. It is further shown that for some systems the dynamic cavity offers a way to compute the stationary state of a non-equilibrium process effectively, which could then be taken an alternative starting point of causal analysis.

1. Causality in Philosophy, Physics and Statistics

Causality formalizes the universal human experience of agents (causes) taking actions leading to results (effects) [1]. In the Western Philosophical tradition Aristotle postulated four kinds of causes: the material, the formal, the efficient and the final, out of which Bacon later retained the material and the efficient. The Third Law of Newton [1] however states that in Nature there is no separation between cause and effect in the Aristotelian or Baconian sense; Physics fundamentally knows only interactions, and these are always mutual, a state of affairs unchanged since that time and the replacement of the Classical Physics by Quantum Physics. This objection to philosophical causality was alluded to by Russell [2] as

> "[Hume] supposes the law to state that there are propositions 'A causes B' where A and B are classes of events; the fact that such laws do not appear in any well-developed science appears unknown to philosophers."
>
> "History of Western Philosophy", chapter "Hume", page 638

We note that when term causal is used to describe an interaction in modern high-energy Physics it means only that the influence cannot propagate faster than light so that object A at time t_A only depends on what happened at object B at times t_B early enough that a signal from B can reach A at time t_A, and vice versa [5].

The everyday and the philosophical notions of causality are in Physics instead intertwined with reversibility and irreversibility; Nature's laws are time-reversal invariant on the fundamental

[1] The literature on this topic is too vast and variegated to be meaningfully referenced; two classic studies from the perspective of belief systems in traditional societies can be found in [3] and [4].

level, but most ordinarily encountered processes are overwhelmingly likely to only flow in one direction [6, 7]. We say that dropping a glass vase on the floor is the cause of it breaking because it is exceedingly unlikely that the glass pieces would jump back together and fuse into a vase. Similarly, we say an enzyme causes a chemical reaction in one direction when the concentrations of the reactants are such that the opposite reaction is very unlikely. Although the details of this complex process are not fully known, we can also say that smoking causes cancer because the DNA in living cells is mostly that of one and the same genome for each individual – an extremely small subset of all possible DNA sequences of the same length – and cancerogenes in tobacco smoke therefore almost always lead to mutations away from the healthy genotype and into one out of very many deficient genotypes. In Nature cause-effect relationships are thus but abbreviations for processes in physical systems so strongly driven out of thermal equilibrium that they mostly only go one way.

Nevertheless, causal analysis is an important branch of statistics, describing the effects of interventions and answering questions of the "if-so-then-what?" character [8, 9, 10]. Interventions are then taken to be outside Nature, typically ascribed to a human agent, and causality is thus distinct from statistical association studies. If person X is holding a glass vase and person Y trips him over, then person X is quite likely to fall and break the vase. However, we cannot know this for sure without observing the event as person X might for instance be much larger and stronger than Y. Likewise, if we can deactivate enzyme E then we can observe that a catalyzed reaction $S_1 \to S_2$ ceases, while if we can over-activate E then the catalyzed reaction goes faster. This is the paradigm for how molecular biologists identify interactions experimentally; good research practice and common sense hold that observing such direct responses is a more reliable means to acquire knowledge than observing the variations of E and the speed of the reaction $S_1 \to S_2$ in natural undisturbed conditions. For example, the catalyzed reaction may be one in a series of reactions $S_1 \to S_2 \to S_3 \to S_4 \to \cdots$, catalyzed by enzymes $E, E_2, E_3, \ldots$ and the living cell may regulate the production of all these enzymes by the availability of the first substrate S_1 [11]. In this case $E, E_2, E_3, \ldots$ would all vary positively with the speed of the reaction $S_1 \to S_2$ and only a direct experiment can determine which of them actually does the job.

The first purpose of this paper is to show that there is a conceptual parallelism between causal analysis in statistics and long-time response functions in physics. A main difference is interventions in causal analysis are assumed to have an immediate effect while response acts only over time. This means that causal analysis is (comparatively) easier to work with, but is also further removed from physical reality. We point out that various computational methods in statistical physics give access to physical response, and which can therefore serve as basis for alternatives to causal analysis. The second purpose is to show how the tools of message-passing/Belief Propagation, developed to analyze complex static interdependence, have been generalized to describe dynamics with complex interactions, and then known as dynamic cavity. Under suitable assumptions the dynamic cavity simplifies considerably the determination long-time responses, which can hence be considered one specific alternative starting point of a causal analysis. The paper is organized as follows. In Section 2 we describe in simple terms causal analysis, following mainly [12]. In Section 3 we give a short summary of response theory using Markov chains (synchronously updated spin systems) as our main example, and show the parallelism to causal analysis. In Section 4 we describe dynamic cavity as a generalization of message-passing/Belief Propagation to dynamic phenomena, and show that it can be used to turn correlation-response into an alternative to causal analysis, for some systems. In Section 5 we sum up and discuss our results.

2. Causal analysis

Statistical physicists are nowadays conversant with graphical models to describe probability distributions [13, 14]. To explain causal analysis we will start with two very simple Bayesian belief networks

$$(a) \qquad ⓐ\leftarrow ⓑ \rightarrow ⓒ \quad \text{and} \quad (b) \qquad ⓐ\rightarrow ⓑ \rightarrow ⓒ \tag{1}$$

where by Ⓐ → Ⓑ we mean that random variable B is dependent on random variable A in the ordinary sense of probability, and also that A (somehow) causes B. The dependency is encoded in conditional probabilities $P_{A|B}(a|b)$ where a and b are values of A and B. The joint probabilities of the three variables are in the two cases

$$\text{Case } (a) \quad P_B(b)P_{A|B}(a|b)P_{C|B}(c|b) = \frac{P_{A,B}(a,b)P_{C,B}(c,b)}{P_B(b)}$$

$$\text{Case } (b) \quad P_A(a)P_{B|A}(b|a)P_{C|B}(c|b) = \frac{P_{A,B}(a,b)P_{C,B}(c,b)}{P_B(b)}$$

where $P_B(b)$ is the marginal probability of variable B to take value b while $P_{A,B}(a,b)$ is the marginal probability of the pair of variables A and B to take values a and b, and so on. We assume for simplicity that $P(b)$ is different from zero for all values b of B. Both Bayesian belief networks encode the same joint probability; we cannot distinguish by co-variation whether A causes B or B causes A. In the language of factor graphs this joint probability can alternatively be described by a factor graph [15]

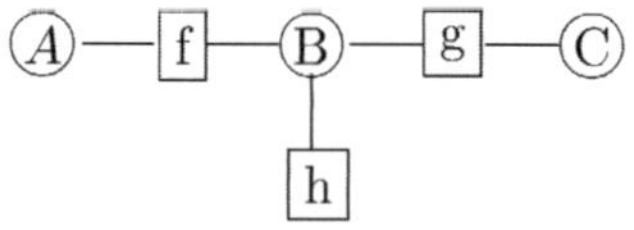

(2)

with factors

$$f(a,b) = P(A=a, B=b) \qquad g(c,b) = P(C=c, B=b) \qquad h(b) = (P(B=b))^{-1} \tag{3}$$

and

$$P_{A,B,C}(a,b,c) = \frac{1}{Z} f(a,b)g(c,b)h(b) \tag{4}$$

The partition function Z is determined by the interactions encoded by the factors of the factor graph, and is in general difficult to compute for large models, but in the simple example considered here it is obviously equal to one.

In both cases described above the joint probability of A and C conditioned on B is then

$$P_{A,C|B}(a,c|\text{see}(B=b)) = \frac{P_{A,B}(a,b)P_{C,B}(c,b)}{(P_B(b))^2} = P_{A|B}(a|b)P_{C|B}(c|b) \tag{5}$$

where we have introduced Pearl's "see" notation [12]. To avoid confusion, let us note again that $P_{A|B}(a|b)$ in above has its ordinary probabilistic meaning of $\frac{\sum_c P_{A,B,C}(a,b,c)}{\sum_{a,c} P_{A,B,C}(a,b,c)}$ and is the same in both models.

If we intervene on B and set its value to b the two Belief networks lead to new joint probabilities on the two remaining variables (A and C). In both cases A and C become independent with probabilities depending parametrically on the set value b, which we can call $P_A^{(B)}(a;b)$ ("the probability distribution of random variable A in the modified model where variable B has been set to constant b") and $P_C^{(B)}(c;b)$ ("the probability distribution of random variable C in the modified model where variable B has been set to constant b"). Introducing Pearl's "do" notation [12] we then have

$$P_{A,C}(a,c|\text{do}(B=b)) = P_A^{(B)}(a;b)P_C^{(B)}(c;b) \tag{6}$$

and the dependencies can be illustrated as a (trivial) Bayesian network

(7)

or as an (equally trivial) factor graph

(8)

The two probability distributions are however not the same in the two cases.

In case *(a)* we have as numerical values $P_A^{(B)}(a;b) = P_{A|B}(a|b)$ and $P_C^{(B)}(c;b) = P_{C|B}(c|b)$, because A and C are then both assumed to be caused by B; the factors in the factor graph (8) are $e = P_{A|B}(a|b)$ and $d = P_{C|B}(c|b)$. For this case (5) and (6) hence describe the same distribution.

In case *(b)* we also have $P_C^{(B)}(c;b) = P_{C|B}(c|b)$, but for the other probability instead $P_A^{(B)}(a;b) = P_A(a)$ corresponding to a factor $e = P_A(a)$ in (8). This difference is ultimately what it means to interpret the arrows in (1) as causes: if A is a cause and B is an effect then A should be unaffected by B, and in particular unaffected by any outside intervention on B. Therefore, whether or not there is any intervention on B, in case *(b)* the marginal probability of A is and remains $P_A(a)$ and the "do" (6) is different from the "see" (5). Expanding on the same point, in case *(b)* the "do-probability" $P_{A,C}(a,c|\text{do}(B=b))$ can be expressed in terms of probabilities observable before the intervention, namely as $P_A(a)P_{C|B}(c|b)$, but is not the same as the "see-probability" $P_{A,C}(a,c|\text{see}(B=b))$ which is, for both cases, $P_{A|B}(a|b)P_{C|B}(c|b)$. The Kullback-Leibler distance between the two is (for this simple example) $KL(\text{see}(B=b)|\text{do}(B=b)) = \sum_a P_{A|B}(a|b)\log\frac{P_{A|B}(a|b)}{P_A(a)}$, which generally is not zero.

We will now raise the abstraction level and following [12] define a general causal model M, also known as *Structural Equation Model*, as a set of exogenous variables U, a set of endogenous variables $V_1,\ldots,V_N$, located in nodes $1,\ldots,N$ in a graph G, for each node i a set of parent nodes $PA_i \subset \{\{1,\ldots,N\}\setminus i\}$ and conditional probabilities $F_i(V_i|V_{PA_i},U)$. The structure of G is determined by there being a link $i \to j$ iff $i \in PA_j$. Additionally one may include in the model specifications a distribution $P(U)$ over the exogenous variables [12]. Each such model defines a joint probability distribution of the endogenous variables as

$$P_M(V_1,\ldots,V_N|U) = \frac{1}{Z_M(U)}\prod_i F_i(V_i|V_{PA_i},U) \tag{9}$$

If G is a Directed Acyclic Graph (DAG), so that dependencies cannot propagate in a loop, clearly $Z_M(U) = 1$. The *do operator* is introduced by Pearl as:

> Interventions and counterfactuals are defined through a mathematical operator called *do*(x), which simulates physical interventions by deleting certain functions from the model, replacing them with a constant $X = x$, while keeping the rest of the model unchanged. The resulting model is denoted M_x. The post-intervention distribution resulting from the action *do*$(X = x)$ is given by the equation
>
> $$P_M(y|do(x)) = P_{M_x}(y)$$
>
> Judea Pearl, "The Do-Calculus Revisited" (2012) [12]

It is useful to compare and contrast the do operation with the cavity method to be discussed in more detail below in Section 4. Both modify a probabilistic model by eliminating one or more variables, figuratively opening a hole (or cavity) in the factor graph. A first difference is that

Journal of Physics: Conference Series **699** (2016) 012002 doi:10.1088/1742-6596/699/1/012002

in the cavity method the variable and all its interactions are eliminated as if they were never there, while in a do operation the variable is set to a constant and the value of that constant matters. Instead of (9) we thus have, taking $X = V_k$ for some k,

$$P_{M_{v_k}} = \frac{1}{Z_{M_{v_k}}(U)} \prod_{\substack{i \neq k \\ k \notin V_{PA_i}}} F_i(V_i|V_{PA_i}, U) \prod_{\substack{i \neq k \\ k \in V_{PA_i}}} F_i^{V_k = v_k}(V_i|\{V_{PA_i}\} \setminus V_k, U) \tag{10}$$

where $Z_{M_{v_k}}$ is a new normalization constant and $F_i^{V_k=v_k}$ is a new function obtained from F_i by setting the variable V_k to the constant v_k. Hence we can express the "see" and the "do" as

$$P(V_i|\text{see}(V_k = v_k)) = \frac{P_M(V_i, V_k)}{P_M(V_k)} \qquad\qquad P(V_i|\text{do}(V_k = v_k)) = P_{M_{v_k}}(V_i) \tag{11}$$

which again shows how and why the two concepts differ. A graphical illustration of the do operation is given in Fig. 1.

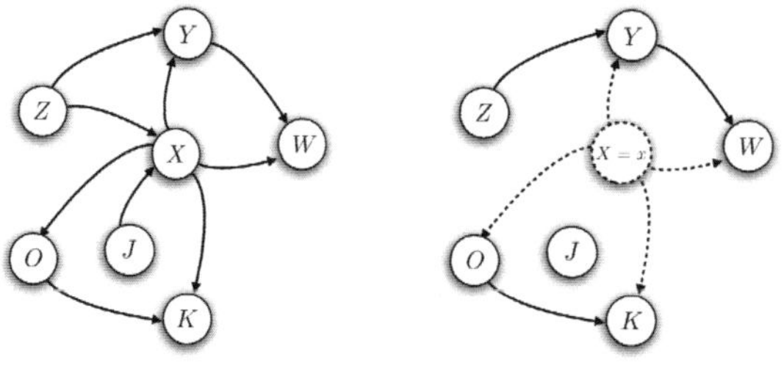

Figure 1. Illustration of the do operation. Left panel: a Bayesian belief network with a central node containing variable X. Right panel: reduced Bayesian belief network after intervening on variable X setting it to value x. Node containing X and outgoing links are indicated by dashed lines symbolizing that (10) depends parametrically on x. Ingoing links to node containing X are eliminated together with random variable X which does not appear in (10).

A second and more important difference is that the do operation is formulated for Bayesian belief networks of which (as we have seen) there can be many corresponding to the same joint probability distribution. Under the operations do(X) for different X, each Bayesian belief network (*i.e.* each direction of the arrows in Eq. (1)) hence specifies a different set of changes of the joint probability distribution encoded in a factor graph.

An important question in causal analysis has been whether probabilities after an intervention, *i.e.* $P_M(y|do(x))$, y standing for any subset of the endogenous variables, can be determined from observations before the intervention, *i.e.* from the set $P_M(z)$, z standing for some other subset. When this is so one says that a causal effect query is *identifiable* because it can be decided (the probability $P_M(y|do(x))$ estimated) from data obtained before an intervention. In both the simple examples above this was the case, only the $P_M(y|do(x))$'s were not the same. More generally, a causal effect query is always identifiable from passively observed $P_M(z)$, provided that all variables in M are observed and G is known [10]. In less technical terms this last statement means nothing else than given sufficient data one can in principle estimate conditional probabilities, and given a direction of the arrows in a Bayesian belief network one can translate this information into what the conditional probabilities will be in a modified model; the situation is more complicated when some variables are unobserved (un-measured). The *Do-Calculus* of Pearl consists of three rules for deciding identifiability when G is known and is a DAG, and some of the variables are unobserved. The Do-Calculus can hence be used to determine (in perhaps quite complex settings) whether a separate experiment is necessary, or if an hypothetical question can be answered with the data already at hand.

3. Response Theory

At the basis of scientific mathematical philosophy is the idea that the regularities of the world are best expressed by how it changes in time, famously stated by Newton to Leibniz as the anagram *6accdae13eff7i319n4o4qrr4s8t12ux* [16]. As discussed in Introduction, cause-effect relationships in Physics are only short-hand descriptions of situations where some object or process A partly (or wholly) determines the (deterministic or probabilistic) rate of change of some other object or process B. Such time-ordered relationships are often assumed in Econometrics and then (given quite strong technical restrictions) referred to as Granger-causation or G-causation [17]. They also appear in informal discussions of causal analysis as *e.g.* in "the current causes the voltage to drop across the resistor" – in physical terms the current is a response to a non-equilibrium initial state (a voltage difference across a capacitor) and the rate of change of the voltage is proportional to the current, the proportionality being the capacitance. A bit more abstractly one can say "reckless driving causes accidents" if one takes accidents to be random events the frequency of which depend positively on "recklessness"; when that changes over time one expects the aggregate number of accidents to follow, with some delay. Other phrases expressing causality such as "you will fail this course because of your laziness" have a clear time separation in their grammatical structure indicating that the effect is understood to come after the cause [2].

Let us therefore substitute the Bayesian belief network in Section 2 by a minimal model encoding the same dependencies as a probabilistic evolution law:

$$P(V_1(t), \dots, V_N(t)) = \prod_{i=1}^{N} F_i(V_i(t)|V_{PA_i}(t-1))P(V_1(t-1), \dots, V_N(t-1)) \tag{12}$$

The notation is here the same as in (9) except that the variables are now indexed by time (t) and a possible dependence on exogenous variables has been suppressed. Up to the technical simplification of synchronous dynamics, (12) is a prototype for a physically realistic mutual dependency. It could be realized in a biological regulatory system, say in a signal transduction network, where the cause-effect relationship between PA_i and V_i would have the underlying mechanistic interpretation of PA_i being the kinases, phosphatases and other enzymes catalyzing the phosphorylation, de-phosphorylation and other modifications to unit i. The endogenous variables U are then concentrations of molecules at constant concentrations, which could be sugars and other carbon sources for bacteria, or hormones and other signaling molecules in multi-cellular organisms. For long times the probability distribution in (12) would then reach stationary state which we will denote

$$P^*(V_1, \dots, V_N|U) = \lim_{t\to\infty} P(V_1(t), \dots, V_N(t)|U) \tag{13}$$

Since P^* in (12) is at least as realistic as P in (9) as a representation of how the endogenous variables depend on the exogenous variables we could also use it to define an analogy of the do operation. We can thus set

$$P^*_M(y|\mathrm{do}(x = X)) = P^*_{M_x}(y) \tag{14}$$

where the right-hand side is interpreted as the long-time response of the system to intervention on X.

In Physics a response function related to a generic quantity $V_i(t)$ is normally defined in the linear regime as $R_{ij}(t,t') = \langle \partial V_i(t)/\partial H_j(t')\rangle$ where $H_j(t')$ is a general parameter which can be varied within the system. An example, in ferromagnetic systems, is the susceptibility function $\chi_{ij}(t,t') = \langle \partial M_i(t)/\partial H_j(t')\rangle$ which gives the change in the local magnetization $M_i(t)$ on site i at

[2] All these three example phrases are taken from the introduction to [8], one of the first modern papers on causal analysis in statistics.

time t due to an impulse change of an external field $H_j(t')$ acting on a different site j at an earlier time t'. The response is then proportional to the (small) change $\Delta H_j(t')$, the proportionality being the response kernel $K(t,t')$. One may also consider the response to a finite step-like change [18] and the *do* operation defined by (14) is clearly of this more general type.

We can therefore now state the first result of this paper: any means to efficiently solve for the stationary state of (12), before and after intervention, can be the basis for an alternative to standard causal analysis. We will below in Section 4 highlight the possibilities recently opened by dynamic cavity, but the statement is more general. Monte Carlo methods [19, 20], mean-field methods [21], exact results on the SSEP model and analogous systems [22], macroscopic fluctuation theory [23], generating functions expansions [24], and any other general or specialized method to analyze (12) can be used as building blocks for a causal reasoning which includes the notion of time and time delays, and which is thus more natural from the physical point of view, and closer to common sense.

We end by noting that a great deal is known about response functions for systems near thermodynamic equilibrium where they are related to correlation functions through the Fluctuation-Dissipation-Theorem, which generically takes the form [25, 26]

$$\frac{1}{T}[C(\tau=0)-C(\tau)] = \int_0^{\tau} R(\tau')d\tau' \tag{15}$$

where $\tau = t - t'$, time translational invariance is assumed, and T is temperature. The left-hand side of above is measured in the unperturbed system and the right hand side in the perturbed system. Causal effect queries are therefore always identifiable in systems at or near thermodynamic equilibrium from observing no more than the correlation between the variable which is set and the variable one wants to predict. This relation between correlation and response has been used to improve network inference [27]. A formula analogous to (15) also exists far from equilibrium [18] but as it requires as input the gradient of the stationary state with respect to the parameter it is only useful when that stationary state can be determined.

4. Dynamic cavity

In this section we describe how the techniques now generally called message-passing or Belief Propagation can be generalized to analyze evolution laws like (12). Message-passing techniques have been invented independently in different fields [13, 15]. In Physics they are also known as the cavity method [14], and usually traced back to [28]. Their purpose is to compute marginal probabilities over some (usually small) subset of variables in a probabilistic model described by a factor graph which is done by storing partial computations in nodes representing the variables and then forwarding such partial results to neighbors in the graph for further processing. Message-passing converges and is exact if the underlying graph is a tree but also often converges and is a very good approximation if the underlying graph has only long loops, a fact that has many theoretical and practical applications in coding theory and elsewhere [14]. The fixed points of the algorithms correspond to stationary points under variation of the Bethe approximation to the free energy in the corresponding statistical mechanics problem [13, 14]. Situations where the message-passing equations have more than one fixed point are outside the scope of this brief presentation and their analogues have (to our knowledge) not been studied for the dynamic cavity described below.

Our point of departure is now the observation that the dynamics (12) naturally leads to a probability distribution on variable histories

$$P(X_1,\ldots,X_N) = P(V_1(0),\ldots,V_N(0))\prod_{t=1}^{T}\prod_{i=1}^{N} F_i(V_i(t)|V_{PA_i}(t-1)) \tag{16}$$

where $X_i = \{V_i(0), \ldots, V_i(T)\}$ for $i = 1, \ldots, N$. Before continuing, let us note that if the variables are Boolean and take values $\{-1, 1\}$ ("spins") then (12) specifies a dynamics of a spin system under synchronous updates, and if further all the transition probabilities of the type $F_i(V_i(t)|V_{PA_i}(t-1)) \propto \exp\left(V_i(t)\left(h_i + \sum_j J_{ij}V_j(t-1)\right)\right)$ are known as the Kinetic Ising model [29]. The parent set V_{PA_i} is then comprised of the variables V_j for which J_{ij} is non-zero. When $J_{ij} = J_{ji}$ for all pairs (i,j) the system has a stationary state $P(V_1, \ldots, V_N) \propto \exp\left(\sum_i h_i V_i + \sum_{ij} J_{ij} V_i V_j\right)$ and (12) then simulates a system in thermal equilibrium, albeit under the somewhat unphysical synchronous update rule. In the more general case when $J_{ij} \neq J_{ji}$, and in particular for fully asymmetric models where J_{ij} can only be non-zero when J_{ji} equals to zero, (12) on the other hand simulates a non-equilibrium system. The stationary probability distribution of such a physical system naturally depends on the exact update rule; we will briefly comment on this issue in Section 5 below.

The first result on reducing the complexity of (16) dates back almost thirty years [30] and pertains to fully asymmetric models. For these an influence $X_j \to X_i$ must traverse a loop in G to get back to X_j, and when there are no loops, or when these can otherwise be disregarded, the marginal probability of X_j is independent of X_i. This leads to simple equation for the marginalization over a single variable and and single time, namely

$$P_i(V_i, t) = \sum_{V_j \in PA_i} F_i(V_i|V_{PA_i}) \prod_j P_j(V_j, t-1) \qquad \text{(Fully asymmetric)} \qquad (17)$$

We now generalize a bit and assume that the dependency graph G has the property associated with the effectiveness of standard message-passing *i.e.* that it is a tree, or at least locally tree-like. That is, we assume that one cannot form circular dependency chains $i \to j \to k \to \cdots \to i$ where $V_i \in PA_j, V_j \in PA_k, \ldots \in PA_i$ unless either somewhere the chain backtracks as $\cdots j \to k \to j \to \cdots$ or the chain is long, on the order of the graph diameter of G. The term "dynamic cavity" was introduced in [31] for such situations where it was used to obtain rigorous bounds on the consensus threshold for the majority dynamics. Methods have later been developed to treat, in principle exactly, such problems when the dynamical law is modified to only allow transitions in one direction [32, 33]. An important step was taken in [34] where marginals in a stationary state were computed approximately based on an ansatz, recently extended to also cover transient phenomena [35, 36]. The main problem is then that even when the dependency graph G of the dynamics in (12) is locally tree-like this is not the case for dependencies in (16) due to "loops-in-time". These dependencies have been resolved by a graph expansion technique [32, 33, 35] as we will now explain.

First, as for the Kinetic Ising model it is often convenient to define transition functions only up to a normalization $F_i(V_i|V_{PA_i}) \propto \exp\left(r_i(V_i, V_{PA_i})\right)$. The normalization constant is then $N_i(V_{PA_i}) = \sum_{V_i} \exp\left(r_i(V_i, V_{PA_i})\right)$, a function that does not depend explicitly on V_i. Assuming further for simplicity that interaction functions r_i are only pair-wise the dependency graph can be illustrated as in Fig. 2. The model defined on variable histories, (16), now has short-loop dependencies even if the graph G itself does not. This can be seen by tracing the dependency of one of these variable, say $X_i = \{V_i(0), \ldots, V_i(T)\}$. Pick a time t and note that $V_i(t)$ depends on $V_j(t-1)$ for all $j \in PA_i$. Then pick two of these variables X_j and X_k such that $i \in PA_j$ and $i \in PA_k$, then $V_j(t-1)$ and $V_k(t-1)$ both depend on $V_i(t-2)$. At the same time $V_j(t-1)$ and $V_k(t-1)$ are however also dependent through the normalization $N_i(V_{PA_i})$, and X_i, X_j and X_k are therefore connected in a dependency loop of length three.

While there are many approaches to get rid of loops in factor graphs we will use one which is well adapted to the dynamics. For every pair i and j such that $V_i \in PA_j$ (whether or not also $V_j \in PA_i$) we introduce a new compound variable $(X_i^{(ij)}, X_j^{(ij)})$ interpreted as "variable $X_i^{(ij)}$ of

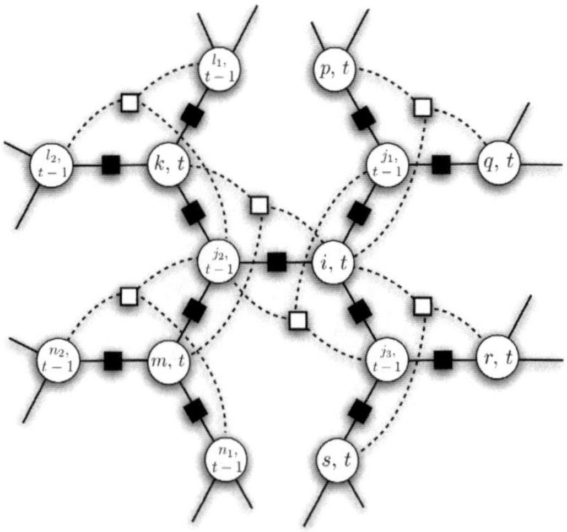

Figure 2. A tree-like dependency graph with the normalization constants in the transition functions split off as separate factor nodes (boxes). It has been assumed that the dependencies are not fully asymmetric so that when node i depends on node j, node j in general also depends on node i. Dependencies between nodes (in general mutual) are indicated by (undirected) lines. In the kinetic model loops emerge from variables at different times in (16).

type X_i belonging to link (i,j) and $X_j^{(ij)}$ of type X_j also belonging to link (i,j)" [3]. Introducing now the consistency requirement that the variables $X_i^{(ij)}$ take the same value for all the links (i,j) where this type of variable is found we can rewrite (16) as

$$P(\{X_i^{(ij)}, X_j^{(ij)}\}) = P_{\text{init}} \cdot \prod_{t=1}^{T} \prod_i F_i(\overline{V_i(t)} | \{V_j^{(ij)}(t-1)\}_{j \in PA_i}) \cdot \prod_{t=1}^{T} \prod_i \mathbf{1}_{V_i^{(ij_1)}(t) = V_i^{(ij_2)}(t) = \dots} \quad (18)$$

where P_{init} is the probability distribution on the initial conditions translated to the new variables, $V_j^{(ij)}(t)$ are the restrictions of the variables $X_j^{(ij)}$ to a single time t and $\overline{V_i(t)}$ is any suitable average of the $V_i^{(ij)}(t)$ for different j [35]. The graph expansion is illustrated in Fig. 3. Introducing messages in the standard way and summing out the consistency conditions we thus arrive at [35]

$$m_{i \to (ij)}(X_i^{(ij)}, X_j^{(ij)}) \propto \sum_{\{X_k^{(ik)}\}} \Phi_i(X_i^{(ij)}, X_j^{(ij)}, \{X_k^{(ik)}\}) \prod_{k \in \partial i \setminus j} m_{k \to (ik)}(X_k^{(ik)}, X_i^{(ij)}) \quad (19)$$

where $\Phi_i(X_i, X_j, \{X_k\}) = \prod_{t=1}^{T} F_i(V_i(t) | V_j(t-1), \{V_k(t-1)\}_{k \in PA_i \setminus j})$. Equation (19) are the dynamic cavity update equations corresponding to the ordinary cavity update equations applied to the model (16) on variable histories. A trace of the dynamic origin remains

[3] To be precise the two parts of the compound variable are distinguished by their index (i or j) and not by their order in the pair. When a message is to be transmitted from i to j they are naturally read in the order $(X_i^{(ij)}, X_j^{(ij)})$ while if the message is transmitted in the opposite direction the natural order is $(X_j^{(ij)}, X_i^{(ij)})$.

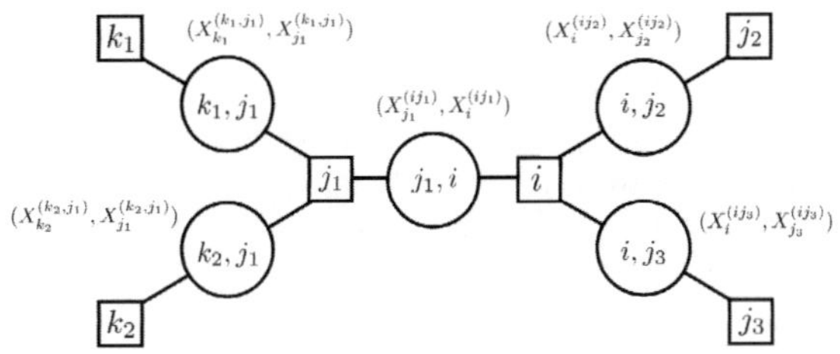

Figure 3. Auxiliary graph obtained from a part of Fig. 2 where loops have been removed by a graph expansion procedure. The new variable nodes contain histories of two variables that were neighbors in the original graph while new factor nodes (one per each old variable node) contain both the transition functions and the consistency conditions.

in that the probability $m_{i\to(ij)}(X_i^{(ij)}, X_j^{(ij)})$ can be taken to depend on the full history $X_i^{(ij)} = \{V_i^{(ij)}(0), \ldots, V_i^{(ij)}(T)\}$ of the first argument, but only a one unit shorter history $X_j^{(ij)\prime} = \{V_j^{(ij)}(0), \ldots, V_j^{(ij)}(T-1)\}$ of the second argument. For a discussion as well as a description of the analogous dynamic cavity output equations, see [35]. To make (19) practical further assumptions are needed, to close the iterations in a low-dimensional subspace of the functions $m_{i\to(ij)}(X_i^{(ij)}, X_j^{(ij)})$. In [35] good results were reported based on closure in the class of 1-step Markov processes, leading to schemes not much more complicated than (17) while in [36] even better results were reported from a more involved procedure. The field is in active development and likely even better approximations will appear in the near future.

We will now take the point of view that the probabilities P^* and $P^*_{M_x}$ in (13) and (14) are efficiently computable and ask what are the implications for causal analysis. First, the assumption of synchronous updates is unrealistic in most natural systems but certainly no more than the assumption of instantaneous dependence made in (9). In most problems where an underlying mechanistic explanation is conceivable "causes" are ultimately to be interpreted as variables influencing transition rates, and the simplest example of such dynamics is (12). In stationary state an underlying explanation, which one could call "mechanistic causes", leads to a joint probability distribution P^* with generally many more dependencies. That is, there will be one (directed) dependency graph G describing the probabilistic evolution law (12) and another (undirected) factor graph F describing the probability P^* in (13), and F will almost always be (much) larger and (much) richer than G. For a worked-out example of such an effect, in the relaxation towards equilibrium of the Kinetic Ising model on a 1D lattice [29], see [37].

5. Summary and discussion

We have given a brief introduction to causal analysis and discussed how it extends the tools of factor graphs and probabilistic models to describe outside interventions that change the models themselves. We have compared and contrasted causal analysis to the analysis of dynamic processes by physical response theory and pointed out the possibilities recently opened up through dynamic cavity.

Causal analysis considers causal relationships to be the fundamental building blocks of reality [10] and aims to discover which are these casual dependencies in the system under investigation (usually DAG networks). Although this certainly is a fascinating goal, it is at odds with physical theory which does not admit causes and effects in the philosophical sense on the fundamental level, but only for macroscopic (irreversible) processes. For such processes the flow of time is however essential, and causes are thus naturally understood as variables influencing transition rates between various states in a system. The stationary states of such processes are normally quite complicated reflecting not only dependencies in the transition rates, but also chains of such dependencies of arbitrary length, the only major exception being systems in thermodynamic equilibrium. Therefore, great caution is called for when interpreting the results of causal analysis as causes in an everyday sense. The mechanisms identified by causal analysis include (except in thermal equilibrium) both underlying direct effects and many kinds of indirect effects where the setting of one variable influences the behavior of another at a later time through one or many intermediaries.

The major advantage of causal analysis is instead in its relative simplicity of its basic ansatz. Up to recent times few methods except Monte Carlo were available to analyze the dynamics of non-equilibrium systems, and determining their stationary states was therefore laborious. Several techniques may however now give access to non-equilibrium stationary states including improved Monte Carlo harnessing advances in algorithms and hardware, and mean-field methods and other analytical or semi-analytical methods as discussed above. We have discussed that when the interactions are arranged on a tree, but are not strictly one-way, the dynamic cavity method has emerged as a new alternative yielding quite accurate estimates at comparatively low computational costs. Many major issues however remain to be solved in that approach, the most important one perhaps being how to extend the dynamic cavity (if this is possible) to continuous-time processes.

Acknowledgments

E.A. acknowledges valuable discussions with Dr. Antti Hyttinen and Profs. Timo Koski and Jukka Corander. This work has been supported by the European Union through Marie Curie ITN "NETADIS" (GDF), by Swedish Science Council through grant 621-2012-2982 (EA), and by the Academy of Finland through its Center of Excellence COIN (EA).

References

[1] Newton I 1687 *Principia* "Lex III: Actioni contrariam semper et æqualem esse reactionem: sive corporum duorum actiones in se mutuo semper esse æquales et in partes contrarias dirigi" [Law III: To every action there is always opposed an equal reaction: or the mutual actions of two bodies upon each other are always equal, and directed to contrary parts]

[2] Russel B, *History of Western Philosophy* (Simon & Schuster/Touchstone, 1945:1967)

[3] Lévi-Strauss C 1962 *La pensée sauvage* (Plon, Paris)

[4] Bourdieu P 1980 *Le sens pratique* (Les Éditions de Minuit, Collection "Le sens commun", Paris)

[5] Weinberg S 1985 *The Quantum Theory of Fields: volume 1 Foundations* (Cambridge University Press, Cambridge UK)

[6] Feynman R P 1964 The Distinction of Past and Future In *The Character of Physical Law* (The MIT Press) chapter 5

[7] Chibbaro S, Rondoni L and Vulpiani 2014 From Microscopic Reversibility to Macroscopic Irreversibility In *Reductionism, Emergence and Levels of Reality* (Springer International Publishing) chapter 4

[8] Pearl J and Verma T 1994 A theory of inferred causation In *Proceedings of Logic, Methodology and Philosophy of Science IX* (Elsevier Science B.V.)

[9] Pearl J 2009 Causal inference in statistics: An overview *Statistics Surveys* **3** 96–146

[10] Pearl J 2000 *Causality* (Cambridge University Press)

[11] Alon U 2007 *An introduction to Systems Biology* (Chapman & Hall/CRC)

[12] Pearl J 2012 The Do-Calculus Revisited In *Proceedings of the Twenty-Eighth Conference on Uncertainty in Artificial Intelligence (UAI2012) Preprint* arXiv:1210.4852

[13] Yedidia J S, Freeman W T and Weiss Y 2003 Understanding belief propagation and its generalizations In *Exploring artificial intelligence in the new millennium* **8** (Morgan Kaufmann Publishers Inc. San Francisco, CA, USA) pp. 236–239

[14] Mézard M and Montanari A 2009 *Information, physics, and computation* (Oxford University Press)

[15] Kschischang F R, Frey B J and Loeliger H-A 2001 Factor Graphs and the Sum-Product Algorithm *IEEE transactions on Information Theory* **47** 498–519

[16] Newton I, Letter to Leibniz on October 24 1676, *Opera omnia IV*, 540 ff, to be rendered as "Data æquatione quotcunque fluentes quantitae involvente fluxiones invenire et vice versa" [Given in an equation the fluents

of any number of quantities, to find the fluxions and vice versa]. Quoted by V.I. Arnold in *Geometrical Methods in the Theory of Ordinary Differential Equations*, Preface to First Edition, Springer (1983:1989) and there translated as "It is useful to solve differential equations".

[17] Granger C 1969 Investigating Causal Relations by Econometric Models and Cross-spectral Methods *Econometrica* **37** 424–438

[18] Marconi U M B, Puglisi A, Rondoni L and Vulpiani A 2008 Fluctuation-Dissipation: Response Theory in Statistical Physics *Physics Reports* **461** 111–195

[19] Newman M E J and Barkema G T 1999 *Monte Carlo Methods in Statistical Physics* (Clarendon Press)

[20] Robert D and Casella G 2004 *Monte Carlo Statistical Methods* (Springer-Verlag, New York)

[21] Kappen H J and Spanjers J J 2000 Mean field theory for asymmetric neural networks *Phys. Rev.* E **61** 5658

[22] Derrida B 2007 Non-equilibrium steady states: fluctuations and large deviations of the density and of the current *J. Stat. Mech.* P07023

[23] Bertini L, Sole A D, Gabrielli D, Jona-Lasinio G and Landim C 2015 Macroscopic fluctuation theory *Reviews of Modern Physics* **87** 593–636

[24] Mimura K and Coolen A C C 2009 Parallel dynamics of disordered Ising spin systems on finitely connected directed random graphs with arbitrary degree distributions *J. Phys. A: Math. Theor.* **42** 415001

[25] Kubo R 1966 The fluctuation-dissipation theorem *Reports on progress in physics*, **29** 255

[26] Villamaina D, Baldassarri A, Puglisi A and Vulpiani A 2009 The fluctuation-dissipation relation: how does one compare correlation functions and responses? *J. Stat. Mech.* P07024

[27] Raymond J and Ricci-Tersenghi F 2013 Correcting beliefs in the mean-field and Bethe approximations using linear response In *2013 IEEE International Conference on Communications Workshops (ICC)* pp. 1429–1433

[28] Bethe H A 1935 Statistical theory of superlattices *Proc. Roy. Soc. London* A **150** 552–558

[29] Glauber R J 1963 Time-Dependent Statistics of the Ising Model *J. Math. Phys.* **4** 294

[30] Derrida B, Gardner E and Zippelius A 1987 An exactly solvable asymmetric neural network model *Euro. Phys. Lett.* **4** 167

[31] Kanoria Y and Montanari A 2011 Majority dynamics on trees and the dynamic cavity method *The Annals of Applied Probability* **21** 1694–1748

[32] Lokhov A Y, Mézard M and Zdeborová L 2015 Dynamic message-passing equations for models with unidirectional dynamics *Phys. Rev.* E **91** 012811

[33] Altarelli F, Braunstein A, Dall'Asta L, Ingrosso A and Zecchina R 2014 The zero-patient problem with noisy observations *J. Stat. Mech.* P10016

[34] Neri I and Bollé D 2009 The cavity approach to parallel dynamics of Ising spins on a graph *J. Stat. Mech.* P08009

[35] Ferraro G D and Aurell E 2015 Dynamic message-passing approach for kinetic spin models with reversible dynamics *Phys. Rev.* E **92**, 010102

[36] Barthel T, Bacco C D and Franz S 2015 A matrix product algorithm for stochastic dynamics on locally tree-like graphs *Preprint* arXiv:1508.03295

[37] Ferraro G D and Aurell E 2014 Perturbative large deviation analysis of non-equilibrium dynamics *J. Phys. Soc. Japan* **83** 084001

NMR spectral analysis using prior knowledge

Takuma Kasai[1,2,5], Kenji Nagata[3], Masato Okada[3] and Takanori Kigawa[1,2,4,5]

[1] Laboratory for Biomolecular Structure and Dynamics, Cell Dynamics Research Core, RIKEN Quantitative Biology Center, Yokohama, Japan
[2] JST CREST, Yokohama, Japan
[3] Department of Complexity Science and Engineering, Graduate School of Frontier Sciences, The University of Tokyo, Kashiwa, Japan
[4] Department of Computational Intelligence and Systems Science, Interdisciplinary Graduate School of Science and Engineering, Tokyo Institute of Technology, Yokohama, Japan

E-mail: takuma.kasai@riken.jp, kigawa@riken.jp

Abstract. Signal assignment is a fundamental step for analyses of protein structure and dynamics with nuclear magnetic resonance (NMR). Main-chain signal assignment is achieved with a sequential assignment method and/or an amino-acid selective stable isotope labeling (AASIL) method. Combinatorial selective labeling (CSL) methods, as well as our labeling strategy, stable isotope encoding (SiCode), were developed to reduce the required number of labeled samples, since one of the drawbacks of AASIL is that many samples are needed. Signal overlapping in NMR spectra interferes with amino-acid determination by CSL and SiCode. Since spectral deconvolution by peak fitting with a gradient method cannot resolve closely overlapped signals, we developed a new method to perform both peak fitting and amino acid determination simultaneously, with a replica exchange Monte Carlo method, incorporating prior knowledge of stable-isotope labeling ratios and the amino-acid sequence of the protein.

1. Introduction

Nuclear magnetic resonance (NMR) is a widely used method for protein analysis. In NMR spectra, each observable atom in the protein gives a signal at its specific frequency, called a chemical shift, which is determined by the chemical environment of the atom. Moreover, the characteristics of the signals, such as appearances, intensities, and/or line widths, reflect the chemical bonds, distances, mobilities, and/or environments of the atoms, depending on the types of measurements. Therefore, by combining various NMR measurements, the three-dimensional structures and dynamics of the protein can be analyzed.

The determination of the chemical shifts of the atoms, which is called the signal assignment, is usually the first step of the analysis. Sequential assignment with triple resonance measurements is a commonly used technique for the assignment of main chain atoms, by searching neighboring amino acid residues [1]. Additionally or alternatively, amino-acid selective stable isotope labeling (AASIL) is used, especially for challenging targets such as large proteins [2], low-solubility proteins [3] or for protein analyses by in-cell NMR [4], because in such cases, the fast decay of NMR signals, the low

[5] To whom any correspondence should be addressed.

Journal of Physics: Conference Series **699** (2016) 012003 doi:10.1088/1742-6596/699/1/012003

signal-to-noise ratio and/or the signal overlapping may interfere with triple resonance measurements. Naturally abundant ^{12}C and ^{14}N are unobservable with NMR, while ^{13}C and ^{15}N are observable stable isotopes (SIs). For the simple AASIL method, each protein sample corresponds to one amino acid that is selectively ^{15}N labeled, while the others retain ^{14}N. Using these sets of protein samples, one can know the amino acid from which each signal is derived. Furthermore, in the dual selective approach, using a combination of two amino acids in which one is selectively ^{15}N labeled and the other is ^{13}C labeled, one can also know the amino acid of the preceding (N-terminal side) residue by a measurement using the amide nitrogen and the carbonyl carbon [5, 6]. This information greatly narrows down the assignment possibilities, and therefore is useful for main chain assignments. However, because the number of standard amino acids is 20, the simple AASIL method requires a large number of labeled protein samples, and thus generates arduous sample preparation workloads and consumes NMR machine time.

To reduce the number of labeled samples required in AASIL, various combinatorial selective labeling (CSL) methods were proposed [7-18]. In CSL, each amino acid is represented by a combination of multiple samples. We developed a new strategy, named "stable isotope encoding" (SiCode) [19], to further reduce the number of samples, by regarding AASIL as a "encoding-and-decoding" process. Specifically, the amino-acid information is encoded into the SI-labeling ratio pattern of the samples, and decoded from the signal intensity ratio of the NMR spectra. With this strategy, using ternary digits as codewords, 19 amino acids are represented by only 3 labeled samples [19]. However, for both CSL and SiCode, occasional signal overlapping may lead to misinterpretation of the amino acid information. Although the original SiCode procedure successfully decoded some overlapped signals [19], further improvement is needed.

In this paper, we report an improved decoding procedure for SiCode, which utilizes the prior knowledge of the SI labeling pattern and the amino acid sequence of the protein for signal deconvolution. Nagata et al. [20] reported that the Bayesian spectral deconvolution and model selection problems were simultaneously solved, by finding both the fitting parameters and free energy for model selection with the replica exchange Monte Carlo (REMC) method. This method has been applied to SiCode, by regarding the decoding process as a model selection problem.

2. Theory

2.1. *The SiCode decoding problem*

In the SiCode strategy, the SI labeling pattern is regarded as a codeword table and thus is predefined, as shown in table 1, for example. $\boldsymbol{c}(a) \in [0,1]^S$ and $\boldsymbol{n}(a) \in [n_{\min}, 1]^S$ are S-dimensional vectors that represent the ^{13}C and ^{15}N labeling ratios of amino acid a, respectively, where S is the number of labeled samples, and $n_{\min}$ is a minimum ^{15}N labeling ratio and is set to be greater than zero to avoid loss of information for the ^{13}C labeling ratio, as discussed later. For example, in the labeling pattern shown in table 1, $n_{\min}$ is 0.5. To estimate the signal amplitude, at least one sample is 100% labeled, therefore

$$\forall a, \max \boldsymbol{n}(a) = 1, \max \boldsymbol{c}(a) = 1. \tag{1}$$

^{1}H-^{15}N HSQC (hereafter "HSQC") is a fundamental two-dimensional NMR spectrum. The intensities of the HSQC spectrum are assumed to be

$$\boldsymbol{I}_{\text{HSQC}}(x,y) = \sum_i A_{\text{HSQC}}^i \, \boldsymbol{n}(a_i) \exp\left(-\frac{\left(x - x_0^i\right)^2}{2{\sigma_x^i}^2} - \frac{\left(y - y_0^i\right)^2}{2{\sigma_y^i}^2} \right) \tag{2}$$

where $\boldsymbol{I}_{\text{HSQC}}(x,y)$ is an S-dimensional vector that represents the intensities of HSQC, x and y are the chemical shifts of the ^{1}H and ^{15}N axes, respectively, A_{HSQC}^i is the amplitude of the HSQC signal of residue i, a_i is the amino acid of residue i, x_0^i and y_0^i are the ^{1}H and ^{15}N chemical shifts of the center of the signal of residue i, respectively, and σ_x^i and σ_y^i are the ^{1}H and ^{15}N line widths of the signal of residue i, respectively. Note that A^i_{HSQC} is 0 if a_i is proline, because of the absence of the amide hydrogen.

The two-dimensional version of HN(CO) (hereafter "HNCO") is another spectrum used in the SiCode strategy. As in HSQC, the two axes of HNCO correspond to the amide hydrogen and nitrogen. Therefore, their signals should appear at the same chemical shifts as in HSQC, while their intensities are affected by both the ^{15}N labeling ratio of the amide nitrogen of residue i and ^{13}C labeling ratio of the carbonyl carbon of residue $i-1$, as follows:

$$\boldsymbol{I}_{\mathrm{HNCO}}(x,y)=\sum_i A^i_{\mathrm{HNCO}}\ \boldsymbol{n}(a_i)\circ \boldsymbol{c}(a_{i-1})\exp\left(-\frac{\left(x-x_0^i\right)^2}{2{\sigma_x^i}^2}-\frac{\left(y-y_0^i\right)^2}{2{\sigma_y^i}^2}\right) \quad (3)$$

where $\boldsymbol{I}_{\mathrm{HNCO}}(x,y)$ is an S-dimensional vector that represents the intensities of HNCO, A^i_{HNCO} is the amplitude of the HNCO signal of residue i, and $\circ$ denotes element-wise multiplication. As in HSQC, A^i_{HNCO} is 0 if a_i is proline. Note that if the ^{15}N labeling ratio of one of the samples is a very small value, then we cannot obtain the ^{13}C labeling ratio information for the sample, due to the weak HNCO signal. To avoid this situation, each element of the $\boldsymbol{n}(a)$ value is set to $n_{\min}$ or larger. The decoding problem of SiCode is to determine a_i and a_{i-1} for each signal appearing in the spectra.

Table 1. SI labeling patterns used in this study

amino acid	sample 1		sample 2		sample 3		corresponding codeword
	^{13}C	^{15}N	^{13}C	^{15}N	^{13}C	^{15}N	
G	100%	100%	100%	100%	100%	100%	222
F	100%	100%	100%	100%	50%	75%	221
N	100%	100%	100%	100%	0%	50%	220
L	100%	100%	50%	75%	100%	100%	212
S	100%	100%	50%	75%	50%	75%	211
D	100%	100%	50%	75%	0%	50%	210
M	100%	100%	0%	50%	100%	100%	202
K	100%	100%	0%	50%	50%	75%	201
R	100%	100%	0%	50%	0%	50%	200
A	50%	75%	100%	100%	100%	100%	122
I	50%	75%	100%	100%	50%	75%	121
C	50%	75%	100%	100%	0%	50%	120
V	50%	75%	50%	75%	100%	100%	112
Y	50%	75%	0%	50%	100%	100%	102
Q	0%	50%	100%	100%	100%	100%	022
E	0%	50%	100%	100%	50%	75%	021
H	0%	50%	100%	100%	0%	50%	020
T	0%	50%	50%	75%	100%	100%	012
W	0%	50%	0%	50%	100%	100%	002
P	0%	0%	0%	0%	0%	0%	

2.2. The original version of the decoding procedure (sequential decoding)

In the original version of SiCode, the decoding procedure consisted of three sequential steps: peak fitting, calculation of SI-labeling ratios, and determination of amino acids. Hereafter, we refer to this as "sequential decoding". Since it is difficult to analyze the full region of the spectra at the same time, the spectra are divided into small regions. The least square fitting to the following function is then performed for each region:

$$\begin{pmatrix} \boldsymbol{I}_{\mathrm{HSQC}}(x,y) \\ \boldsymbol{I}_{\mathrm{HNCO}}(x,y) \end{pmatrix} = \sum_{k=1}^{K} \begin{pmatrix} \boldsymbol{B}^k_{\mathrm{HSQC}} \\ \boldsymbol{B}^k_{\mathrm{HNCO}} \end{pmatrix} \exp\left(-\frac{\left(x - x_0^k\right)^2}{2{\sigma_x^k}^2} - \frac{\left(y - y_0^k\right)^2}{2{\sigma_y^k}^2} \right) \tag{4}$$

where K is the total number of signals in the region and is assumed to be given, and $\boldsymbol{B}^k_{\mathrm{HSQC}}$ and $\boldsymbol{B}^k_{\mathrm{HNCO}}$ are the HSQC and HNCO intensities of the k-th signal, respectively. $\boldsymbol{B}^k_{\mathrm{HSQC}}$, $\boldsymbol{B}^k_{\mathrm{HNCO}}$, x_0^k, y_0^k, σ_x^k, and σ_y^k are fitting parameters. As described, at least one sample is 100% labeled in the SiCode labeling pattern for each amino acid. The amplitude of each signal is then estimated by

$$\hat{A}^k_{\mathrm{HSQC}} = \max\left(\boldsymbol{B}^k_{\mathrm{HSQC}}\right) \tag{5}$$

where $\hat{A}^k_{\mathrm{HSQC}}$ is the estimated HSQC amplitude of the k-th signal, assuming that all signals in the region are positive. The ^{15}N labeling ratios are estimated by

$$\widehat{\boldsymbol{n}}_k = \boldsymbol{B}^k_{\mathrm{HSQC}} / \hat{A}^k_{\mathrm{HSQC}} \tag{6}$$

where $\widehat{\boldsymbol{n}}_k$ is are the back-calculated SI-labeling ratios (hereafter named "SI-indices") of ^{15}N. Since HNCO intensities are proportional to not only the ^{13}C labeling ratios of residue $i-1$ but also the ^{15}N labeling ratios of residue i, they should be adjusted as follows:

$$\boldsymbol{B'}^k_{\mathrm{HNCO}} = \boldsymbol{B}^k_{\mathrm{HNCO}} \oslash \widehat{\boldsymbol{n}}_k \tag{7}$$

where $\boldsymbol{B'}^k_{\mathrm{HNCO}}$ is the adjusted HNCO intensitiy of the k-th signal, and $\oslash$ denotes element-wise division. The HNCO amplitude and the ^{13}C labeling ratios are similarly estimated by

$$\widehat{A'}^k_{\mathrm{HNCO}} = \max\left(\boldsymbol{B'}^k_{\mathrm{HNCO}}\right) \tag{8}$$

$$\widehat{\boldsymbol{c}}_k = \boldsymbol{B'}^k_{\mathrm{HNCO}} / \widehat{A'}^k_{\mathrm{HNCO}} \tag{9}$$

where $\hat{A'}^k_{\mathrm{HNCO}}$ is the estimated adjusted HNCO amplitude of the k-th signal, and $\widehat{\boldsymbol{c}}_k$ is the ^{13}C SI-indices. SI indices can be converted to amino acids, using the given labeling pattern:

$$a_i^k = \operatorname*{argmin}_a d(\widehat{\boldsymbol{n}}_k, \boldsymbol{n}(a)) \tag{10}$$

$$a_{i-1}^k = \operatorname*{argmin}_a d(\widehat{\boldsymbol{c}}_k, \boldsymbol{c}(a)) \tag{11}$$

where a_i^k is the amino acid of the residue from which the k-th signal is derived, a_{i-1}^k is the amino acid of its preceeding (N-terminal side) residue, and d denotes the Euclidean distance. Since the codeword table shown in table 1 uses three ternary digits as a codeword (0 for 0%, 1 for 50%, 2 for 100% labeling of ^{13}C; 0 for 50%, 1 for 75%, 2 for 100% labeling of ^{15}N) and all of the 19 possible codewords are occupied, the simple conversion of the SI indices to the nearest ternary digits (i.e., for $\widehat{\boldsymbol{c}}_k$, -0.25 to 0.25 is 0, 0.25 to 0.75 is 1, 0.75 to 1.25 is 2; for $\widehat{\boldsymbol{n}}_k$, 0.375 to 0.625 is 0, 0.625 to 0.875 is 1, 0.875 to 1.125 is 2) provides the same result as equations (10) and (11). As the proline residue is not SI-labeled, a small $\widehat{A'}^k_{\mathrm{HNCO}}$ value suggests that a_{i-1}^k is a proline.

This sequential decoding procedure has two problems. Firstly, it is difficult to find the global optimum of the peak fitting (equation (4)). We used the Levenberg-Marquardt method with the initial parameters determined manually [19]; however, the failure to set appropriate initial parameters leads to trapping in a local optimum. Secondly, since we perform the peak fitting (equation (4)) first and then interpret the resulting amplitude using the codeword table (equations (5) to (11)), peak fitting may fail in cases with close overlapping (as discussed in the Results section).

2.3. Improved decoding procedure with REMC

For a given amino acid set $M = (a_i^k, a_{i-1}^k;\ k = 1 \ldots K)$ and peak parameter set $\Theta = \left(A^k_{\mathrm{HSQC}}, A^k_{\mathrm{HNCO}}, x_0^k, y_0^k, \sigma_x^k, \sigma_y^k;\ k = 1 \ldots K\right)$, the residual error of each spectrum is calculated as:

$$\boldsymbol{R}_{\mathrm{HSQC}}(x,y;M,\Theta) = \boldsymbol{I}_{\mathrm{HSQC}}(x,y) - \sum_{k=1}^{K} A^k_{\mathrm{HSQC}}\, \boldsymbol{n}(a_i^k) \exp\left(-\frac{\left(x - x_0^k\right)^2}{2{\sigma_x^k}^2} - \frac{\left(y - y_0^k\right)^2}{2{\sigma_y^k}^2} \right) \tag{12}$$

$$\boldsymbol{R}_{\mathrm{HNCO}}(x,y;M,\Theta) = \boldsymbol{I}_{\mathrm{HNCO}}(x,y) - \sum_{k=1}^{K} A_{\mathrm{HNCO}}^{k}\,\boldsymbol{n}(a_i^k)\circ \boldsymbol{c}(a_{i-1}^k)\exp\left(-\frac{\left(x-x_0^k\right)^2}{2{\sigma_x^k}^2}-\frac{\left(y-y_0^k\right)^2}{2{\sigma_y^k}^2}\right) \tag{13}$$

where $\boldsymbol{R}_{\mathrm{HSQC}}$ and $\boldsymbol{R}_{\mathrm{HNCO}}$ are the residual errors of HSQC and HNCO, respectively, and A_{HSQC}^{k} and A_{HNCO}^{k} are the HSQC and HNCO amplitudes of the k-th signal, respectively. Assuming that the spectral noise is Gaussian distributed and its magnitude in each spectrum of each sample is known, the sum of the normalized squared residual error $E(M,\Theta)$ is defined as:

$$E(M,\Theta) = \frac{1}{2}\sum_{x,y}\left(\left|\boldsymbol{R}_{\mathrm{HSQC}}(x,y;M,\Theta)\oslash\boldsymbol{\sigma}_{\mathrm{HSQC}}\right|^2 + \left|\boldsymbol{R}_{\mathrm{HNCO}}(x,y;M,\Theta)\oslash\boldsymbol{\sigma}_{\mathrm{HNCO}}\right|^2\right) \tag{14}$$

where $\boldsymbol{\sigma}_{\mathrm{HSQC}}$ and $\boldsymbol{\sigma}_{\mathrm{HNCO}}$ are the standard deviations of the HSQC and HNCO noises, respectively. For Bayesian spectral deconvolution and model selection, regarding the amino acid set M as a model, the marginal likelihood, which is the posterior probability of the observed spectra with the given amino acid set M, is calculated as [20]:

$$L(M) = \int \exp\left(-E(M,\Theta)\right)\varphi(\Theta)\mathrm{d}\Theta \tag{15}$$

where $\varphi(\Theta)$ is the prior probability density of Θ, which in this report we assume has a uniform distribution. Since it is difficult to calculate $L(M)$ analytically, we numerically calculate this value with the REMC method, as described [20]. The amino-acid set M should be selected according to the value $-\log L(M)$, which is called the free energy [20]. For every Markov chain Monte Carlo (MCMC) step, a parameter set Θ is sampled and the corresponding $E(M,\Theta)$ is calculated. From these sampling results, we can find the best fit parameter Θ with the smallest $E(M,\Theta)$ value, the posterior distribution of Θ with the low temperature replica, and $L(M)$ from all replicas, as described [20]. The use of REMC avoids trapping in a local optimum.

If the amino acid sequence of the protein is not known, then the number of possible M is large; namely, $(19\times 20)^K$, since a_i^k is one of all 19 non-proline amino acids and a_{i-1}^k is one of all 20 amino acids. Excluding an impossible combination of amino acids, using the prior knowledge of the amino acid sequence, may improve the decoding correctness in challenging cases. To achieve this, M should comply with

$$\forall p, C(p,M) \leq T(p) \tag{16}$$

where $p = (a_{i-1}, a_i)$ is an adjacent pair of amino acids, $C(p,M)$ is the number of amino-acid pairs p found in M, and $T(p)$ is the number of amino-acid pairs p found in the amino-acid sequence of the protein, assuming that one residue gives no more than one signal in a spectrum. Hereafter, we call this procedure "model-selection decoding".

As discussed, we can obtain $L(M)$ for all possible M by running REMC for each M; however, a combinatorial explosion of the number of possible M occurs as K increases. To find the best fit M and Θ or the posterior distribution of M and Θ, we can alternatively run REMC once, by sampling both the discrete variables M and the continuous variables Θ as parameters. In this report, we assume the uniform distribution of M. Each amino acid pair of M, as well as each parameter of Θ, is updated in MCMC step. This REMC run gives us the best fit M or posterior distribution of M, which can be used as a decoding result for SiCode. Hereafter, we call this procedure "model-optimization decoding".

3. Materials and methods

3.1. Protein preparation and NMR measurement

Three selectively labeled samples of the SH2 domain (residues 291 to 393) of the human BMX protein (UniProt: P51813), with an N-terminal 7-residue cloning artifact (GSSGSSG), were prepared according to the SI-labeling pattern shown in table 1, using an *Escherichia coli* based cell-free protein synthesis system [21-25] with metabolic inhibitors to suppress SI-labeling scrambles [26], as described [19]. To evaluate the decoding procedures in various signal-to-noise ratios (SNRs), NMR spectra were acquired under three conditions: at a 0.35 mM concentration with 8 scans for HSQC and 32 scans for HNCO (hereafter, high-SNR conditions), at a 0.05 mM concentration with 64 scans for HSQC and 256 scans for HNCO (hereafter, medium-SNR conditions), and at a 0.05 mM concentration with 16 scans for HSQC and 64 scans for HNCO (hereafter, low-SNR conditions), as described [19].

Since it is important for SiCode to compensate for variances in concentrations among samples and/or other reasons that affect intensities, the spectra should be normalized using glycine residues [19]. In this report, prior to the REMC analyses, the spectra acquired for samples 1, 2, and 3 were divided by the values of 0.937, 0.911, and 1.000, respectively, in the high-SNR conditions, and 0.997, 0.896, and 0.946, respectively, in the medium- and low-SNR conditions.

3.2. Model-selection decoding

All possible amino-acid combinations M were generated from the amino-acid sequence of BMX SH2, according to equation (16). For each M, REMC was performed essentially as described [20]. The standard deviations of the spectral noise were evaluated using the region without signals shown in figure 1, for each spectrum and each sample. The spectral regions used for analysis were manually determined, as shown in figure 1. The upper limits of A^k_{HSQC} and A^k_{HNCO} were set to 1.2-fold of the maximum of each spectrum in the region, while their lower limits were set to 3-fold of the average of $\boldsymbol{\sigma}_{\mathrm{HSQC}}$ and $\boldsymbol{\sigma}_{\mathrm{HNCO}}$, respectively. The upper and lower limits of x^k_0 and y^k_0 were set to be the ends of the spectral region. The upper and lower limits of σ^k_x were 0.1 ppm and 0.001 ppm, respectively. The upper and lower limits of σ^k_y were 0.75 ppm and 0.01 ppm, respectively. The number of temperatures L was 96, and the inverse temperature β^l was

$$\beta^l = \begin{cases} 0, & (\text{if } l = 1) \\ r^{l-L}, & (\text{otherwise}) \end{cases} \qquad (17)$$

where $r = 1.15$, $l = 1 \ldots L$. The MCMC steps were set to 5,000 for burn-in and 1,000 for sampling. In each MCMC step, the sequential updates of the parameters A^1_{HSQC}, A^1_{HNCO}, ..., A^K_{HSQC}, A^K_{HNCO}, x^1_0, y^1_0, ..., x^K_0, y^K_0, σ^1_x, σ^1_y, ..., σ^K_x, σ^K_y were performed with the Metropolis algorithm, followed by replica exchange between two adjacent temperatures, as described [20].

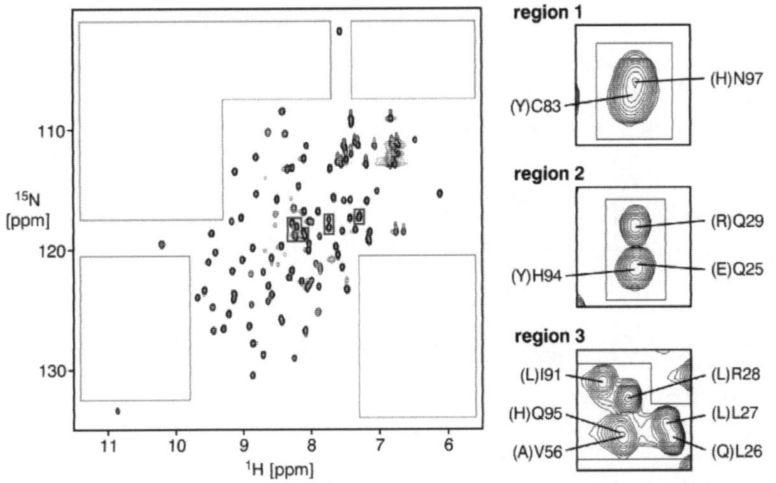

Figure 1. HSQC spectrum of sample 1 of BMX SH2 protein under the high-SNR conditions. Black and red contours represent positive and negative signals, respectively. Regions used for estimation of background noise are shown by red boxes. Regions used in this study for deconvolution of overlapped signals are shown by blue boxes, which are expanded in the right panel. Each peak label consists of a^k_{i-1} (in parentheses), a^k_i, and residue number i.

3.3. Model-optimization decoding

Simultaneous optimizations of both M and Θ were performed essentially as described in the above model-selection decoding, except for the following. In each MCMC step, the amino acid combinations $p_1, \ldots, p_K$ were updated in the sequence before the update of Θ, where the amino-acid pair $p_k = (a_{i-1}^k, a_i^k)$. If M violated equation (16) as a result of the update, this update was rejected, regardless of the change in the residual error. The number of temperatures L was 192 and $r = 1.1$. The number of MCMC steps were set to be 50,000 or 200,000 for burn-in, in the case of $K < 6$ or $K \geq 6$, respectively, and 10,000 for sampling.

4. Results

4.1. Sequential decoding

In this report, we evaluated the decoding procedures with overlapped signals appearing in the spectra. As shown in figure 1, region 1 contains two signals, (Y)C83 and (H)N97, with overlapping that makes the decoding difficult. However, even the sequential decoding successfully decodes these two signals under the high-SNR conditions (figure 2). Decoding the signals in region 2, which contains three signals, (E)Q25, (R)Q29, and (Y)H94, is a more difficult problem because two of them are very closely overlapped (figure 1). The sequential decoding failed to find the correct answer in the analysis of this region under the low-SNR conditions (figure 3).

As discussed in the Theory section, we first perform the peak fitting in the sequential decoding. The resulting amplitude of each signal is then converted to the amino acid information, using the predefined SI-labeling pattern. Therefore, a failure in the peak fitting step leads to the misinterpretation of the amino acid information. As shown in table 1, the SI-labeling ratio of ^{15}N is either 50%, 75%, or 100%; however, SI indices of ^{15}N that differ from these values were obtained; for example, 33.0% and 62.8% in the analysis of region 2 (figure 3). These values clearly indicate the failure of the peak fitting, as the signal deconvolution of closely overlapped signals is difficult. Moreover, one of the decoding results of three signals in region 2 (figure 3), (Y)I, is an amino acid pair that does not appear in the amino-acid sequence of BMX SH2, also indicating failure. Hence, using prior knowledge of the SI-labeling pattern and the amino-acid sequence in the peak fitting step may prevent such failures.

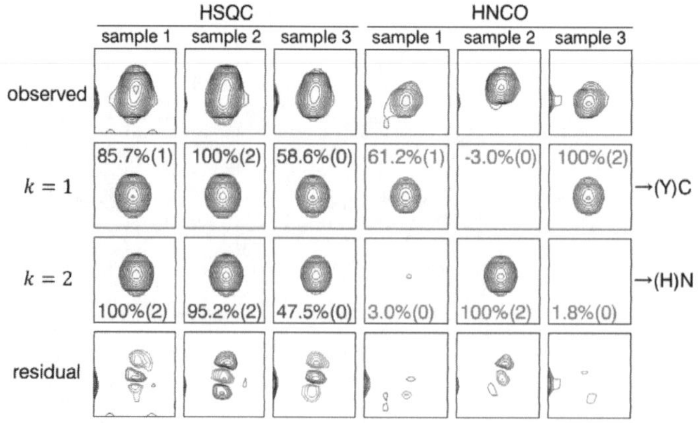

Figure 2. Sequential decoding of region 1 under high-SNR conditions. Observed spectrum ("observed"), deconvoluted signals (k = 1 and 2), and residual error ("residual") are shown. SI indices of ^{15}N ($\hat{\boldsymbol{n}}_k$) and ^{13}C ($\hat{\boldsymbol{c}}_k$) are shown with blue and red numbers, respectively. Each corresponding ternary digit is shown in parentheses immediately following the SI index. Decoding results are shown on the right of the spectra.

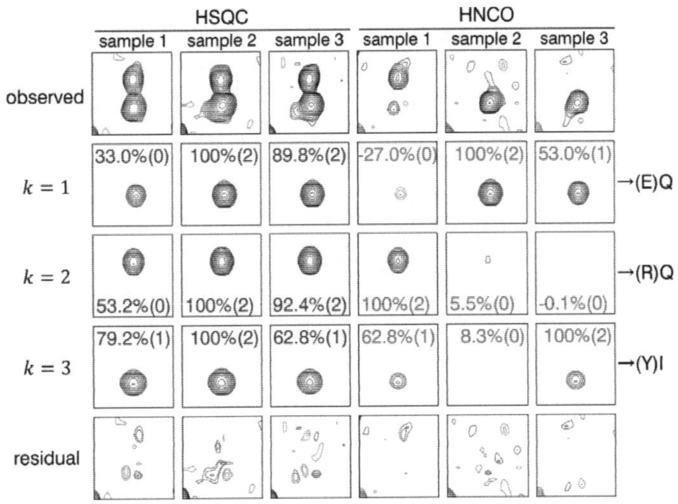

Figure 3. Sequential decoding of region 2 under low-SNR conditions. Spectra are shown as in figure 2.

4.2. Model-selection decoding

Using prior knowledge of the amino acid sequence is achieved by regarding decoding as a model selection problem. All of the possible models M; i.e., combinations of amino acid pairs, can be listed according to equation (16), provided the total signal number K is given. The sampling results of REMC for each model allow us to calculate the free energy, which shows the probability of each model. For each REMC run, the amino acids a_i^k and a_{i-1}^k are given, and as a consequence the SI-labeling ratios $\boldsymbol{n}(a_i^k)$ and $\boldsymbol{c}(a_{i-1}^k)$ are also given. Therefore, the model-selection decoding uses the SI-labeling pattern as prior knowledge for the peak fitting.

We applied this method to analyze region 1, under high-SNR conditions. The number of possible models is 90 or 4,021, for $K = 1$ or $K = 2$, respectively. We assumed that K is either 1 or 2; therefore, 4,111 models in total were tested. Numerical calculations of free energies by REMC for all of the models revealed that the model (H)N+(Y)C had the smallest free energy (figure 4), and was the correct model.

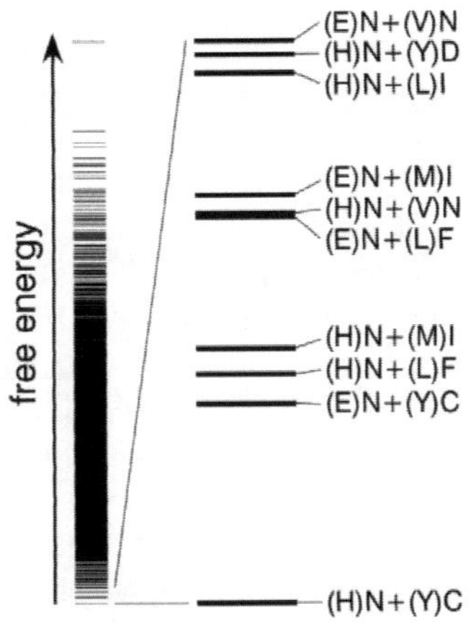

Figure 4. Free energies of each model (combination of amino acid pair). Signals in region 1 were analyzed under high-SNR conditions. The calculated free energy of each model is shown with a black line. All 4,111 models are shown on the left, while 10 models with the lowest free energies are expanded on the right.

4.3. Model-optimization decoding

In the previous section, the model M is given in each REMC run. In case of BMX SH2, the number of possible models is 90, 4,021, or 118,905, for $K = 1, 2$, or 3, respectively. In general, the number of possible models increases exponentially with K. Since model-selection decoding requires REMC runs for all of possible models, it is not feasible in case of $K \geq 3$. Alternatively, the model M and fitting parameters Θ can both be optimized in a single REMC run, to reduce the computation time. This method allows us to obtain the posterior probability distributions of M and Θ, as well as the best fitting parameters M and Θ.

We applied this method to decode the signals in region 2 under the low-SNR conditions, for which the sequential decoding failed to obtain the correct answer (figure 2). As shown in figure 5, the model-optimization decoding successfully obtained the correct answer.

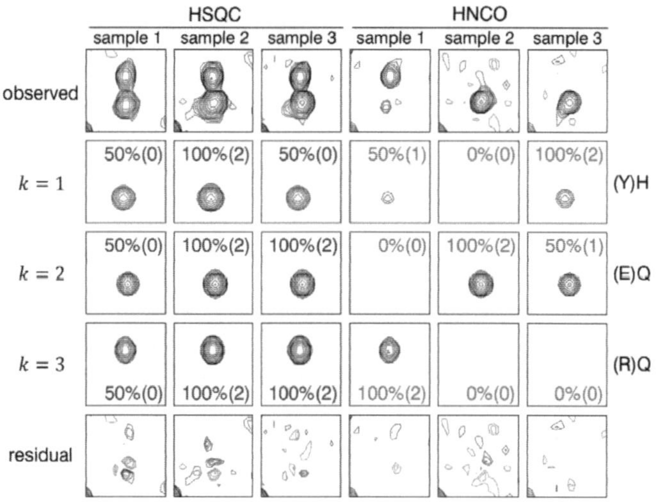

Figure 5. Model-optimization decoding of region 2 under low-SNR conditions. Observed spectrum ("observed"), generated spectra with the best fit parameters M and Θ for each signal ($k = 1$ to 3), and residual error ("residual") are shown. The amino acid pair of each signal is shown on the right, and the corresponding ^{15}N and ^{13}C SI labeling ratios are shown with blue and red numbers, respectively. Each corresponding ternary digit is shown in parentheses immediately following the SI labeling ratio.

The frequency of the appearance of M at the lowest temperature replica corresponds to the estimated posterior probability distribution of M. By model-optimization decoding of the signals in region 1 under the low-SNR conditions, only one model, (H)N+Y(C), was sampled, indicating that the posterior probabilities of the other models were too low to be detected. In contrast, by the same analysis without the prior knowledge of the amino acid sequence; i.e., without limiting models M by equation (16), four models, (H)N+(Y)C, (H)N+(Y)F, (H)N+(Y)N, and (H)N+(M)F, were sampled. Their sampled frequencies; namely, their estimated posterior probabilities, were 51.0%, 24.9%, 24.1%, and 0.02%, respectively. As amino acid pairs (Y)F, (Y)N, and (M)F do not appear in the amino-acid sequence, the models with these pairs are eliminated in the analysis with equation (16). These results show that the prior knowledge of not only the SI-labeling pattern but also the amino-acid sequence certainly facilitates correct decoding.

For the above analyses, the number of signals K is assumed to be given. In the real situation, K is not known, especially in closely overlapped cases. We analyzed region 3 (figure 1), which contains 6 signals, under the medium-SNR conditions with $K = 5, 6, \text{or } 7$ (figure 6). The amino acids were correctly decoded in the $K = 6$ case. However, in the case of $K = 5$, the closely overlapped (A)V56 and (H)Q95 were considered to be a single broad signal (I)T (k = 4), and in the case of K = 7, an additional very broad signal (k = 4) appeared, while the other 6 signals revealed the correct answers. The appearance of the broad signal is due to the wider range of the line width parameter than its actual range in our REMC settings, and thus may be suppressed by narrowing the range. However, in some cases, the appearance of a broad signal may help us to become aware that K is incorrect.

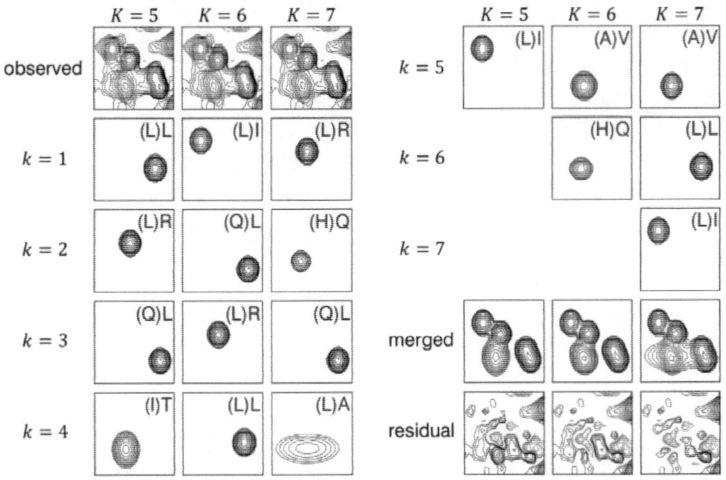

Figure 6. Model-optimization decoding of region 3 under the medium SNR-conditions with various given numbers of signals. HSQC spectra of sample 1 of observed spectra ("observed"), generated spectra with the best fit parameters M and Θ for each signal (k = 1 to 7), sum of the generated spectra ("merged"), and residual error ("residual") are shown. The amino acid pair of each signal is shown with blue letters.

5. Discussion

In this report, we showed that spectral deconvolution with prior knowledge, such as the SI-labeling ratios and the amino acid sequence, improves the decoding reliability in the case of signal overlapping, which interferes with obtaining amino-acid information in SiCode and other CSLs. The previously reported [19] sequential decoding was a kind of heuristics, in which the peak fitting result was assumed to be correct in the following steps. Such heuristics worked well in relatively easier cases. However, in cases with a low SNR and/or close overlapping, the intensities of each signal may contain errors due to inaccurate signal deconvolution. In such cases, using prior knowledge in the early stage of analysis improves the result.

We previously proposed different SI-labeling patterns optimized for various numbers of samples and amino acids [19]. Some of them are more complicated than that shown in table 1, and use more than three SI-labeling levels. The proposed methods; namely, the model-selection decoding and the model-optimization decoding, can also deal with such patterns. Moreover, the NMR spectra of other CSLs can be analyzed with the proposed methods, to improve the tolerance to signal overlapping.

The proposed methods are based on assumptions, such as the correctness of the SI-labeling ratio of the prepared protein samples, the same chemical shifts and line widths among spectra and samples, the two-dimensional Gaussian line shapes, the Gaussian background noise, the absence of minor conformers, and the prior uniform distribution of parameters. Some of these assumptions are actually incorrect to some extent, and result in residual errors of peak fitting. Careful modifications of these assumptions, according to the real phenomena, may further improve the correctness and reliability of the methods. However, as discussed above, even with these assumptions, the proposed methods certainly improved the decoding accuracy.

As shown in equation (16), the prior knowledge of the amino-acid sequence is used as the upper limits of the frequency of amino acid pairs. Using it as the lower limits may further improve the decoding, if the number of signals derived from minor conformers is limited. However, to use it as the lower limits, the full region of the spectra must be analyzed simultaneously. The computation times in this study were 3 min 26 sec for region 1 ($K = 2$), 3 min 15 sec for region 2 ($K = 3$), and 62 min 12 sec for region 3 ($K = 6$), with a computer equipped with an Intel Core i7 4930K CPU (6 cores, 3.4 GHz). The computation time depends largely on the number of signals (K), the number of data points in the region, and the number of MCMC steps. Although the number of MCMC steps in this study is excessive, and thus can be reduced, applying the method to the full region of the spectra ($K > 100$) is not feasible so far, and therefore further improvement is expected.

For small proteins, most of signals are not overlapped, and thus they can be analyzed with the previous sequential decoding. However, close signal overlapping is often observed in large proteins, helical proteins, and/or intrinsically disordered proteins. In addition, a low SNR, caused by low solubility, high molecular weight of the protein and/or challenging situations, such as in-cell NMR, makes the deconvolution difficult. Failures in the signal deconvolution and the amino acid determination lead to mistakes in the assignment, which impedes the NMR analysis of the structures and dynamics of the protein. Spectral analyses using prior knowledge with the help of computation may promote NMR research of difficult proteins in challenging situations.

Acknowledgements

We thank the lab members at RIKEN for their help in sample preparation and data acquisition. We also thank Professors Toshiyuki Tanaka, Shiro Ikeda, and Koji Hukushima for valuable discussions. This work was supported in part by Grants-in-Aid for Scientific Research on Innovative Areas (Grant Nos. 25120003 and 25120009) and a Grant-in-Aid for Challenging Exploratory Research (Grant No. 26650027) from the Ministry of Education, Culture, Sports, Science and Technology (MEXT) of Japan and the Japan Society for the Promotion of Science (JSPS).

References

[1] Grzesiek S and Bax A 1993 Amino acid type determination in the sequential assignment procedure of uniformly ^{13}C/^{15}N-enriched proteins *J. Biomol. NMR* **3** 185-204

[2] Bertelsen E B, Chang L, Gestwicki J E and Zuiderweg E R 2009 Solution conformation of wild-type *E. coli* Hsp70 (DnaK) chaperone complexed with ADP and substrate *Proc. Natl. Acad. Sci. U. S. A.* **106** 8471-6

[3] Cervantes C F, Handley L D, Sue S C, Dyson H J and Komives E A 2013 Long-range effects and functional consequences of stabilizing mutations in the ankyrin repeat domain of IκBα *J. Mol. Biol.* **425** 902-13

[4] Hembram D S, Haremaki T, Hamatsu J, Inoue J, Kamoshida H, Ikeya T, Mishima M, Mikawa T, Hayashi N, Shirakawa M and Ito Y 2013 An in-cell NMR study of monitoring stress-induced increase of cytosolic Ca^{2+} concentration in HeLa cells *Biochem. Biophys. Res. Commun.* **438** 653-9

[5] Kainosho M and Tsuji T 1982 Assignment of the three methionyl carbonyl carbon resonances in *Streptomyces* subtilisin inhibitor by a carbon-13 and nitrogen-15 double-labeling technique. A new strategy for structural studies of proteins in solution *Biochemistry* **21** 6273-9

[6] Yabuki T, Kigawa T, Dohmae N, Takio K, Terada T, Ito Y, Laue E D, Cooper J A, Kainosho M and Yokoyama S 1998 Dual amino acid-selective and site-directed stable-isotope labeling of the human c-Ha-Ras protein by cell-free synthesis *J. Biomol. NMR* **11** 295-306

[7] Hefke F, Bagaria A, Reckel S, Ullrich S J, Dötsch V, Glaubitz C and Güntert P 2011 Optimization of amino acid type-specific ^{13}C and ^{15}N labeling for the backbone assignment of membrane proteins by solution- and solid-state NMR with the UPLABEL algorithm *J. Biomol. NMR* **49** 75-84

[8] Jaipuria G, Krishnarjuna B, Mondal S, Dubey A and Atreya H S 2012 Amino acid selective

labeling and unlabeling for protein resonance assignments *Adv. Exp. Med. Biol.* **992** 95-118

[9] Krishnarjuna B, Jaipuria G, Thakur A, D'Silva P and Atreya H S 2011 Amino acid selective unlabeling for sequence specific resonance assignments in proteins *J. Biomol. NMR* **49** 39-51

[10] Löhr F, Reckel S, Karbyshev M, Connolly P J, Abdul-Manan N, Bernhard F, Moore J M and Dötsch V 2012 Combinatorial triple-selective labeling as a tool to assist membrane protein backbone resonance assignment *J. Biomol. NMR* **52** 197-210

[11] Maslennikov I and Choe S 2013 Advances in NMR structures of integral membrane proteins *Curr. Opin. Struct. Biol.* **23** 555-62

[12] Maslennikov I, Klammt C, Hwang E, Kefala G, Okamura M, Esquivies L, Mörs K, Glaubitz C, Kwiatkowski W, Jeon Y H and Choe S 2010 Membrane domain structures of three classes of histidine kinase receptors by cell-free expression and rapid NMR analysis *Proc. Natl. Acad. Sci. U. S. A.* **107** 10902-7

[13] Parker M J, Aulton-Jones M, Hounslow A M and Craven C J 2004 A combinatorial selective labeling method for the assignment of backbone amide NMR resonances *J. Am. Chem. Soc.* **126** 5020-1

[14] Shi J, Pelton J G, Cho H S and Wemmer D E 2004 Protein signal assignments using specific labeling and cell-free synthesis *J. Biomol. NMR* **28** 235-47

[15] Sobhanifar S, Reckel S, Junge F, Schwarz D, Kai L, Karbyshev M, Löhr F, Bernhard F and Dötsch V 2010 Cell-free expression and stable isotope labelling strategies for membrane proteins *J. Biomol. NMR* **46** 33-43

[16] Staunton D, Schlinkert R, Zanetti G, Colebrook S A and Campbell I D 2006 Cell-free expression and selective isotope labelling in protein NMR *Magn. Reson. Chem.* **44** S2-9

[17] Trbovic N, Klammt C, Koglin A, Löhr F, Bernhard F and Dötsch V 2005 Efficient strategy for the rapid backbone assignment of membrane proteins *J. Am. Chem. Soc.* **127** 13504-5

[18] Wu P S, Ozawa K, Jergic S, Su X C, Dixon N E and Otting G 2006 Amino-acid type identification in ^{15}N-HSQC spectra by combinatorial selective ^{15}N-labelling *J. Biomol. NMR* **34** 13-21

[19] Kasai T, Koshiba S, Yokoyama J and Kigawa T 2015 Stable isotope labeling strategy based on coding theory *J. Biomol. NMR* **63** 213-21

[20] Nagata K, Sugita S and Okada M 2012 Bayesian spectral deconvolution with the exchange Monte Carlo method *Neural Netw* **28** 82-9

[21] Kigawa T, Yabuki T, Matsuda N, Matsuda T, Nakajima R, Tanaka A and Yokoyama S 2004 Preparation of *Escherichia coli* cell extract for highly productive cell-free protein expression *J. Struct. Funct. Genomics* **5** 63-8

[22] Kigawa T, Yabuki T, Yoshida Y, Tsutsui M, Ito Y, Shibata T and Yokoyama S 1999 Cell-free production and stable-isotope labeling of milligram quantities of proteins *FEBS Lett.* **442** 15-9

[23] Matsuda T, Koshiba S, Tochio N, Seki E, Iwasaki N, Yabuki T, Inoue M, Yokoyama S and Kigawa T 2007 Improving cell-free protein synthesis for stable-isotope labeling *J. Biomol. NMR* **37** 225-9

[24] Seki E, Matsuda N, Yokoyama S and Kigawa T 2008 Cell-free protein synthesis system from *Escherichia coli* cells cultured at decreased temperatures improves productivity by decreasing DNA template degradation *Anal. Biochem.* **377** 156-61

[25] Yabuki T, Motoda Y, Hanada K, Nunokawa E, Saito M, Seki E, Inoue M, Kigawa T and Yokoyama S 2007 A robust two-step PCR method of template DNA production for high-throughput cell-free protein synthesis *J. Struct. Funct. Genomics* **8** 173-91

[26] Yokoyama J, Matsuda T, Koshiba S, Tochio N and Kigawa T 2011 A practical method for cell-free protein synthesis to avoid stable isotope scrambling and dilution *Anal. Biochem.* **411** 223-9

Journal of Physics: Conference Series **699** (2016) 012004 doi:10.1088/1742-6596/699/1/012004

Application of random coherence order selection in gradient-enhanced multidimensional NMR

Mark J. Bostock and Daniel Nietlispach

Department of Biochemistry, University of Cambridge, 80 Tennis Court Road, Old Addenbrooke's Site, Cambridge CB2 1GA

E-mail: dn206@cam.ac.uk

Abstract. Development of multidimensional NMR is essential to many applications, for example in high resolution structural studies of biomolecules. Multidimensional techniques enable separation of NMR signals over several dimensions, improving signal resolution, whilst also allowing identification of new connectivities. However, these advantages come at a significant cost. The Fourier transform theorem requires acquisition of a grid of regularly spaced points to satisfy the Nyquist criterion, while frequency discrimination and acquisition of a pure phase spectrum require acquisition of both quadrature components for each time point in every indirect (non-acquisition) dimension, adding a factor of 2^{N-1} to the number of free-induction decays which must be acquired, where N is the number of dimensions. Compressed sensing (CS) ℓ_1-norm minimisation in combination with non-uniform sampling (NUS) has been shown to be extremely successful in overcoming the Nyquist criterion. Previously, maximum entropy reconstruction has also been used to overcome the limitation of frequency discrimination, processing data acquired with only one quadrature component at a given time interval, known as random phase detection (RPD), allowing a factor of two reduction in the number of points for each indirect dimension (Maciejewski et al. 2011 *PNAS* **108** 16640). However, whilst this approach can be easily applied in situations where the quadrature components are acquired as amplitude modulated data, the same principle is not easily extended to phase modulated (P-/N-type) experiments where data is acquired in the form $\exp(i\omega t)$ or $\exp(-i\omega t)$, and which make up many of the multidimensional experiments used in modern NMR. Here we demonstrate a modification of the CS ℓ_1-norm approach to allow random coherence order selection (RCS) for phase modulated experiments; we generalise the nomenclature for RCS and RPD as random quadrature detection (RQD). With this method, the power of RQD can be extended to the full suite of experiments available to modern NMR spectroscopy, allowing resolution enhancements for all indirect dimensions; alone or in combination with NUS, RQD can be used to improve experimental resolution, or shorten experiment times, of considerable benefit to the challenging applications undertaken by modern NMR.

1. Introduction

NMR spectroscopy and MRI both suffer limitations due to the measurement time required. In the context of high-resolution biomolecular NMR, this requires samples to be highly concentrated and stable over the length of the measurement, properties which may be mutually exclusive. NMR experiments may be defined as "sampling limited" or "sensitivity limited" [1]. In the former, the length of the experiment is determined by the need to sample multiple indirect dimensions, often to high resolution; in this situation, signal-to-noise is ample, and so time is spent acquiring points on a Nyquist grid to achieve appropriate resolution in the various indirect (non-acquisition) dimensions. In the sensitivity limited regime, it is the sample concentration which is limiting and so many repetitions of the experiment are needed to achieve appropriate signal-to-noise. In fact, the two regimes are often overlapping; for a given amount of available measurement time (typically a maximum of one week on most high resolution spectrometers due to the need to replace cryogens after this time) the requirement to acquire high resolution and multiple indirect dimensions limits the maximum number of scans and hence the signal-to-noise, while the need to achieve sufficient signal-to-noise for assignment may lead to reductions in resolution. In this region, which is typically the case for studies of large biomolecular systems, such as membrane proteins [2, 3, 4] and protein complexes [5, 6], NMR measurements are usually a trade-off between resolution and sensitivity leading to poor quality spectra, which are hard to assign. Higher dimensional experiments, which could otherwise be extremely beneficial for large molecular systems, are usually inaccessible.

A significant step towards alleviating such problems can be achieved by using time-domain undersampling. Such sparse or non-uniform sampling approaches (NUS) were demonstrated in the early days of multidimensional NMR as an alternative to acquisition of the full Nyquist grid, which requires points to be acquired at $\frac{1}{SW(\mathrm{Hz})}$ where SW is the maximum spectral width in each indirect dimension [7, 8]. In the NUS regime it is no longer possible to use the Fourier transform (FT) to reconstruct frequency domain data and as such, a range of different reconstruction approaches have been developed which include, amongst others, non-uniform Fourier transform [9, 10, 11, 12], maximum entropy (MaxEnt) reconstruction [7, 13, 14], multi-dimensional decomposition [15, 16], SIFT [17] and a range of techniques which reconstruct projections through the various indirect dimensions for example G-matrix Fourier transform NMR [18] and projection-reconstruction NMR [19].

Recently, a new class of reconstructions, inspired by compressed sensing (CS) theory [20, 21, 22], has been developed [23, 24]. Such methods may be convex, based on ℓ_1-norm reconstruction and typically implemented using algorithms such as iterative hard (IHT) or soft (IST) thresholding [24, 25, 26] or non-convex methods such as iteratively-reweighted least squares (IRLS) reconstruction which reduces the norm during the reconstruction i.e. $\ell_{p\to 0}$-norm reconstruction [24, 27]. More recent developments may allow the use of a range of minimisation functions within a given algorithmic framework [28].

To date CS reconstruction has been applied to NUS data allowing improvements in resolution or sensitivity whilst maintaining accurate intensity reconstruction even in high dynamic range spectra [23, 24, 25, 26]. In addition to the requirement to sample to high resolution, which may be alleviated by NUS based approaches, the other major contribution to the length of NMR experiments is the requirement for frequency discrimination. This is typically achieved via quadrature detection; acquiring two orthogonal phases for each time point. An alternative approach, time-proportional phase incrementation (TPPI) [29], requires oversampling by a factor of two in each indirect dimension. In either case, the effect of frequency discrimination is to increase the number of acquired free induction decays (FIDs) by 2^{N-1} where N is the dimensionality.

Reducing the number of frequency discrimination points would reduce the sampling limitation, allowing sampling to higher resolution for a given experiment time or, in cases where there is ample sensitivity, reduce the experiment time. This can be achieved using the property of randomness, allowing single phase detection, with random selection of either the real or imaginary component. Like NUS data, spectra recorded with random frequency discrimination can be processed with the discrete Fourier transform (DFT) if missing points are replaced with zeros, however this introduces significant artifacts. Consequently, alternative reconstruction approaches may be used to suppress such artifacts. Application of MaxEnt has been demonstrated where the data is recorded according to the States protocol [30], which results in the two quadrature components represented by cosine and sine modulated datasets, each of which generates an absorption mode spectrum. This approach is known as random phase detection (RPD) and is equivalent to recording 0° or 90° phase components [31]. CS processing of such spectra suppresses artifacts in a similar manner to MaxEnt.

However, many modern NMR experiments use gradient coherence order selection to improve data quality e.g. for improved water suppression or to increase signal-to-noise via sensitivity enhanced and TROSY implementations. The two quadrature points are P- and N-type data of the form (ignoring relaxation)

$$\begin{aligned}\exp(i\Omega_S t_1)\exp(i\Omega_I t_2) \quad &\text{(P-type)}\\ \exp(-i\Omega_S t_1)\exp(i\Omega_I t_2) \quad &\text{(N-type)}\end{aligned}$$

where I and S represent proton and heteronuclear magnetisation and t_2 and t_1 represent the direct and indirect dimensions respectively.

Unlike amplitude modulated data, FT of each individual echo/anti-echo dataset generates signals with the unfavourable phase-twist lineshape [32]. The desired pure phase absorption mode lineshape is achieved by a combination of the two datasets to generate cosine and sine modulated data, which can then be combined via the States routine to generate a purely absorptive spectrum [33].

Clearly for a P-/N-type dataset, randomly acquiring either the P- or N- component, which we name random coherence selection (RCS), cannot be handled by standard reconstruction methods due to the intense artifacts arising from the phase-twist lineshape. Consequently, a modification of the processing method is required. In this paper we demonstrate adaption of the CS-IHT algorithm [25] to process P-/N-type RCS-data (CS-RCS). Given the prevalence of gradient-selected experiments in modern NMR this is a decisive improvement. For both forms of frequency discriminated data, amplitude and phase modulated, RCS and/or RPD may be combined with NUS allowing optimisation of the experimental setup for the high resolution required in modern NMR.

2. Theory

Consider a system of linear equations

$$\mathbf{Ax} = \mathbf{b} \tag{1}$$

where $\mathbf{A}$ is an $M \times N$ matrix and $\mathbf{x}$ is a vector of length N to be recovered from measurements $\mathbf{b}$ where $M < N$. In the context of NMR, $\mathbf{A}$ is the inverse Fourier transform, $\mathbf{b}$ is the time-domain data and $\mathbf{x}$ is the frequency domain spectrum. Where $M < N$, i.e. the data is undersampled, equation (1) has infinitely many solutions. However, CS theory states that $\mathbf{x}$ can be recovered almost exactly if $\mathbf{x}$ is sufficiently sparse by minimising the ℓ_1-norm

$$\min_{\mathbf{x}} ||\mathbf{x}||_1 \text{ subject to } \mathbf{Ax} = \mathbf{b} \tag{2}$$

where

$$||\mathbf{x}||_1 = \sum_i |x_i| \tag{3}$$

For NMR spectra which contain noise, equation (2) is relaxed to

$$\min_{\mathbf{x}} ||\mathbf{x}||_1 \text{ subject to } ||\mathbf{Ax} - \mathbf{b}||_2 \leq \delta \tag{4}$$

CS requires (i) sparse representation of the signal (in a given basis, not necessarily the frequency domain) and (ii) incoherent sampling with respect to that basis. This translates into randomising the sampling schedule, which has been shown to benefit other sampling approaches [34].

Gradient coherence selection experiments produce P- and N-type data of the form

$$\begin{aligned} &\exp(i\Omega_S t_1)\exp(i\Omega_I t_2) \quad \text{(P-type)} \\ &\exp(-i\Omega_S t_1)\exp(i\Omega_I t_2) \quad \text{(N-type)} \end{aligned}$$

In contrast to amplitude modulated data, P- and N-type data shows frequency discrimination, but produces an undesirable phase-twist lineshape [32]. In order to produce an absorption mode lineshape, P- and N-type data (S_P and S_N) is combined to give cosine and sine modulated data ($S_{\cos}$ and $S_{\sin}$) as follows

$$\begin{aligned} S_{\cos}(t_1,t_2) &= (S_P + S_N)/2 \\ S_{\sin}(t_1,t_2) &= (S_P - S_N)/2i \end{aligned}$$

For amplitude modulated data, in order to remove the 2^{N-1} factor resulting from the requirement for frequency discrimination, RPD may be used [31]. Although such methods can be demonstrated for 1D spectra, the benefit of both RPD and NUS-based methods is clearly only realised when recording more than one dimension as such higher or indirect dimensions require a separate FID to be recorded for each time-increment or quadrature-detection point. For a two dimensional experiment, a matrix of hypercomplex points, $\mathbf{z}$, is recorded, represented according to hypercomplex notation as [31, 35]

$$z_{k_1,k_2} = x_{k_1,k_2} + i_1 y_{k_1,k_2} + i_2 r_{k_1,k_2} + i_1 i_2 s_{k_1,k_2} \tag{5}$$

where

$$\begin{aligned} i_1^2 = i_2^2 &= -1 \\ i_1 \cdot i_2 &= i_2 \cdot i_1 \end{aligned} \tag{6}$$

Typically in a 2D experiment, only the indirect dimension would be undersampled giving

$$z_k^{\text{RPD}} = \begin{cases} x_{k_1,k_2} + i_1 y_{k_1,k_2} & \text{if } p_k = 0, \\ i_2 r_{k_1,k_2} + i_1 i_2 s_{k_1,k_2} & \text{if } p_k = 1. \end{cases} \tag{7}$$

Although random phase detection assumes collection of only one phase component at each time-point for the indirect dimensions (i.e. a factor of two reduction in the number of points per indirect dimension), both phase components may be detected, or, in combination with an NUS based schedule, neither may be collected.

Amplitude modulated data of this form may be easily reconstructed through existing reconstruction methods such as MaxEnt and CS by replacing missing points with zeros. Clearly each component produces an absorption mode lineshape and selecting cosine or sine components randomly resolves the sign ambiguity, achieving frequency discrimination. In this case the spectrum will contain only one peak but with additional noise-like sampling artifacts. Applying CS (or other reconstructions methods such as MaxEnt [31]) to such time-domain data can recover a largely artifact-free spectrum.

In contrast RCS of P-/N-type data is non-trivial to reconstruct. In a fully-sampled experiment P-/N- data is combined as follows

$$\frac{1}{2}\left(\exp(i\Omega_S t_1)\exp(i\Omega_I t_2) + \exp(-i\Omega_S t_1)\exp(i\Omega_I t_2)\right) = \cos(\Omega_S t_1)\exp(i\Omega_I t_2) \tag{8}$$

$$\frac{1}{2i}\left(\exp(i\Omega_S t_1)\exp(i\Omega_I t_2) - \exp(-i\Omega_S t_1)\exp(i\Omega_I t_2)\right) = \sin(\Omega_S t_1)\exp(i\Omega_I t_2) \tag{9}$$

allowing processing via the States method [30]. However, this clearly requires both P and N points to be present, which in an RCS experiment would not be the case. In order to combine RCS acquisition with CS, it is important that the data matching step in equation (4) $||\mathbf{Ax} - \mathbf{b}||_2$ uses the raw undersampled data i.e. P-/N-points. If converting the data from cosine-/sine-type to P/N-type is represented by a function $\mathbf{I}$ equation (4) may be reformulated as

$$\min_{\mathbf{x}} ||\mathbf{x}|| \text{ subject to } ||\mathbf{AIx} - \mathbf{b}||_2 \leq \delta \tag{10}$$

As with RPD, the P- and N- points must be sampled randomly for each time-increment to prevent aliasing artifacts from the missing points.

3. Results

Figure 1 demonstrates CS reconstruction (Figure 1B) of an RCS version of a 2D [^{1}H,^{15}N]-SOFAST TROSY for the chicken α-spectrin SRC homology 3 (SH3) domain, a 7 kDa, 62 amino acid protein [36] in comparison to the fast Fourier transform (FFT) reconstruction (Figure 1A) of the fully-sampled data. The reconstruction is of high quality, with good recovery of the peak positions as well as the peak shapes and intensities. RCS data for the CS reconstruction was recorded for half the time of the full reconstruction due to the selection of only one of each quadrature detection point in the ^{15}N dimension (i.e. 56 points for RCS, compared to 56 complex points (112 total points) for the full reconstruction). This reduced the experiment time by 50% to 3 min. Consequently, for a given length of experiment, the time-saving from RCS could be used to improve the resolution. Nevertheless, although a 2D provides a useful illustration of the principle, clearly the benefits of this approach come at higher dimensions. Since CS relies on sparsity in the frequency domain, extending a spectrum into higher dimensions increases the sparsity reducing the number of points required to reconstruct the spectrum.

In Figure 2 the benefit of RQD for higher dimensional experiments is evident. Figure 2 shows a ^{1}H,^{13}C-plane from a 3D [^{1}H,^{15}N]-TROSY HNCA for S195A-human factor IX, a 297 amino acid, 33 kDa protein [37]. In Figure 2A, a plane from the fully-sampled FT-reconstructed spectrum is shown recorded with 384 ^{15}N, ^{13}C complex pairs, equivalent to 1536 total points or 16×24 complex points in the ^{15}N and ^{13}C dimensions respectively. This correlates to t_{max} for ^{15}N and ^{13}C of 9.22 ms and 5.28 ms respectively. In Figure 2B data is acquired with RCS in the constant time ^{15}N dimension recorded using gradient selection (P-/N-type data), and RPD in the ^{13}C dimension recorded as States-TPPI (cosine-/sine- data). Dimensions 2 and 3 are reconstructed with CS-RCS and normal CS respectively. This allows a factor of two reduction in points per indirect dimension, i.e. a total factor of four saving. In this example, the time-saving is used to sample to higher resolution, in this case equivalent to sampling a matrix of 32×48 complex points for ^{15}N and ^{13}C respectively. This results in t_{max} of 18.44 ms and 10.56 ms for ^{15}N and ^{13}C respectively. The two sampling schedules are compared in Figures 2C and D, with the axes representing total points in each dimension. The considerable improvement in ^{13}C resolution is evident in Figure 2B, for example by the peaks indicated with an arrow.

With sufficient sparsity NUS and RQD may be combined for significant undersampling. In Figure 3, a ^{1}H,^{15}N-plane from a 3D [^{15}N,^{1}H]-TROSY HN(CO)CA for S195A-human factor

Journal of Physics: Conference Series **699** (2016) 012004 doi:10.1088/1742-6596/699/1/012004

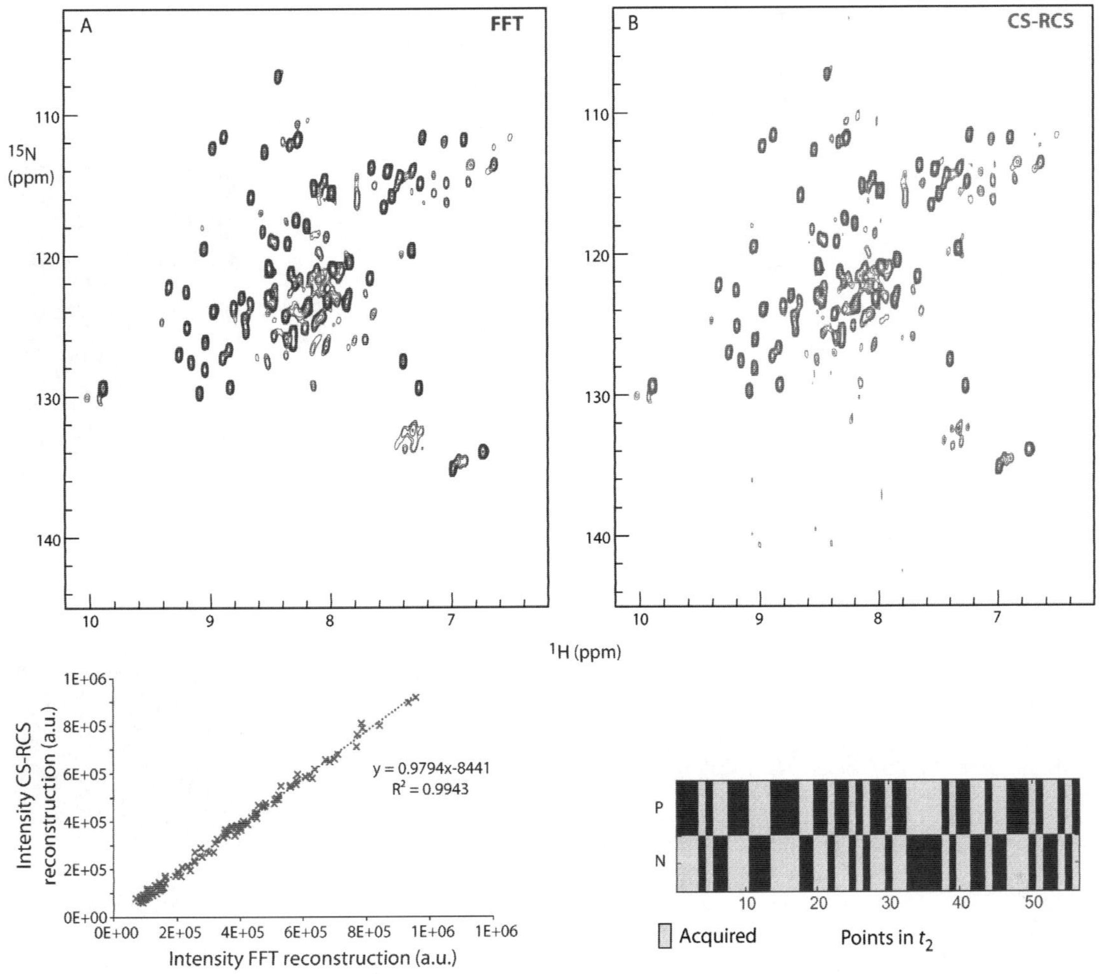

Figure 1. Reconstruction of a 2D [^{1}H,^{15}N]-SOFAST TROSY spectrum for an α-spectrin SH3 domain using FFT reconstruction of fully-sampled data (A) and CS-IHT reconstruction of RCS data (B). The sampling schedule for the RCS data is shown below, with acquired points indicated. 56 points were recorded from total 112 points (56 complex points), i.e. a factor of two reduction in the number of points. The sampled points were selected randomly. Intensities are compared for equivalent peaks picked in the two spectra showing good reproduction of intensities in the CS-RCS reconstruction. The increased noise level in the CS-RCS spectrum results from running the experiment for only half the amount of time of the fully-sampled spectrum. Data was recorded for a 0.3 mM SH3-domain sample at 298 K on a Bruker DRX500 equipped with a 5 mm TCI cryoprobe, with 4 scans, ^{15}N spectral width = 2530.4 Hz and $t_{1\max}$ = 22.1 ms. Recording time for the fully-sampled data was 6 min.

IX is shown. This experiment is recorded with P-/N-type selection in the ^{15}N dimension and amplitude modulation in the ^{13}C dimension. The experiment was acquired using an exponentially weighted NUS schedule recording 350 ^{15}N, ^{13}C pairs from a fully-sampled matrix of 1536 complex pairs, i.e. 22.8% sampling with $t_{1\max}(^{15}N) = 11.52$ ms and $t_{1\max}(^{13}C) = 8.21$ ms. CS reconstruction of the NUS experiment is shown in Figure 3A. Incorporating RCS acquisition with CS-RCS processing in dimension 2 (^{15}N) and RPD acquisition with normal CS-processing in dimension 3 (^{13}C) is shown in Figure 3B. In this case the experiment was undersampled by a further factor of four, equivalent to 5.7% sampling. This equates to an experiment time of approximately 4.5 h. The two sampling schedules for the indirect dimensions are shown below.

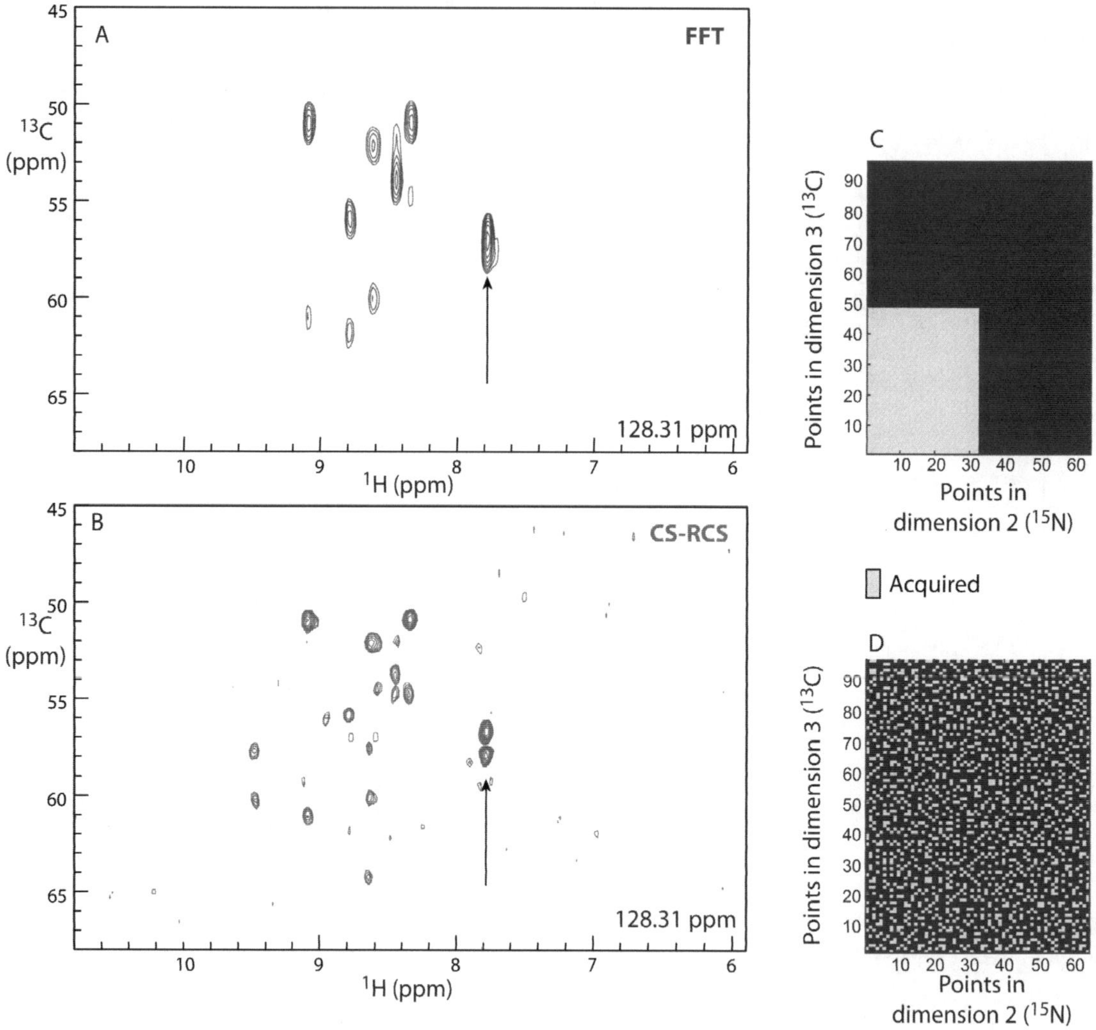

Figure 2. A ^{1}H,^{13}C-plane from a 3D [^{1}H,^{15}N]-TROSY HNCA for U-[^{2}H,^{15}N,^{13}C]-labelled S195A-human factor IX, taken at 128.31 ppm in the ^{15}N dimension (dimension 2). In A, FFT reconstruction of fully-sampled data, equivalent to 384 ^{15}N, ^{13}C pairs (1536 total points) or 16 × 24 complex points in the ^{15}N and ^{13}C dimensions respectively is shown. This corresponds to $t_{\max}$ for ^{15}N and ^{13}C of 9.22 ms and 5.28 ms respectively. In B, CS reconstruction with RCS in the constant-time ^{15}N dimension and RPD in the ^{13}C dimension is shown. Since this allows a factor of two undersampling in each indirect dimension, i.e. equivalent to 25% sampling, this allows sampling to higher resolution for the same experiment time giving $t_{\max}$ of 18.44 ms and 10.56 ms for ^{15}N and ^{13}C respectively. This is equivalent to sampling points from a matrix of 32 × 48 complex points. The reconstruction uses CS-RCS in dimension 2 (^{15}N) and normal CS in dimension 3 (^{13}C). An example of the improved resolution in the CS-RCS spectrum is indicated by a black arrow in A and B. Adjacent to each spectrum, C and D show the region of sampled points, with acquired points indicated. The axes represent total points in each dimension. Data was recorded on a 0.25 mM U-[^{2}H,^{15}N,^{13}C]-labelled S195A-human factor IX using a Bruker DRX500 equipped with a 5 mm TCI cryoprobe at 298 K, with 96 scans. The spectral widths in the indirect dimensions were 1736.1 Hz (^{15}N) and 4545.5 Hz (^{13}C).

As before, peak shapes and positions are well recovered.

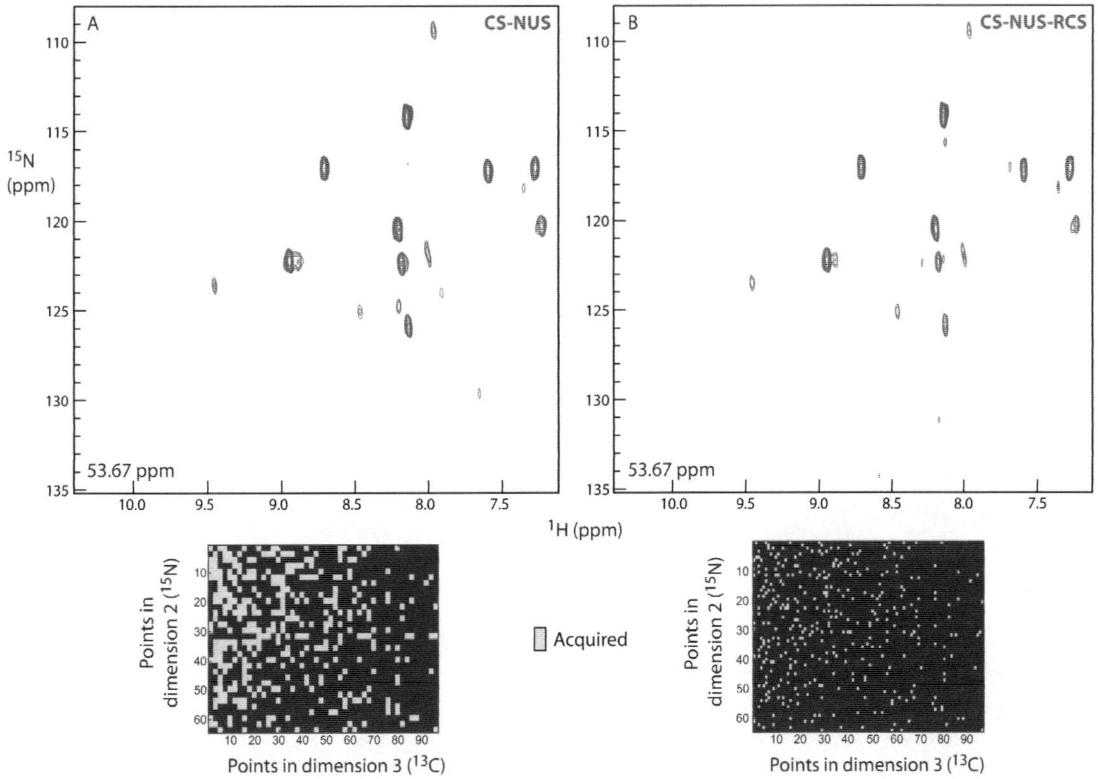

Figure 3. A ^{1}H,^{15}N-plane from the reconstruction of a 3D [^{1}H,^{15}N]-TROSY HN(CO)CA for U-[^{2}H,^{15}N,^{13}C]-labelled S195A-human factor IX, taken at 53.67 ppm in the ^{13}C dimension (dimension 3). In A, CS reconstruction of 22.8% NUS data, equivalent to 350 ^{15}N, ^{13}C complex pairs (1400 total points), giving $t_{1\max}(^{15}N) = 11.52$ ms and $t_{1\max}(^{13}C) = 8.21$ ms, from a fully-sampled matrix of 1536 complex pairs (6144 total points) is shown. In B, CS reconstruction of 5.7% NUS-RPD data, equivalent to 350 points (from 6144 total points) is shown, using CS-RCS reconstruction in dimension 2 (^{15}N) and normal CS-processing in dimension 3 (^{13}C). Sampling schedules for the indirect dimensions are shown below each spectrum, with acquired points indicated. In each case, the resolution is the same, defined by the NUS schedule. RCS and RPD are used here for time-reduction. Data was recorded on a 0.25 mM U-[^{2}H,^{15}N,^{13}C]-labelled S195A-human factor IX using a Bruker AV800 equipped with a 5 mm TXI HCN cryoprobe at 298 K, with 320 scans. The spectral widths in the indirect dimensions were 2777.8 Hz (^{15}N) and 5848.0 Hz (^{13}C).

4. Conclusions

The previously introduced RPD reduces the number of points required for frequency discrimination by a factor of two in each indirect dimension where the data is recorded as cosine and sine components [31]. RCS, introduced in this paper, allows this factor of two saving to be extended into indirect dimensions recorded with gradient selection (P- and N-type data). Both RPD and RCS generate artifacts, similar to NUS. However, processing with CS or MaxEnt can minimise the artifacts from RPD spectra. The adaption of the CS algorithm in this paper, CS-RCS, allows this approach to be applied to minimise the artifacts present in RCS data. This extends the power of RQD to the full suite of experiments available to modern NMR which typically combine indirect dimensions with cosine-/sine- and P-/N-type data. The time savings generated by RQD can be used to enhance resolution, by sampling to higher $t_{\max}$ values in indirect dimensions, or to reduce experiment time in sampling limited regimes. The data shown in this paper demonstrates successful application of these ideas to 2D and 3D experiments.

Since CS processing relies on spectral sparsity, further improvements can be expected for 4D and higher dimensional experiments.

5. Methods

Spectra for the SH3 domain were recorded for a 0.3 mM SH3-domain sample using a 2D [^{1}H,^{15}N]-SOFAST TROSY sequence on a Bruker DRX500 equipped with a 5 mm TCI cryoprobe at 298 K, with 56 complex points in the ^{15}N dimension, $t_{1\max} = 22.1$ ms, and 4 scans. The ^{15}N spectral width was 2530.4 Hz. Recording time for the fully-sampled data was 6 min.

A 0.25 mM U-[^{2}H,^{15}N,^{13}C]-labelled S195A-human factor IX sample was used for 3D experiments. The HNCA was recorded on a Bruker DRX500 equipped with a 5 mm TCI cryoprobe with 96 scans and the HN(CO)CA was recorded on a Bruker AV800 spectrometer with 320 scans equipped with a 5 mm TXI HCN cryoprobe, both at 298 K. The full HNCA spectrum was recorded with 16×24 complex pairs in the ^{15}N and ^{13}C dimensions respectively, with spectral widths in the indirect dimensions of 1736.1 Hz (^{15}N) and 4545.5 Hz (^{13}C) and $t_{\max}$ for ^{15}N and ^{13}C of 9.22 ms and 5.28 ms respectively. The HN(CO)CA was recorded as an NUS experiment with 350 complex pairs according to an exponential schedule with spectral widths in the indirect dimensions of 2777.8 Hz (^{15}N) and 5848.0 Hz (^{13}C) and $t_{1\max}$ for ^{15}N and ^{13}C of 11.52 ms and 8.21 ms respectively.

Spectra were reconstructed with Azara (W. Boucher, unpublished) or MATLAB for FT reconstructions and with custom-written MATLAB and Python scripts for CS-processing, using a modified form of the IHT algorithm [25]. NUS schedules were generated using ScheduleTool [38] and RQD schedules using custom-written MATLAB code. Spectra were analysed in Azara and CCPN Analysis [39].

Acknowledgments

Thanks to Dr. Daniel Holland for useful discussions and to Dr. Jennifer Kopanic for the S195A-Factor IX sample.

References

[1] Szyperski T, Yeh D C, Sukumaran D K, Moseley H N B and Montelione G T 2002 *PNAS* **99** 8009
[2] Gautier A, Mott H R, Bostock M J, Kirkpatrick J P and Nietlispach D 2010 *Nat. Struct. Mol. Biol.* **17** 768
[3] Hiller S, Garces R G, Malia T J, Orekhov V Y, Colombini M and Wagner G 2008 *Science* **321** 1206
[4] Kim H J, Howell S C, Van Horn W D, Jeon Y H and Sanders C R 2009 *Prog. Nucl. Magn. Reson. Spectrosc.* **55** 335
[5] Fiaux J, Bertelsen E B, Horwich A L and Wüthrich K 2002 *Nature* **418** 207
[6] Sprangers R, Velyvis A and Kay L E 2007 *Nat. Methods* **4** 697
[7] Barna J C J, Laue E D, Mayger M R and Worrall S J P 1987 *J. Magn. Reson.* **73** 69
[8] Schmieder P, Stern A S, Wagner G and Hoch J C 1994 *J. Biomol. NMR* **4** 483
[9] Coggins B E and Zhou P 2008 *J. Biomol. NMR* **42** 225
[10] Kazimierczuk K, Koźmiński W and Zhukov I 2006 *J. Magn. Reson.* **179** 323
[11] Marion D 2005 *J. Biomol. NMR* **32** 141
[12] Marion D 2006 *J. Biomol. NMR* **36** 45
[13] Hoch J C, Stern A S, Donoho D L and Johnstone I M 1990 *J. Magn. Reson.* **86** 236
[14] Rovnyak D, Frueh D P, Sastry M, Sun Z-Y J, Stern A S, Hoch J C and Wagner G 2004 *J. Magn. Reson.* **170** 15
[15] Orekhov V Y, Ibraghimov I V and Billeter M 2001 *J. Biomol. NMR* **20** 49
[16] Tugarinov V, Kay L E, Ibraghimov I V and Orekhov V Y 2005 *J. Am. Chem. Soc.* **127** 2767
[17] Matsuki Y, Eddy M T and Herzfeld J 2009 *J. Am. Chem. Soc.* **131** 4648
[18] Atreya H S and Szyperski T 2004 *PNAS* **101** 9642
[19] Kupče E and Freeman R 2003 *J. Am. Chem. Soc.* **125** 13958
[20] Candes E J, Romberg J and Tao T 2006 *IEEE Trans. Inf. Theory* **52** 489
[21] Donoho D L 2006 *IEEE Trans. Inf. Theory* **52** 1289
[22] Logan B F 1965 Properties of high-pass signals. PhD thesis, Columbia University, New York.

[23] Holland D J, Bostock M J, Gladden L F and Nietlispach D 2011 *Angew. Chem. Int. Ed.* **50** 6548
[24] Kazimierczuk K and Orekhov V Y 2011 *Angew. Chem. Int. Ed.* **50** 5556
[25] Bostock M J, Holland D J and Nietlispach D 2012 *J. Biomol. NMR* **54** 15
[26] Hyberts S G, Milbradt A G, Wagner A B, Arthanari H and Wagner G 2012 *J. Biomol. NMR* **52** 315
[27] Kazimierczuk K and Orekhov V Y 2012 *J. Magn. Reson.* **223** 1
[28] Sun S, Gill M, Li Y, Huang M and Byrd R A 2015 *J. Biomol. NMR* **62** 105
[29] Marion D and Wüthrich K 1983 *Biochem. Biophys. Res. Commun.* **113** 967
[30] States D, Haberkorn R and Ruben D 1982 *J. Magn. Reson.* **48** 286
[31] Maciejewski M W, Fenwick M, Schuyler A D, Stern A S, Gorbatyuk V and Hoch J C 2011 *PNAS* **108** 16640
[32] Keeler J and Neuhaus D 1985 *J. Magn. Reson.* **63** 454
[33] Davis A L, Keeler J, Laue E D and Moskau D 1992 *J. Magn. Reson.* **98** 207
[34] Hoch J, Maciejewski M and Filipovic B 2008 *J. Magn. Reson.* **193** 317
[35] Delsuc M A 1988 *J. Magn. Reson.* **77** 119
[36] Musacchio A, Noble M, Pauptit R, Wierenga R and Saraste M 1992 *Nature* **359** 851
[37] Johnson D J D, Langdown J and Huntington J A 2010 *PNAS* **107** 645
[38] Maciejewski M W and Gorbatyuk V Sample Scheduler for Non-Uniform Data Sets `http://sbtools.uchc.edu/nmr/sample_scheduler/`
[39] Vranken W F, Boucher W, Stevens T J, Fogh R H, Pajon A, Llinas M, Ulrich E L, Markley J L, Ionides J and Laue E D 2005 *Proteins* **59** 687

Journal of Physics: Conference Series **699** (2016) 012005 doi:10.1088/1742-6596/699/1/012005

Protein NMR Structure Refinement based on Bayesian Inference

Teppei Ikeya[1,2*], Shiro Ikeda[3], Takanori Kigawa[2,4], Yutaka Ito[1,2] and Peter Güntert[1,5,6*]

[1]Tokyo Metropolitan University, 1-1 Minami-ohsawa, Hachioji, Tokyo 192-0397, Japan
[2]CREST/Japan Science and Technology Agency (JST), Japan
[3]The Institute of Statistical Mathematics, 10-3 Midori-cho, Tachikawa, Tokyo 190-8562, Japan
[4]Laboratory for Biomolecular Structure and Dynamics, RIKEN Quantitative Biology Center (QBiC), 1-7-22 Suehiro-cho, Tsurumi-ku, Yokohama, Kanagawa 230-0045, Japan
[5]Institute of Biophysical Chemistry and Center for Biomolecular Magnetic Resonance, Goethe University Frankfurt, Max-von-Laue-Str. 9, 60438 Frankfurt am Main, Germany
[6]Laboratory of Physical Chemistry, ETH Zürich, Vladimir-Prelog-Weg 2, 8093 Zurich, Switzerland

E-mail: tikeya@tmu.ac.jp, guentert@em.uni-frankfurt.de

Abstract. Nuclear Magnetic Resonance (NMR) spectroscopy is a tool to investigate three-dimensional (3D) structures and dynamics of biomacromolecules at atomic resolution in solution or more natural environments such as living cells. Since NMR data are principally only spectra with peak signals, it is required to properly deduce structural information from the sparse experimental data with their imperfections and uncertainty, and to visualize 3D conformations by NMR structure calculation. In order to efficiently analyse the data, Rieping et al. proposed a new structure calculation method based on Bayes' theorem. We implemented a similar approach into the program CYANA with some modifications. It allows us to handle automatic NOE cross peak assignments in unambiguous and ambiguous usages, and to create a prior distribution based on a physical force field with the generalized Born implicit water model. The sampling scheme for obtaining the posterior is performed by a hybrid Monte Carlo algorithm combined with Markov chain Monte Carlo (MCMC) by the Gibbs sampler, and molecular dynamics simulation (MD) for obtaining a canonical ensemble of conformations. Since it is not trivial to search the entire function space particularly for exploring the conformational prior due to the extraordinarily large conformation space of proteins, the replica exchange method is performed, in which several MCMC calculations with different temperatures run in parallel as replicas. It is shown with simulated data or randomly deleted experimental peaks that the new structure calculation method can provide accurate structures even with less peaks, especially compared with the conventional method. In particular, it dramatically improves in-cell structures of the proteins GB1 and TTHA1718 using exclusively information obtained in living *Escherichia coli* (*E. coli*) cells.

1. Introduction

Biomacromolecules such as proteins, DNA and RNA have three-dimensional (3D) structures and dynamical properties, and provide essential functions contributing to almost all biological events and phenomena. It is notably interesting that the structures and dynamics of proteins have been designed exquisitely during their long evolution process in order to accurately perform individual vital tasks. To elucidate the molecular structures and properties, therefore, allows to understand mechanisms of biological activities and human diseases at atomic level, and to promote drug discovery and protein engineering.

There are currently three main methods to determine 3D structures of biomacromolecules at atomic resolution, X-ray crystallography, cryo-electron microscopy (cryo-EM) and nuclear magnetic resonance (NMR) spectroscopy. X-ray crystallography is the most popular and powerful as a 3D structure determination tool for biomacromolecules. It can investigate large molecules in detail routinely once a crystal of the target molecule has been obtained. However, it requires to produce a crystal of the molecule, which is usually not trivial, and can observe only the crystal state in which the molecule is packed, immobilized and occasionally skewed by surrounding molecules in the crystal lattice despite of the importance of the molecular dynamics for the functions of proteins. Cryo-EM currently becomes popular to observe large macromolecular assemblies because it has virtually no limitation on upper molecular size. Its drawback of inherently low resolution has recently ameliorated considerably and cryo-EM is approaching atomic resolution [1]. NMR spectroscopy is the second most commonly used tool to investigate the biomacromolecules. It observes the precession of nuclear spins as nuclear resonance of emitted electromagnetic radiation under strong magnetic fields, which is obtained by a superconducting magnet and modulated by the surrounding electronic environments. By NMR, one can determine, with atomic resolution, the chemical configurations, structures, and dynamics of a molecule.

A principal advantage of NMR is that biological macromolecules can be studied non-invasively in their native form in nearly natural environment, and that besides static structural information also dynamic aspects of protein function can be investigated on timescales ranging from nanoseconds to months. It does not require preparing crystals of the molecules like X-ray, and can obtain structural information at atomic resolution. A large number (thousands) of signals can be recorded and resolved in a single NMR spectrum by spreading them in multiple dimensions. Furthermore, a method to investigate protein structures in living cells by NMR [2, 3], so-called in-cell NMR, has recently been developed, which uncovers new pictures of protein behaviors in the cells. A principal limitation of solution NMR spectroscopy relates to the size of the macromolecule. Due to their slow molecular tumbling larger molecules exhibit rapid transverse relaxation that leads to rapidly decaying signals and broad resonance lines. These are manifested in the spectra by weak, broad and overlapping peaks, which eventually impedes their detection and assignment. Therefore, for structure determination, the NMR method is in general limited to soluble proteins below 30 kDa of molecular weight, or to even smaller masses for proteins that have a large effective size because of their environment, e.g. proteins within living cells. It is of high interest to overcome the molecular size limitation and to establish more accurate structure determination of such difficult molecule samples.

Since the original NMR data are principally only peak intensities and positions in a spectrum, it is necessary to properly deduce the structural information from the data with its imperfections and uncertainties, and to visualize 3D conformations by NMR structure calculation. The structure calculation can be a model fitting approach starting from a model structure that embodies the information of chemical composition and bonds of a molecule, which is generated by some other method. In the case of proteins, the initial model is a conformation of an extended linear amino acid chain, which is optimized to find atomic coordinates that satisfy expected distances and dihedral angles derived from the experimental NMR data. Conventional structure determination is typically achieved by a molecular dynamics simulation (MD) approach with distance and dihedral angle restraints derived from the data, in which a target function is defined as the sum of the (suitably simplified) physical potential energy of a molecule and the sum of squared errors between experimental data and molecular conformations, with a weight factor to keep a balance between them.

Various NMR structure calculation methods have been proposed to determine more accurate 3D structures for proteins of lager molecular size or in cells with sparse and noisy information. Nilges and coworkers presented in 2005 a NMR structure calculation method based on Bayes theorem, so-called Inferential Structure Determination (ISD) [4, 5]. Since, as noted above, the raw NMR experimental data are not direct structural information, it is necessary to convert the NMR data into restraints on distances and dihedral angles among atoms by relations that are based on physical principals but contain also some nuisance parameters. Whereas these parameters are fixed to user-defined values in the conventional methods, the Bayesian method can deduce them with a statistical model by parameter sampling schemes. In a Bayesian NMR structure calculation, the target function is replaced by the posterior, and the contributions of prior and likelihood are computed from the data without a predefined explicit weight factor.

The program CYANA [6, 7] is widely used for NMR structure calculations as well as occasionally for molecular modeling in *de novo* design of proteins. As an optimization method for the target function, it adopts torsion angle molecular dynamics (TAMD) that enables to obtain converged structures quickly with longer step-sizes than MD in Cartesian coordinate space. The potential energy of CYANA is a physical force field optimized for torsion angle space derived from the Cartesian space force field of the MD program Amber [8] and with a Generalized Born (GB) implicit water model [9]. The physical force field and water model can achieve a more accurate estimation of the prior distribution of the structure ensemble, in which TAMD can reduce the computational cost for obtaining the marginal likelihood as well as molecular conformations. Here, we present a new NMR structure refinement method based on Bayesian inference implemented into CYANA. We employed the Bayesian modeling to Nuclear Overhauser Effect (NOE) data, which has the form of restraints on distances between two hydrogen atoms that are shorter than approximately 5 Å. The method is composed of a MD calculation to quickly obtain rough global structures with automatic NOE assignment, and subsequent structure refinement by Bayesian inference addressing both unambiguously and ambiguously assigned NOE peak lists. The final conformational ensembles are evaluated by principal component analysis (PCA).

Three different types of peak lists were prepared as input data for the validation of the new method. (1) Fully simulated data reconstructed from known structures. (2) Sparse experimental data obtained by randomly deleting a given amount of peaks from the original experimental data set. (3) Complete in-cell NMR data sets that are intrinsically sparse and noisy and one of the main targets of this method. Using fully simulated and experimental data allow us in-depth analyses and verification for practical non-ideal data, respectively.

2. Theory

The target function (T) of a conventional NMR structure calculation usually consists of the χ^2-term between predicted and observed data, the physical potential energy (E), and a weight factor (w):

$$T(\theta) = \chi^2(\theta) + wE(\theta), \tag{1}$$

where θ denotes the set of torsion angles used to describe the protein structure. In Bayes' theorem, the target function can be replaced by the posterior probability for the evaluation of an ensemble of conformations:

$$P(\theta|D) \propto P(\theta)P(D|\theta) \tag{2}$$

where θ is coordinates in torsion angle space, and D is the experimental data. In modern NMR structure determination, the most essential experimental data yielding 3D structure information is the NOE peak volume V_{kl} derived from a dipolar interaction between spin k and l, and hence in this study the NOE data is described in the framework of Bayesian inference, while the other experimental data are treated in conventional forms, e.g. dihedral angle restraints. The likelihood function is described by a lognormal distribution

$$V_{kl} = \frac{\gamma}{r_{kl}^{6}} \tag{3}$$

$$P(V|\theta) = \prod_{i=1}^{n} L(V_{k_i l_i}|\theta) \tag{4}$$

$$L(V_{kl}|\theta) = \frac{1}{\sqrt{2\pi\sigma^2}V_{kl}} \exp\left(-\frac{1}{2\sigma^2}\ln^2\left(\frac{V_{kl}}{\mu}\right)\right) \tag{5}$$

where n is the number of peaks σ is the standard deviation and μ is an expected peak volume. It is necessary to convert the peak volume into distance information between two spins r_{kl}. In the conventional approach this is achieved using the simple equation (3) with a calibration constant γ. Thus, the prior of the Bayesian modeling is

$$P(\sigma,\theta,\gamma) = P(\sigma)P(\theta)P(\gamma) \tag{6}$$

$$P(\theta) = \frac{1}{Z(\beta)} \exp(-\beta E(\theta)) \tag{7}$$

$$\sigma \sim G[a,b]$$
$$\gamma \sim LN[\mu_\gamma, \sigma_\gamma]$$

where a and b are shape and scale parameters of the Gamma function G, respectively, LN is the lognormal distribution, and $P(\theta)$ is described as the canonical ensemble of molecular structures with partition function Z and inverse temperature β.

The sampling algorithm for obtaining the posterior is based on Markov chain Monte Carlo (MCMC) with the Gibbs sampler for the σ and γ parameters, and MD for sampling the canonical ensemble of conformations. In a general sense, this MC method along with MD is known as hybrid MC or Hamiltonian MC (HMC) in Bayesian statistics, by which high acceptance rates on Metropolis criteria can be achieved even for large parameter dimensions. While the general HMC method introduces an independent auxiliary variable describing a pseudo kinetic energy term for effective sampling in MC, the method here is a more direct combination of MC and MD. The MD method is used for sampling molecular conformations because it is known that its sampling of protein conformations is far superior to MC due to covalent structure restrictions and the tightly packed globular shape [10, 11]. The target function $T(\theta)$ in MD is composed of the physical potential $E(\theta)$ and a pseudo energy $L(\theta, \gamma, \sigma)$ with γ and σ, which are sampled by the Gibbs sampler,

$$T(\theta) = \beta E(\theta) + L(\theta,\gamma,\sigma) \tag{8}$$

$$E(\theta) = \sum E_{dihedral} + \sum E_{vdw} + \sum E_{vdw14} + \sum E_{electro} + \sum E_{electro14} + \sum E_{GB}$$

$$L(\theta,\gamma,\sigma) = \frac{1}{2\sigma^2}\sum_{i=1}^{n}\ln\left(\frac{V_i r_i^6}{\gamma}\right)$$

where $E_{dihderal}$, E_{vdw}, E_{vdw14}, $E_{electro}$, $E_{electro14}$, and E_{GB} are energy terms for dihedral angles, van der Waals, 1-4 van der Waals, electrostatic, 1-4 electrostatic, and Generalized Born (GB) implicit water model interactions.

In the case of an ambiguously assigned distance restraint with m assignment possibilities, the distance r_i is replaced by

$$\bar{r} = \left(\sum_{j=1}^{m} r_{k_j l_j}^{-6} \right)^{-1/6} \tag{9}$$

3. Material and Methods

Figure 1 shows a flow chart of this method. The initial step of the structure calculation is principally identical with the conventional structure calculation with automatic Nuclear Overhauser Effect SpectroscopY (NOESY) cross peak assignments by the program CYANA [12]. The calibration constant, γ, is determined such that the median value of all peak volumes or intensities in a given NOESY spectrum corresponds to a predefined distance, usually 4.0 Å. All upper distance bound restraints are calculated from the corresponding peak intensity in the spectrum using the calibration constant. The conformation with the lowest target function value is selected from the 20 conformers calculated with the distance restraints obtained in the 7th cycle, and regularized for the molecular geometry of the Amber ff03 force field which is slightly different from the original CYANA geometry used in the initial step. The regularization was achieved by recalculating the structures in CYANA using distance and torsion angle restraints. Distance restraints with an upper bound of 0.1 Å were used for the distances to the corresponding N, C$^\alpha$, and C' atoms in the structured regions of the conformers after the initial 7th cycle of CYANA. Torsion angle restraints were set with a width of 20° centered around the value of each torsion angle in that structure [13]. The regularized coordinates were used for the next refinement stage based on Bayesian inference and as reference structures for the calculation of root-mean-square deviations (RMSDs).

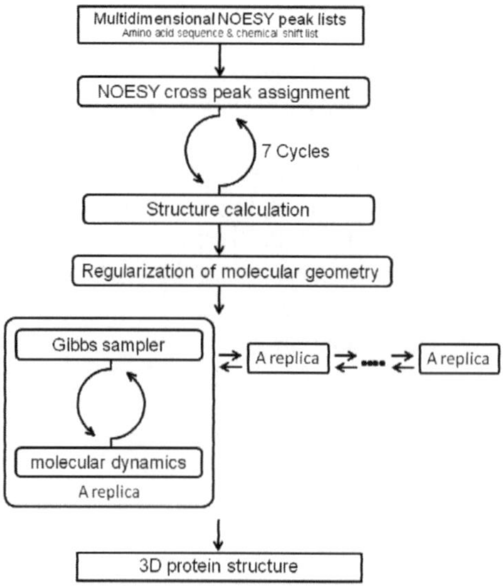

Figure 1. Structure refinement by Bayesian inference.

A replica in Fig. 1 denotes a hybrid Monte Carlo (HMC) calculation for sampling θ, γ, σ, and evaluating the posterior. The torsion angle coordinates, θ, are sampled by molecular dynamics simulation with the current γ and σ, in which the Amber physical force field with the GB model is modified for torsion angle space and used for sampling the canonical ensemble. The calibration constant and standard deviation of NOE volumes, γ and σ, are sampled with the current θ by the Gibbs sampler. These three parameters are optimized by alternatively performing MD and the Gibbs sampler in iterative steps. In the practical sampling of these parameters, it is not trivial to search the entire functional space particularly on exploring the conformational prior due to the extraordinarily large conformational space of proteins. Hence, the replica exchange method is employed, in which several MCMC calculations with different temperatures are performed in parallel as replicas, and each is exchanged at certain intervals.

The accuracy of structures was quantified by the RMSD value to the reference structure for the backbone atoms N, C$^\alpha$, and C' in the structured regions of the proteins. Bayesian modelling allows to obtain a posterior probability for all parameters including atomic coordinates and NOE calibration constants, and final structure ensembles can be validated based on the posterior distribution. Since the parameter space is high-dimensional, principal component analysis (PCA) were performed to visualize the distributions of structure ensembles. Cartesian coordinates of the atoms in the final structure ensembles were used as input variables for PCA. The posterior distribution on the principal components elucidates structure ensembles from the estimated distribution due to wider search of the conformational space satisfying current experimental data by MCMC. In here, we show all conformers within 1 standard deviation of the maximum of the posterior. Structures were visualized using the program MOLMOL [14].

3.1. Conventional structure calculation as reference

Conventional structure calculations were done using the standard protocol implemented in CYANA [7, 12]. 100 initial conformers with random torsion angle values were subjected to simulated annealing using 10000 torsion angle dynamics steps, and the 20 lowest target function conformers were selected to represent the NMR structure of the protein.

For comparison, we alternatively refined the CYANA structures with the program OPALp [15], which we routinely calculate to obtain final structures deposited to Protein Data Bank (PDB). The program can perform the conjugate gradient minimization in Cartesian space with Amber ff94 and TIP3P implicit water model. For comparison of the conventional and Bayesian algorithms, the same peak lists were given to both approaches.

3.2. Data set generation for structure calculation

Two types of simulation data were prepared to validate this method: (1) Fully simulated data that are reconstructed from known structures. (2) Experimental data that were modified by randomly deleting peaks. The first data sets permit to evaluate the predictive accuracy of the method, and the second ones can assess its performance in more realistic cases.

The fully simulated peak lists were derived from 3D structures of the proteins TTHA1718 (66 aa, PDB 2ROE) [2] and GB1 (57 aa, PDB 2J52) [16]. The peak intensities were calculated from the structures according to the relationship $V_{kl} = \gamma\, r_{kl}^{-6}$ for all atom pairs within the range 2.4–5.0 Å for ^{13}C-separated, ^{13}C-separated aromatic and ^{15}N-separated NOESYs. The individual peaks were given slightly different calibration constant values taken from a Gaussian distribution centered at the given γ. Table 1 shows the numbers of simulated peaks for TTHA1718 and GB1 that were randomly selected from atom pairs fulfilling the above criteria. Other restraints like hydrogen bond and torsion angle were not used. The data sets were designed so as to reflect challenging conditions for the conventional structure determination by CYANA, e.g. cases with a severely limited amount of structural information such as in-cell NMR. The peak lists were assigned with the original atom pairs in order to validate the performance of the conventional and new methods exclusively in terms of the quality and amount of distance information, but excluding contributions from the accuracy of the CYANA automatic NOE assignment algorithm.

Reduced experimental peak lists were generated by randomly deleting given amounts of the NMR distance restraint data for three different proteins, i.e. TTHA1718, the 140-residue ENTH-VHS domain At3g16270(9–135) from *Arabidopsis thaliana* (ENTH) [17, 18], and the 114-residue Src homology domain 2 from the human feline sarcoma oncogene Fes (SH2) [19]. The amounts of deleted peaks were chosen such that structures calculated by the conventional CYANA method were disturbed by 1.0–4.0 Å RMSD from the original ones, since such data sets are the prime practical targets for the present structure refinement method. Considering the dynamics of molecules and imperfections of experimental data, it does not make sense to attempt obtaining structures with less than 1.0 Å RMSD. On the other hand, it is not trivial to dramatically improve structures with more than 4.0 Å RMSD because the conformational space that should be search for this would become vast. The manually determined chemical shift assignments were used from the BMRB with accession numbers 11035 for TTHA1718, 5928 for ENTH, and 6331 for SH2.

The chemical shifts for GB1 were determined in our group. Peak lists were obtained with the automated peak picking algorithm of the program NMRView [20] and Azara [21] with manual corrections or modifications for ^{15}N-, and ^{13}C-resolved NOESY spectra. Torsion angle and hydrogen bond restraints were not used in any of the calculations.

In-cell NMR experiments and data collection were performed as reported previously [2, 3].

4. Results and Discussion

4.1. Fully simulated peak lists of TTHA1718 and GB1

First, we validated the performance of the method with the fully simulated NOE peak lists from 3D structures of TTHA1718 and GB1. Structure calculations using either the conventional CYANA-OPALp approach or the Bayesian refinement were independently performed with these data sets. Table 1 shows the γ and σ values used for simulating peaks, and the predicted values by CYANA and the Bayesian method. Assuming that the peak intensities include not only distance information but also various other physical properties and noise, they were produced on fluctuations by normal distribution with the standard deviation σ. Figure 2 shows the structures derived from the two methods and distributions of the calibration constants in TTHA1718 generated by the Bayesian method. The fluctuations of the calibration constants and standard deviations are insensitive to the number of iterations in the Gibbs sampler, showing that the number of iterations were sufficient (Fig. 2C). In addition, it was confirmed that the replicas with the highest and lowest temperatures exchanged one or more times during the replica exchange MC process, and all variables as well as the conformational space of the protein were sampled sufficiently (data not shown). The expected γ and σ values were computed from the distributions (Table 1). Although it is not trivial to infer those parameters because the process requires simultaneously distance information among atoms, the results demonstrate that the Bayesian method determined those parameters rather precisely in all spectra of both TTHA1718 and GB1. There is no σ in the CYANA calculations since γ was determined such that the median value of all peak intensities in a given NOESY spectrum corresponds to a predefined distance of 4.0 Å. These γ values by CYANA are relatively accurate despite the fairly naïve presumption for obtaining the distances. Nevertheless, the Bayesian method shows better predictions than CYANA in all cases. The peak list of the ^{13}C-separated aromatic NOESY of TTHA1718 contained only 9 peaks, so that the predicted values differ slightly more from the original than those for the other spectra. The estimated σ are slightly smaller than the original values used to simulate the data, which means that the predicted distributions of γ were narrower and less ambiguous or underestimated.

Table 1. Parameters set for creating simulated NOESY peak lists

	peaks[a]	γ set[b]	CYANA	Bayesian	σ set[b]	CYANA	Bayesian
TTHA1718:							
^{13}C-NOESY	86	20.74	19.51	20.69	2.39	-	2.20
^{13}C-aro NOESY[c]	9	17.05	19.28	17.72	5.71	-	4.70
^{15}N NOESY	61	12.27	11.25	12.23	3.30	-	3.05
GB1:							
^{13}C-NOESY	97	20.53	19.56	20.65	2.33	-	1.23
^{13}C-aro NOESY[c]	15	13.17	11.70	12.96	4.76	-	2.82
^{15}N NOESY	41	12.32	10.95	12.34	3.74	-	2.68

[a] Number of used peaks
[b] γ and σ values used for creating the fully simulated NOESY peak lists.
[c] ^{13}C-separated aromatic NOESY

While the conventional method selects the 20 conformers with the lowest target function values from 100 computed with different random seeds, the Bayesian approach predicts the posterior distribution and allows to choose structure ensembles from the estimated distribution due to wider search on conformational space satisfying current experimental data by MCMC. In here, we exhibit all conformers within 1 standard deviation of the maximum of the posterior. Although Figure 2a shows 564 conformers of TTHA1718, the bundle structures are noticeably more converged than the 20 of the conventional method (Figure 2a), and closer to the reference structure deposited in PDB. The tube models depicting the coordinates of C$^{\alpha}$ atoms clearly demonstrate that the ensemble of the new method is less broadening than those of the conventional. Whereas the backbone RMSDs to the mean and the reference by CYANA were 1.62 and 2.22 Å, respectively, the Bayesian approach improved the accuracy of the structures to 1.45 Å of backbone RMSD to the reference.

Journal of Physics: Conference Series **699** (2016) 012005 doi:10.1088/1742-6596/699/1/012005

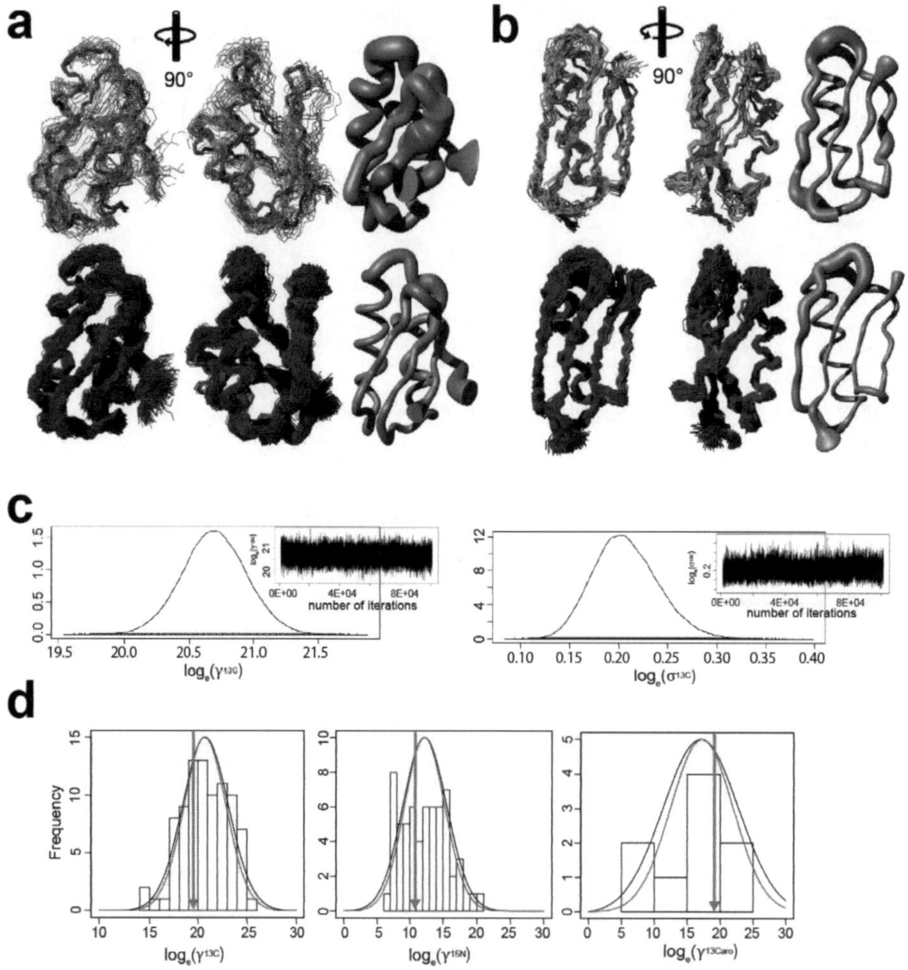

Figure 2. Bayesian structure calculations with fully simulated data of TTHA1718 and GB1. Superpositions of the 20 reference structures from the full data sets (red) conformations yielded by the Bayesian method (blue), showing the backbone (N, C$^\alpha$, C') atoms. Green superpositions are the 20 structures by the conventional method. Deviations of C$^\alpha$ atoms in the Bayesian-refined conformers are shown as tube models. (**a**) TTHA1718. (**b**) GB1. (**c**) Distributions of γ and σ for the ^{13}C-NOESY of TTHA1718. Insets show trajectories of γ and σ in the HMC sampling. (**d**) Histograms and distribution curves of γ for the ^{13}C-, ^{15}N-, and ^{13}C aromatic-NOESY spectra. The distributions of γ in the input peaks (black) and predicted by the Bayesian method (red), as well as the values determined by CYANA (green arrows) are shown.

These results demonstrate that the new Bayesian method for refining protein structures enables a more accurate interpretation of peak intensities than the conventional CYANA approach even in the case of having less structural information in the case that the data are composed of relatively ideal distributions of the peak intensities without additional noise signals. Next, we validated it for more realistic data containing noise signals and non-ideal NOE distributions.

4.2. Imperfect experimental peak lists of four different proteins

Journal of Physics: Conference Series **699** (2016) 012005 doi:10.1088/1742-6596/699/1/012005

The robustness of the algorithm for more realistic data was investigated with respect to missing peaks. Starting from experimental data sets for the proteins TTHA1718, SH2, and ENTH, approximately 55–90% of the peaks were randomly deleted. (see Methods).

Table 2. RMSD values of three proteins

run	1		2		3	
	conv.[a]	Bayes	conv.	Bayes	conv.	Bayes
ENTH:						
Peaks[b]	1229/6823		1010/6823		903/6823	
RMSD to mean (Å)[c]	1.48/2.01	0.69/1.06	1.92/2.46	0.74/1.18	2.13/2.80	1.24/1.70
RMSD to reference (Å)[d]	1.92/2.21	1.25/1.86	2.41/2.74	1.04/1.65	3.51/3.77	1.86/2.47
SH2:						
Peaks	2832/5422		2175/5422		2033/5422	
RMSD to mean (Å)	1.71/2.28	0.65/0.96	2.19/2.76	1.13/1.57	2.92/3.45	1.38/1.80
RMSD to reference (Å)	1.39/1.90	1.27/1.82	2.24/2.76	2.28/2.83	3.10/3.55	2.25/2.65
TTHA1718:						
Peaks	472/3205		348/3205		299/3205	
RMSD to mean (Å)	1.47/2.03	0.84/1.32	2.13/2.83	0.82/1.33	2.43/3.18	1.52/2.06
RMSD to reference (Å)	1.57/1.89	1.55/1.97	2.40/2.77	1.32/2.00	3.21/3.59	2.36/2.85

[a] conventional structure determination by CYANA
[b] Number of used/original peaks
[c] RMSD to the mean structure of the structure ensemble (backbone/sidechain)
[d] RMSD to the reference structure (backbone/all heavy atoms)

Results of the calculations used with three different data sets for ENTH, SH2 and TTHA1718 are summarized in Table 2. The reference structures of the three proteins were obtained by conventional CYANA calculations and OPALp refinement with the original experiment data set. The RMSD to the reference structure depended on the proteins and simulated data set. The Bayesian refinements exquisitely resulted in lower RMSDs to the mean and reference than the conventional structure calculations except for the RMSD of SH2 to the reference. Representative structure ensembles in the three independent calculations, run 3 of ENTH, run 3 of SH2 and run 2 of TTHA1718, clearly demonstrate that the Bayesian-refined structures approach the reference structure more closely than those of the conventional method (Fig. 3). The structures with 2–4 Å RMSD by the conventional method were most improved by the Bayesian method, whereas the improvement was less significant for the ones with approximately 1 Å RMSD to the reference, probably owing to the dynamics of molecules and imperfections of experimental data.

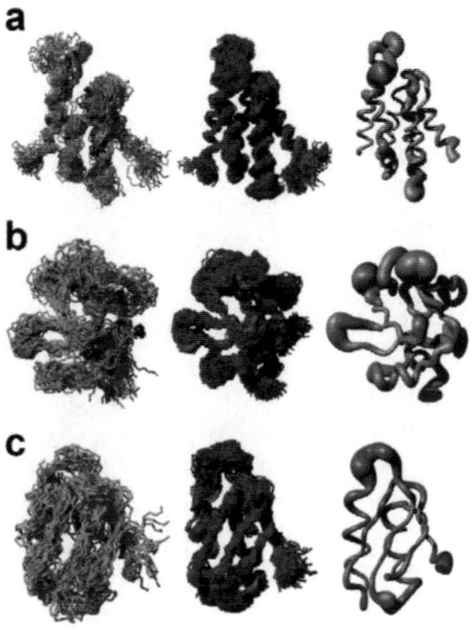

Figure 3. Structures bundles of 3 proteins calculated from reduced data sets by the conventional CYANA method (green) and Bayesian refinement (blue) superimposed on conventionally determined reference structures from the full data set (red), (a) ENTH. (b) SH2. (c) TTHA1718. Deviations of C^{α} atoms in the Bayesian-refined conformers are shown as tube models.

4.3. In-cell NMR structure determination

One of the targets of this method is to determine protein structures in living cells, so called in-cell NMR structure determination. So far, we obtained 3D NOESY spectra of two proteins in living cells, TTHA1718 and GB1, which were of sufficient quality for structure calculation. In the case of the first in-cell structure determination of the protein TTHA1718 [2, 3], we employed backbone hydrogen bond restraints for the β-sheet region where their existence was indicated by inter-strand NOEs. While this approach has been used also for *in vitro* NMR structure determinations, it may obscure deviations from canonical secondary structure manifested in the experimental data because it explicitly fixes standard secondary structure hydrogen bonds for ranges of residues identified by spectroscopists. It is instructive to improve in-cell NMR by our present approach of data-driven structure determination with minimum prior information. Thus, the in-cell TTHA1718 structures were calculated by the Bayesian approach without the hydrogen bond restraints. Since the chemical shift differences between *in vitro* and in-cell were relatively small for these proteins except for the metal binding region of TTHA1718, the accuracy of the in-cell structures could be validated by comparison with the *in vitro* conformers. The Bayesian approach was compared to the conventional refinement method with CYANA and OPALp.

Figure 4a shows the final, Bayesian-refined structures of GB1 in living cells. The Bayesian calculation computes the posterior distribution by replica exchange Monte Carlo simulation, which can efficiently sample the marginal likelihood (Fig. 4b,c). The structure ensemble was validated by PCA, in which the distribution of the first principal component (PC1) was composed of one major and other minor populations in the vicinity of the major region (Fig. 4d). The minor populations include structures that differ from the independently determined *in vitro* structure by approximately 1.6 Å RMSD. This suggests that PCA could efficiently exclude inaccurate ensembles. Thus, we chose from the posterior distribution 1416 conformations within 1 standard deviation from the mean of PC1 (Fig. 4a,d), whereas the 20 structures with the lowest energies are shown as the NMR structure ensemble in the conventional structure determination. The conventional structure ensemble is well converged with an average backbone RMSD of 0.40 Å to the mean structure. The backbone RMSD between the mean structure and the *in vitro* structure is 1.17 Å. Selecting the 20 highest posterior probability structures for comparison with the conventional method, the RMSD of these 20 structures is 0.49 Å to its mean and 1.02 Å to the *in vitro* structure. In addition, Bayesian inference provided the distributions of the calibration constants and their standard deviations for the three NOESY spectra. These distributions are broader for the $^{13}C/^{13}C$-separated NOESY than for the ^{15}N-separated and ^{13}C-separated NOESYs, presumably due to the smaller number of peaks (Fig. 4e,f).

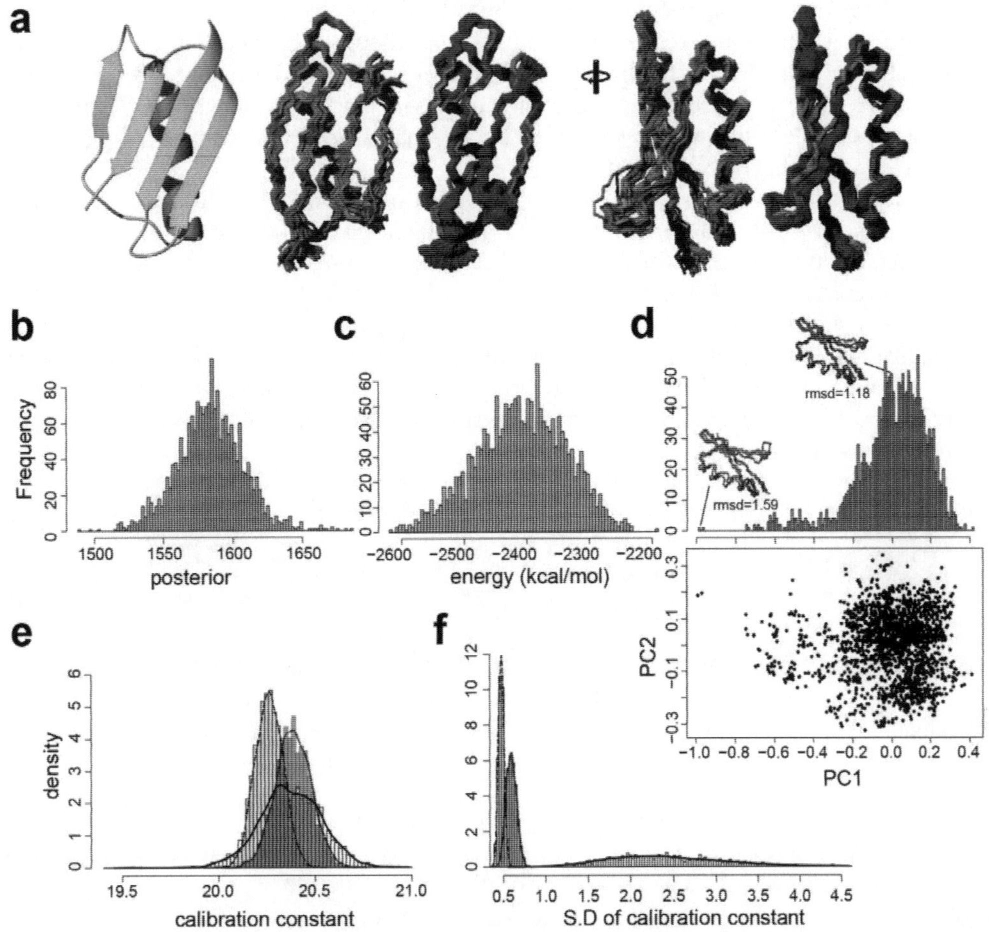

Figure 4. NMR structure of the protein GB1 in living *E. coli* cells. (**a**) Ribbon diagram of the structure of GB1 in living *E. coli* cells with the highest posterior (left). Superpositions of the 20 final structures of purified GB1 in vitro (red) and the 1416 conformations representing the in-cell structure of GB1 within one standard deviation of the first principal component of the posterior distribution (blue). Green superpositions are the conventional CYANA structure calculation. (**b**) Posterior distribution of the in-cell GB1 conformations. (**c**) Potential energy distribution of the in-cell GB1 conformations. (**d**) Distributions of the first principal component (top) and the first and second ones (bottom). (**e**) Distributions of the calibration constants computed by the Bayesian method of $^{13}C/^{13}C$-separated (bold), ^{13}C-separated (dashed), and ^{15}N-separated (dotted) NOESY spectra. (**f**) Corresponding distributions of the standard deviation of the calibration constants.

5. Conclusion

In summary, the results of this paper show that the Bayesian-based structure calculation implemented into CYANA enables to significantly improve structures calculated by the original CYANA/OPALp method. A currently routinely used approach for NMR structure determination using CYANA is to obtain global structures by TAMD and automated NOESY cross peak assignment, and to refine the structures by other software like OPALp, CNS or Amber. However, those structure calculation and subsequent refinement methods have considered the weight factor of the potential energy and the calibration constant describing the relationship between NOE intensities and atom-atom distances exclusively in the initial stage by user-predefined or naive approaches. The weight and calibration constants are closely associated with possible conformational ensembles, and conversely those conformations are restricted by distance restraints derived from those parameters. Thus, it is rational to compute the structures and parameters simultaneously from the experimental data. Since it usually requires substantial calculation time to obtain global folds of proteins and accurate NOE assignments due to the recursive steps of the structure calculation and assignment, our method first performs the conventional structure calculation with a simplified target function to obtain the global fold, and then refines these structures by the Bayesian method.

In addition, it is cumbersome for many spectroscopists to use several software packages in which parameters and data formats are different, due to strong parameter dependence, non-standard format requirements, lack of documentation for conversion from one software to others, or high computation time demands. In contrast, using the CYANA refinement algorithm is straightforward. Moreover, seamless processes without manual intervention are a prerequisite for fully automatic structure determination like the FLYA method. The full automatic calculation is usually composed of automatic peak picking, chemical shift and NOE assignment, structure calculation, and refinement. All the stages should be principally validated by structures and experimental data, and a recursive approach of the stages can be the most reliable.

A principal advantage of the Bayesian refinement algorithm is that it searches a wider conformational space with the other explanatory parameters by the replica exchange and Gibbs sampler methods. It can also be used with solid state NMR data, for structure-based assignment, or resonance assignment based exclusively on NOESY spectra, which will be treated elsewhere.

References

[1] Wang R Y R, Kudryashev M, Li X M, Egelman E H, Basler M, Cheng Y F, Baker D and DiMaio F 2015 *Nat. Methods* **12** 335–338
[2] Sakakibara D, Sasaki A, Ikeya T, Hamatsu J, Hanashima T, Mishima M, Yoshimasu M, Hayashi N, Mikawa T, Wälchli M, Smith B O, Shirakawa M, Güntert P and Ito Y 2009 *Nature* **458** 102–105
[3] Ikeya T, Sasaki A, Sakakibara D, Shigemitsu Y, Hamatsu J, Hanashima T, Mishima M, Yoshimasu M, Hayashi N, Mikawa T, Nietlispach D, Wälchli M, Smith B O, Shirakawa M, Güntert P and Ito Y 2010 *Nat. Protoc.* **5** 1051–1060
[4] Rieping W, Habeck M and Nilges M 2005 *Science* **309** 303–306
[5] Habeck M, Nilges M and Rieping W 2005 *Phys. Rev.* E **72** 031912
[6] Güntert P 2003 *Prog. Nucl. Magn. Reson. Spectrosc.* **43** 105–125
[7] Güntert P, Mumenthaler C and Wüthrich K 1997 *J. Mol. Biol.* **273** 283–298
[8] Duan Y, Wu C, Chowdhury S, Lee M C, Xiong G M, Zhang W, Yang R, Cieplak P, Luo R, Lee T, Caldwell J, Wang J M and Kollman P 2003 *J. Comput. Chem.* **24** 1999–2012
[9] Baker N A 2005 *Curr. Opin. Struct. Biol.* **15** 137–143
[10] Yamashita H, Endo S, Wako H and Kidera A 2001 *Chem. Phys. Lett.* **342** 382–386
[11] Northrup S H and McCammon J A 1980 *Biopolymers* **19** 1001–1016
[12] Güntert P and Buchner L 2015 *J. Biomol. NMR* **62** 453–471
[13] Kirchner D K and Güntert P 2011 *BMC Bioinformatics* **12** 170
[14] Koradi R, Billeter M and Wüthrich K 1996 *J. Mol. Graphics* **14** 51–55
[15] Koradi R, Billeter M and Güntert P 2000 *Comput. Phys. Commun.* **124** 139–147
[16] Frick I M, Wikström M, Forsén S, Drakenberg T, Gomi H, Sjöbring U and Björck L 1992 *Proc. Natl. Acad. Sci. USA* **89** 8532–8536

[17] López-Méndez B, Pantoja-Uceda D, Tomizawa T, Koshiba S, Kigawa T, Shirouzu M, Terada T, Inoue M, Yabuki T, Aoki M, Seki E, Matsuda T, Hirota H, Yoshida M, Tanaka A, Osanai T, Seki M, Shinozaki K, Yokoyama S and Güntert P 2004 *J. Biomol. NMR* **29** 205–206
[18] Scott A, Pantoja-Uceda D, Koshiba S, Inoue M, Kigawa T, Terada T, Shirouzu M, Tanaka A, Sugano S, Yokoyama S and Güntert P 2005 *J. Biomol. NMR* **31** 357–361
[19] Scott A, Pantoja-Uceda D, Koshiba S, Inoue M, Kigawa T, Terada T, Shirouzu M, Tanaka A, Sugano S, Yokoyama S and Güntert P 2004 *J. Biomol. NMR* **30** 463–464
[20] Johnson B A and Blevins R A 1994 *J. Biomol. NMR* **4** 603–614
[21] Boucher W 2010 Azara v. 2.8, in, Cambridge, UK

International Meeting on High-Dimensional Data-Driven Science (HD^3-2015) IOP Publishing
Journal of Physics: Conference Series **699** (2016) 012006 doi:10.1088/1742-6596/699/1/012006

Imaging black holes with sparse modeling

Mareki Honma[1,2], Kazunori Akiyama[1,3,4], Fumie Tazaki[1], Kazuki Kuramochi[1,3], Shiro Ikeda[5,6], Kazuhiro Hada[1] and Makoto Uemura[7]

[1]Mizusawa VLBI Observatory, National Astronomical Observatory of Japan, Hoshigaoka, Ohshu, Iwate 023-0861, Japan
[2]Department of Astronomical Science, SOKENDAI (The Graduate University for Advanced Studies), Hoshigaoka, Ohshu, Iwate 023-0861, Japan
[3]Department of Astronomy, Graduate School of Science, The University of Tokyo, Hongo, Bunkyo-ku, Tokyo 113-0033, Japan
[4]Haystack Observatory, Massachusetts Institute of Technology, Route 40, Westford, MA 01886, USA
[5]Institute of Statistical Mathematics, Tachikawa, Tokyo 190-8562, Japan
[6]Department of Statistical Science, SOKENDAI (The Graduate University for Advanced Studies), Tachikawa, Tokyo 190-8562, Japan
[7]Astrophysical Science Center, Hiroshima University, Higashi-Hiroshima, Hiroshima 739-8526, Japan

E-mail: mareki.honma@nao.ac.jp

Abstract. We introduce a new imaging method for radio interferometry based on sparse-modeling. The direct observables in radio interferometry are visibilities, which are Fourier transformation of an astronomical image on the sky-plane, and incomplete sampling of visibilities in the spatial frequency domain results in an under-determined problem, which has been usually solved with 0 filling to un-sampled grids. In this paper we propose to directly solve this under-determined problem using sparse modeling without 0 filling, which realizes super resolution, i.e., resolution higher than the standard refraction limit. We show simulation results of sparse modeling for the Event Horizon Telescope (EHT) observations of super-massive black holes and demonstrate that our approach has significant merit in observations of black hole shadows expected to be realized in near future. We also present some results with the method applied to real data, and also discuss more advanced techniques for practical observations such as imaging with closure phase as well as treating the effect of interstellar scattering effect.

1. Introduction

Now it is believed that nearly all the galaxies harbor a super-massive black hole at their centers, and it is considered as the central engine of active galactic nuclei (AGN), which emit enormous amount of radiation from a compact region. Exactly speaking, however, the observational evidence obtained so far only reveals the existence of extremely dense objects, and there has been no direct confirmation that these compact and dense objects are real black holes surrounded by an event horizon, which is the critical radius within which even light cannot escape.

In order to obtain the direct confirmation of existence of black holes, one has to resolve the structure around a black hole at a scale of Schwarzschild radius. Detection of the black hole shadow, which is produced by unstable orbits of photons near the black hole, is the most direct confirmation of existence of black holes. Since the black holes are extremely compact, detection of the black hole shadows requires unprecedentedly-high angular resolution. As far as observations with electro-magnetic wave are concerned, the angular resolution of any telescope is roughly given by the ratio of the wavelength to the diameter of the telescope, namely,

$$\theta \approx \lambda/D. \tag{1}$$

Here θ is the angular resolution, λ is the wavelength, and D is the diameter of the telescope (or in case of radio interferometer, the maximum baseline). Currently, the highest angular resolution is given by VLBI (Very Long Baseline Interferometry, [1]), which utilizes the maximum baseline length corresponding to the diameter of the Earth (or even beyond by using an on-board radio telescope orbiting the Earth, e.g, VSOP launched by Japan or Russian Radio Astron satellite).

The largest black hole known so far (in terms of angular size) is Sgr A*, the supper-massive black hole at the center of our own Galaxy (the Milky Way). According to the most recent observations of stellar orbits around Sgr A*, the mass of Sgr A* is determined to be $4\times10^6 M_\odot$, where $M_\odot$ denotes the mass of the Sun. At a distance of 8 kpc, the angular size of Schwarzschild radius of Sgr A* is ~10 μas (note that 1 μas corresponds to 10^{-6} arcsecond). The second largest black hole is the one at the center of a giant elliptical galaxy M87 located in the Virgo cluster. With an expected mass of 6×10^9 $M_\odot$ and an expected distance of 17 Mpc, the angular radius of M87's black hole is about 7 μas. Therefore, for resolving super-massive black holes, one requires an angular resolution at 10 μas level or higher. In order to achieve that high angular resolution, the EHT (Event Horizon Telescope) collaboration has been building a global VLBI array operating at a wavelength of 1.3 mm or shorter [2]. Pioneering observations of the EHT (with only three stations) have already detected compact structures in Sgr A* and M87 black holes corresponding to a scale of 4~5 Schwarzschild radii, demonstrating that mm-VLBI is in fact capable of resolving the structure in the vicinity of super-massive black holes [3, 4, 5].

Currently the EHT collaboration has extended the array to include more stations in US, Europe, Mexico, Chile, the Antarctica, and Green-land. Amongst these stations, the key is ALMA (Atacama Large Millimeter and sub-millimeter Array) located in Chile, because it has extremely-high sensitivity. In order to realize VLBI observations with ALMA, the international collaboration called APP (ALMA Phase-up Project) has recently completed the ALMA's VLBI capability. In fact, VLBI with ALMA will be plausibly in operation from ALMA Cycle-4, starting in early 2017.

However, even when the EHT with ALMA becomes available, one would require higher angular resolution than the array could provide at the initial phase. For instance, the initial observations of the full EHT is planned to be conducted at a wavelength of 1.3 mm, and when combined with the maximum baseline length of ~9000 km, this provides an angular resolution of ~25 μas. Meanwhile, the angular size of the black hole shadow is expected to be a few times of the Schwarzschild radius, and thus the angular resolution provided by the initial EHT may not be high enough to clearly resolve a black hole shadow. Another uncertainty is the mass of the black hole, with which the angular size of a black hole is linearly scaled. For instance, the black hole mass of M87 is still under debate and there exists a factor of two difference among the recent determinations of the black hole mass. This indicates that the angular size of the M87 black hole is also uncertain by a factor of two, and thus in case of smaller mass ($3 \times 10^9 M_\odot$, [6]), it may not be easy to resolve the black hole shadow even with the angular resolution of EHT.

In order to overcome this issue, we have been developing a new method of radio interferometric imaging based on sparse modeling (Honma et al. 2014, [7]), which significantly boost the angular resolution of image compared to the standard method currently used. In this paper we would like to introduce its basic concept as well as recent results of simulation and its application to real observational data.

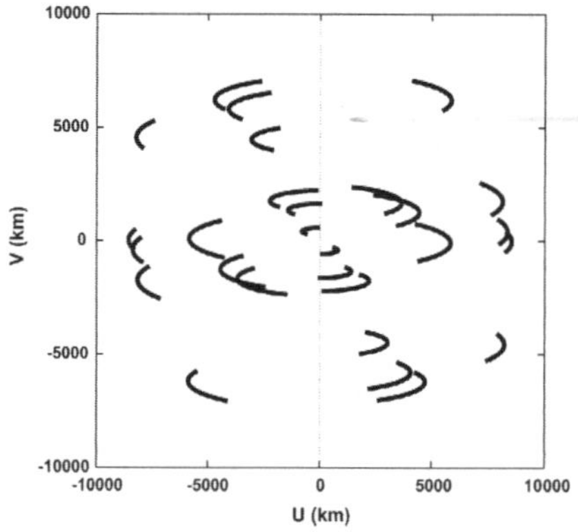

Figure 1. Simulated U-V coverage of the observations of M87 black hole with the six-station EHT array. Here U and V are shown in unit of km.

2. Radio Interferometry Imaging and Sparse Modeling

The basic equation between the direct observables (visibilities $\mathcal{V}$) and the image I is the two-dimensional Fourier transform [1], namely,

$$I(x, y) = \iint \mathcal{V}(u, v)\, e^{-2\pi i(ux+vy)}\, du\, dv. \tag{2}$$

Here the visibility $\mathcal{V}(u, v)$ is a function of spatial frequency $(u,\ v)$, which is projected baseline length of interferometer normalized with the wavelength, i.e., $u \equiv U/\lambda$ and $v \equiv V/\lambda$, where U and V are physical lengths of the projected baselines. In theory, the parameters u and v should run from $-\infty$ to $+\infty$, but in practice this is impossible as the maximum U and V are limited by the longest baseline length. Practically, the Fourier transform in the above equation is executed as a discrete Fourier transformation over a finite range of sampling grids. Since the Fourier transformation is one-to-one conversion from visibility domain to image domain (and vice versa), to obtain an image with N^2 grids, one has to have full visibilities for N^2 grids in the u-v plane. However, sparse locations of VLBI stations make the sampling of the u-v space far from complete (e.g., see figure 1), and a large number of visibility grids in the u-v space remain un-sampled.

In the standard process of interferometric imaging, these un-sampled grids are filled with 0, which we call "0-filling". By doing this, one can conduct the Fourier transform from the visibility domain to the image domain. However, insertion of 0 to un-sampled grids causes severe degradation of the resultant image. In fact, it is this 0-filling process that causes infinite number of side lobes and finite beam size in the synthesized beam. Practically, the degradation of the images is corrected for by a method called CLEAN.

Sparse modeling provides a new approach to overcome above issues by directly solving the under-determined problem. This approach avoids 0-filling and thus consequently provides breakthrough for obtaining better resolution than the standard imaging with CLEAN algorithm.

As an introduction of sparse modeling, here we introduce lp-norm $||\mathbf{x}||_p$ as follows,

$$||\mathbf{x}||_p = ||\mathbf{x}||_0 \ \text{ for } p = 0, \tag{3}$$

$$||\mathbf{x}||_p = (\sum |x_i|^p)^{1/p} \ \text{ for } p > 0. \tag{4}$$

Here $||\mathbf{x}||_0$ describes the number of non-0 components of $\mathbf{x}$. Thus $l0$-norm directly represents the sparseness of the vector $\mathbf{x}$, i.e., if $l0$-norm is smaller, the vector $\mathbf{x}$ has more 0 components

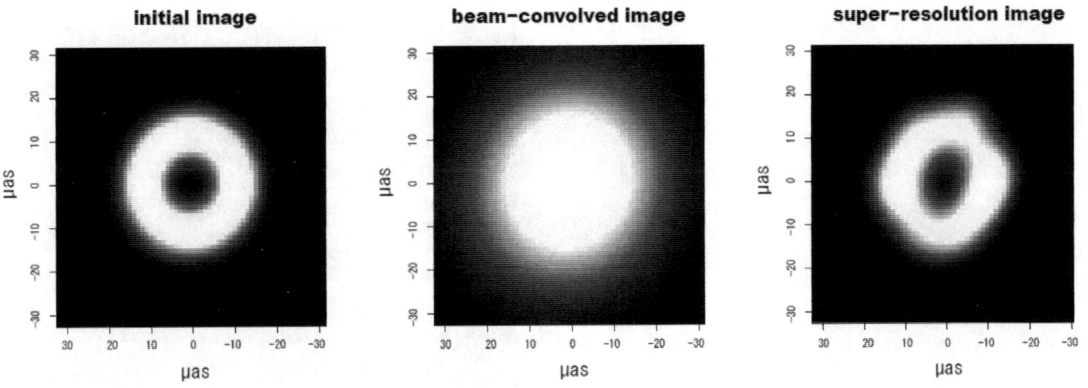

Figure 2. Simulated images of M87. From left to right, the initial model, the image with 0-filling, and the image with LASSO. Improvement of resolution in the LASSO image is significant.

and thus more "sparse". The basic idea of sparse modeling is to solve the under-determined problem by using the sparseness of the solution as regularizer, i.e., by selecting the optimal solution having the smallest $l0$-norm such as,

$$\mathbf{x} = \arg\min ||\mathbf{x}||_0 \text{ subject to } \mathbf{z} = \mathbf{A}\mathbf{x}. \tag{5}$$

Note that $\mathbf{z}$ is the observational quantities (visibilities in the present paper), $\mathbf{x}$ is the solution to be obtained (values of image grids), and $\mathbf{A}$ is the observational matrix (two-dimensional Fourier transform).

Practically speaking, finding the optimal solution with the smallest $l0$-norm requires huge computation cost because one has to try all the possible combinations of 0 components in $\mathbf{x}$. Therefore, obtaining an interferometry image using $l0$ is practically impossible. Instead, in this paper we alternatively use LASSO (Least Absolute Shrinkage and Selection Operator), which was originally developed by Tibshirani (1996) [8]. LASSO uses $l1$-norm to find the optimal solution from infinite number of solutions by,

$$\mathbf{x} = \arg\min \left(||\mathbf{z} - \mathbf{A}\mathbf{x}||_2^2 + \Lambda||\mathbf{x}||_1\right). \tag{6}$$

Here the first term in the argument describes goodness of fit, while the second term represents the penalty term based on the $l1$-norm. We note that Λ is a regularization parameter, which can be used to adjust the degree of sparsity by changing the weight of the $l1$-norm penalty.

Since the Fourier transformation given in equation (2) can be also written in a form of matrix (i.e., simultaneous linear equations), one can use LASSO to solve the under-determined equation in interferometer imaging.

3. Simulation

In order to demonstrate the capability of sparse modeling in interferometry imaging, here we first show the results of simulated observations of the super-massive black hole of M87 with the EHT. Here we consider the EHT array with six stations including CARMA (Combined Array for Research in Millimeter-wave, California, USA), SMTO (Submillimeter Telescope Observatory, Arizona, USA), JCMT (James Clerk Maxwell Telescope, Hawaii, USA), ALMA (Atacama Large Millimeter/submillimeter Array, Chile), LMT (Large Millimeter Telescope, Mexico), and IRAM (Institute de Radio Astronomie Millimétrique, Spain). All these stations are ready or nearly

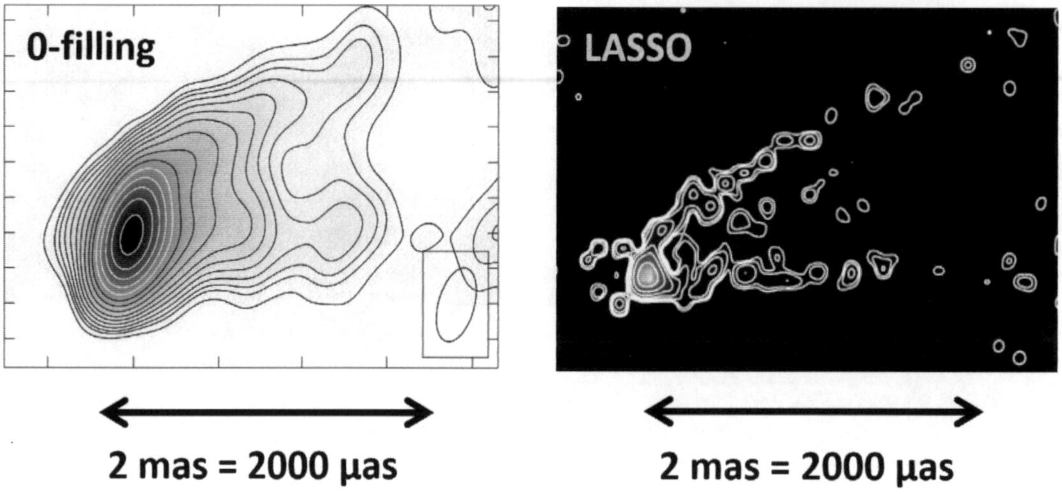

Figure 3. Standard and LASSO images of M87 observed with VLBA at a wavelength of 7 mm. In the two plots, exactly the same data are used. The angular resolution is better in the LASSO image, and the detailed structure of the M87 jet can be traced in more detail.

ready for VLBI observations at 1.3 mm, and the real data will become available within a few years from now on.

For the target source M87, we consider the smaller mass case with a black hole mass of $3 \times 10^9 M_\odot$ [6], for which resolving the black hole shadow would be more difficult. This corresponds to an angular size of the Schwarzschild radius of 4 μas at a source distance of 17 Mpc. The angular size of the black hole shadow is expected to be $\sim 20\ \mu as$, which is comparable or slightly less than the angular resolution (given by λ/D) of the EHT array operating at 1.3 mm. Figure 1 shows the u-v coverage of the simulated EHT array. As seen in the figure, the array provides relatively good sampling of the u-v plane, with maximum baselines of $8000 \sim 9000$ km in both north-south and east-west directions. However, it is obvious that there exist many "holes" in the u-v plane, where no visibility data are sampled. This under-sampling of the u-v plane makes the Fourier transformation in the interferometric imaging an under-determined problem.

In figure 2, we show the simulation results of M87 black hole images with the EHT. Here we assumed a simple doughnut-shape black hole shadow (the left panel), and created the simulated visibilities using the u-v coverage shown in figure 1 with a noise corresponding to $\sim$5% of the peak amplitude. Then we obtained the images of M87 by the standard method with 0-filling to un-sampled grids (the panel in the middle) and by LASSO (in the right). As shown in figure 2, imaging with LASSO provides better angular resolution and one can clearly trace the doughnut-shaped structure. Meanwhile, the standard imaging with 0-filling smears out the structure of the black hole shadow. This clearly demonstrates the merit of LASSO, which directly solves the under-determined problem of interferometric imaging.

4. Application to real data

We have seen that LASSO can provide a better angular resolution in interferometric imaging based on simulated data for the EHT observations. Here, to test if this applies to the real observations, we process the data of real VLBI observations in the same manner described above. For this purpose, here we use VLBA observations of M87 at 43 GHz obtained by Hada et al.(2011) [9]. The observations were conducted toward the same target for the simulated EHT observations described above, but with a different array (10 stations of VLBA) and a different observing wavelength (7 mm). The resolution of the VLBA observations is not high enough to resolve a structure at a scale of the black hole shadow, but instead, the VLBA observations at 7 mm are powerful to trace the relativistic jet launched from the central black hole of M87 (e.g., [9]).

In figure 3, we show the resultant image of M87 obtained by the standard imaging (with 0-filling, left panel) and that by LASSO (right panel). Here the two images are reproduced from exactly the same data. Nevertheless, the LASSO image shows a dramatic improvement of the angular resolution, and it provides detailed structure of M87 jet: two most distinct features are a) clear separation of limb structures and b) obvious detection of the counter-jet components (to the left of the bright radio core in figure 3). The results presented here for the real VLBA observations clearly demonstrate the potential of sparse modeling technique for improving the angular resolution of the images. Hence, interferometric imaging with LASSO is expected to provide significant merit in studying the structure around the black hole and jet.

5. More advanced techniques

5.1. *Imaging with closure phase*

In the previous section, we implicitly assumed that we could obtain the full visibility information, i.e., both the amplitudes and phases of the complex visibilities. However, in practice, it is sometimes difficult to obtain the full visibilities because accurate calibration of phase is still an issue for mm VLBI. In fact, in the pioneering works of 1.3-mm observations of Sgr A* and M87 ([3, 4]), the only observables were amplitudes, and the phase information was totally missed and thus images are not yet available.

Recently there have been attempts to obtain the phases in mm VLBI, but obtaining complete phases (i.e., baseline-based visibility phase) is still difficult. Currently the only available information related with the phase is the closure phase, which is the phase term of the bi-spectrum defined as

$$\mathcal{V}_{123} = |\mathcal{V}_{12}||\mathcal{V}_{23}||\mathcal{V}_{31}|e^{i(\phi_{12}+\phi_{23}+\phi_{31})}. \tag{7}$$

Note that the closure phase $\phi_{123} \equiv \phi_{12}+\phi_{23}+\phi_{31}$ is obtained along a set of triangular baselines. The merit of introduction of the closure phase is that it is significantly easier to obtain compared to the individual baseline-based phase such as ϕ_{12}, ϕ_{23}, and ϕ_{31}. This is due to the fact that the largest error source of phase calibration is the tropospheric fluctuation at each station, and such station-based errors are totally canceled out when the closure is taken along the triangular baselines. In fact, recently Akiyama et al.[10] reports the first detection of closure phase in the VLBI observations of M87 at 1.3 mm. At the early phase of mm VLBI observations with ALMA, it is also expected that the available phase information could be limited to closure phases. The trade-off of closure phase is that one cannot describe the observational equation in a form of simultaneous linear equations, and the equation becomes non-linear. Therefore, it requires an iterative approach to handle the imaging with closure phase, and below we briefly discuss two possible approaches for this issue.

One way to deal with closure phase is to estimate the baseline-based phases from closure phases. Since the number of independent closure phases is always smaller than the number of independent baseline-based phases, this problem always becomes an under-determined problem, and this can be solved by techniques similar to sparse modeling, i.e., by introducing a reasonable regularizer. How to select an optimum regularizer depends on the nature of images. Ikeda et al.(2016)[11] proposes a method called PRECL (Phase REtrieval from CLosure phase), which uses smoothness of phase in the u-v plane as a regularizer: the phases should be similar if two baseline vectors are similar to each other, while the phases can be largely different if two baseline vectors are totally different. By introducing such a regularizer, Ikeda et al.[11] demonstrated

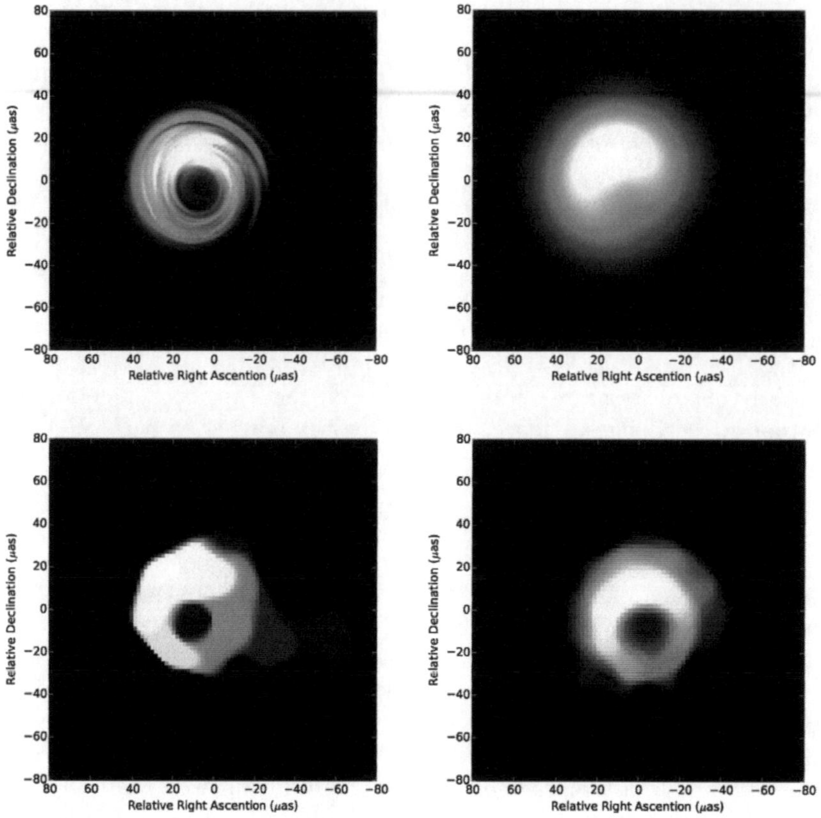

Figure 4. Imaging simulation of M87 using closure phase. The top left panel is the model by Dexter et al.[12], top right is the model convolved with a synthesized beam with 25 μas. The bottom left is the image with phase retrieval from closure (PRECL, [11]), and bottom right is the direct solution of non-linear problem based on iterative approach.

that the baseline-based phases can be well-reproduced from the closure phases using the real full-track interferometric observations with 10 stations. In figure 4, we show an example of more realistic model image of M87 black hole (Dexter et al. 2012, [12]) with the image constructed from closure phase using PRECL (bottom left of figure 4). Note that for the results in figure 4, in addition to the $l1$-norm we introduced another regularization term called the total variation (TV, see the next subsection) to obtain an effective angular resolution of the image. If this kind of method turns out to be applicable to general cases, this will introduce significant merit to the interferometric imaging, as this will simplify the phase calibration process of radio interferometer.

Another approach is to solve the non-linear problem directly by iteration. This can be done by calculating cost function (which includes model errors as well as sparsity term) and its gradient for a given image, and by searching for the image with the minimum cost based on iterative improvements of the image. Akiyama et al.[13] have shown that the practical algorithm such as L-BFGS-B [15, 16] can be used to solve such kind of problem at a practical computational cost. Of course, to obtain an optimum solution by an iterative process, the key is how to select the initial model in the iteration. Selecting a good initial model is still an open issue, but there are a few possible options to handle this problem. For instance, one may start with a low resolution image (small number of the grids) and use the result of low-resolution imaging as the initial model for imaging with higher resolution. The bottom right panel of figure 4 shows the image obtained in such a way. Another possibility is to use other model to find an initial guess of images. The method like PRECL described above would be a good candidate for finding an initial image. The merit of the direct solution of the non-linear problem is that one can use regularizers in any form, such as non-linear cost function.

5.2. Optimal image resolution

A practical question on the super-resolution images obtained in the previous sections is: *what is the effective resolution of the images?* The effective resolution could be dependent of observational conditions such as the *u-v* coverage of the array and signal-to-noise ratio of the data. The most direct way to obtain the effective resolution is to use the data (and also the resultant images) themselves. As one of such approaches, here we propose to introduce the total variation (TV) term to the regularizer of the imaging equation, namely,

$$\mathbf{x} = \arg\min \left(||\mathbf{z} - \mathbf{A}\mathbf{x}||_2^2 + \Lambda||\mathbf{x}||_1 + \Lambda_{\mathrm{TV}} \sum \sqrt{|I_{i+1,j} - I_{i,j}|^2 + |I_{i,j+1} - I_{i,j}|^2} \right), \tag{8}$$

where the subscripts i and j denote the indexes of image grids. The inclusion of the TV term will smoothen the image and by varying the weight of the TV term (Λ_{TV}), one can change the effective resolution of the image. Thus, by searching the optimum value of the parameter, one may determine effective resolution. Note that large Λ_{TV} results in resolution too low to reproduce the observed visibilities while too small Λ_{TV} would not introduce any improvement of goodness of fit in the visibilities. Thus, the optimum resolution may be determined by the image with Λ_{TV} that gives this transition in the goodness of fit.

5.3. Deblurring scattering effect

In the above discussion, we mainly focus on the case for M87. The other promising target for resolving the black hole shadow is Sgr A*, the super-massive black hole in our Galaxy. Since the angular size of Schwarzschild radius is expected to be $\sim$10 μas, the black hole shadow of Sgr A* should be larger than M87. Meanwhile, an issue for observing Sgr A* is the scattering effect due to the interstellar medium lying between the Earth and the Galactic center. In fact, the scattering effect is still non-negligible for VLBI observations at 1.3 mm. The main effect of the scattering can be described by a "Gaussian-kernel", which blurs the image of the black hole shadow. There have been studies to reconstruct a shadow image by deblurring the scattering effect (e.g., Fish et al. 2014, [14]). Since the main effect of the scattering is to worsen signal-to-noise ratio in particular at longer baselines, the effective resolution of the image will be degraded. To overcome this issue, we have been studying if sparse modeling approach is also effective for deblurring the scattered image of Sgr A*, which will be presented in a separate future work.

6. Future prospects

The new imaging techniques described above are all targeted at the forthcoming mm-VLBI observations (EHT) including ALMA (Atacama Large Millimeter/sub-millimeter Array). In order to make ALMA ready to conduct mm VLBI observations, the EHT collaboration and the ALMA Phase-up Project (APP) have brought VLBI capability to ALMA, and it is nearly completed. In fact, intercontinental VLBI fringes were successfully detected at both 4 mm and 1.3 mm with phase-up ALMA. Currently it is planned to open ALMA's VLBI capability for ALMA Cycle-4, which will be called in 2016 spring, and actual observations will be expected to start in early 2017. Therefore, we will obtain real mm-VLBI data in late 2017 or early 2018, and with the techniques described above hopefully we will be able to detect the black hole shadow for the first time.

Acknowledgments

The present study was financially supported by MEXT/JSPS grant-in-aid for innovative area (KAKENHI No 25120007).

References

[1] Thompson A R, Moran J M and Swenson G W Jr 2001 *Interferometry and Synthesis in Radio Astronomy*, 2nd edition (New York: Wiley-Interscience)
[2] Doeleman S S, Fish V L, Broderick A E, Loeb A and Rogers A E E 2009 *Astrophys. J.* **695** 59
[3] Doeleman S S et al. 2008 *Nature* **455** 78
[4] Doeleman S S et al. 2012 *Science* **338** 355
[5] Fish V et al. 2011 *Astrophys. J.* **726** 36
[6] Gebhardt K and Thomas J 2009 *Astrophys. J.* **700** 1690
[7] Honma M, Akiyama K, Uemura M and Ikeda S 2014 *Publ. Astron. Soc. Japan* **66** 95
[8] Tibshirani R 1996 *J. R. Statist. Soc.* B **58** 267
[9] Hada K et al. 2011 *Nature* **477** 185
[10] Akiyama K et al. 2015 *Astrophys. J.* **807** 150
[11] Ikeda S et al. 2016 submitted to *Publ. Astron. Soc. Japan*
[12] Dexter J, McKinney J C and Agol E 2012 *Mon. Not. R. Astron. Soc.* **421** 1517
[13] Akiyama K et al. 2016 in preparation
[14] Fish V et al. 2014 *Astrophys. J.* **795** 134
[15] Byrd R H, Lu P and Nocedal J 1995 *SIAM J. Sci. Stat. Com.* **16** 1190
[16] Zhu C, Byrd R H and Nocedal J 1997 *ACM Trans. Math. Soft.* **23** 550

Future of High-Dimensional Data-Driven Exoplanet Science

Eric B. Ford

Center for Astrostatistics, Institute for CyberScience, Center for Exoplanets and Habitable Worlds, Department of Astronomy & Astrophysics, The Pennsylvania State University, 525 Davey Laboratory, University Park, PA, 16802, USA

eford@psu.edu

Abstract. The detection and characterization of exoplanets has come a long way since the 1990's. For example, instruments specifically designed for Doppler planet surveys feature environmental controls to minimize instrumental effects and advanced calibration systems. Combining these instruments with powerful telescopes, astronomers have detected thousands of exoplanets. The application of Bayesian algorithms has improved the quality and reliability with which astronomers characterize the mass and orbits of exoplanets. Thanks to continued improvements in instrumentation, now the detection of extrasolar low-mass planets is limited primarily by stellar activity, rather than observational uncertainties. This presents a new set of challenges which will require cross-disciplinary research to combine improved statistical algorithms with an astrophysical understanding of stellar activity and the details of astronomical instrumentation. I describe these challenges and outline the roles of parameter estimation over high-dimensional parameter spaces, marginalizing over uncertainties in stellar astrophysics and machine learning for the next generation of Doppler planet searches.

1. Introduction

Historically, the field of exoplanets has shied away from the challenges of high-dimensional data analysis. Typically, astronomers rely on physical intuition to guide data analysis, reducing large and complex astronomical data into a few observable parameters for further analysis. Only recently has it become practical for astronomers to analyze their hard-won data in its full glory. This presents exciting opportunities for statisticians and computer scientists to contribute to advancing humankind's knowledge of planets beyond our solar system. However, doing so will require careful attention to many "real-world" effects that complicate the measurement process.

2. Doppler Planet Surveys

2.1. Doppler Observations

Doppler planet surveys aim to discover planets by repeatedly measuring the "radial velocity" of a star (i.e., stellar velocity projected onto the line-of-sight). For an isolated star, the star's velocity would be constant (relative to the solar system barycenter). For a star hosting one or more planets, the reflex velocity of the star about the planetary system's center-of-mass and its variation with time provides a mechanism for detecting and characterizing its planets. Each observation yields a spectrum of the target star, including many absorption lines which provide information necessary to measure the star's velocity. The raw data from each Echelle spectrograph consists of one or more images measured with CCD chips, each containing several "orders" (rows) of the star's spectrum (see Figure 1). Astronomers take dozens or hundreds of spectra of a given target star, spread over days to decades. Discovering exoplanets requires performing differential radial velocity measurements, i.e., measuring very small shifts of the spectra lines, that correspond to just a few silicon atoms. Developing practical techniques to achieve the required Doppler precision and stability was a major accomplishment that enabled the discovery of exoplanets [1].

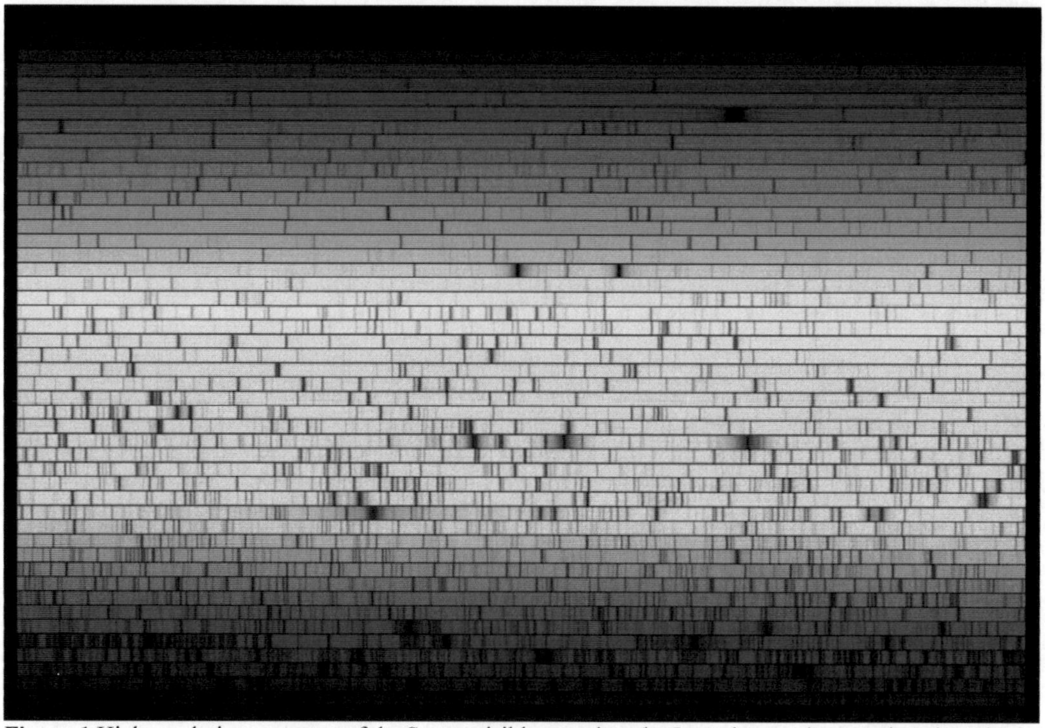

Figure 1 High resolution spectrum of the Sun at visible wavelength. In each row, the wavelength varies by 6 nm, starting from 700 nm at the top and extending to 400nm at the bottom. Dark regions are due to stellar absorption lines, as the effects of Earth's atmosphere have been removed from this simulated image [2]. Real astronomical observations are further complicated by curvature of each row, non-linear wavelength scale, and non-uniform illumination as a function of wavelength. (Image Credit: N.A.Sharp, NOAO/NSO/Kitt Peak FTS/AURA/NSF.)

2.2. Bayesian Analysis of Doppler Observations

Existing Doppler survey rely heavily on astronomical expertise to reduce each spectrum into a few numbers, e.g., the time of observation, a differential velocity measurement of the star's radial velocity, and an estimate of the uncertainty in the velocity measurement. A time-series of such measurements can be analyzed to discover planets and characterize their masses and orbital properties (See Figure 2). The likelihood is derived by combining a physical model (e.g., a single planet traveling on a Keplerian orbit or a system of multiple planets interacting under Newton's laws of motion and gravity) and a simple statistical model (e.g., each observation includes an error modeled as uncorrelated, Gaussian noise). In a Bayesian context, the ideal priors would be based on the intrinsic distribution of exoplanet. Of course, those were completely unknown only a few decades ago. Astronomers typically adopt relatively weak priors, such as those developed during a 2006 program at the Statistical and Applied Mathematical Sciences Institute [3].

Fortunately, the choice of priors for orbital parameters typically has only a weak effect on posterior distributions.

A complete description of a planetary system would require seven parameters per body (mass, three position coordinates and three velocity coordinates at a reference epoch) plus any additional parameters to describe the observational setup. (E.g., if a star was observed by multiple instruments, then there may be offsets between radial velocity scales for each instruments.) However, several parameters are not important due to degeneracies (e.g., translations of the center-of-mass position of the planetary system, rotation of the orbital plane about the line of sight) or near degeneracies (e.g., the mutual inclination of orbital planes only affects the observed radial velocities if the planets' mutual gravitational interactions are detectable). Therefore, the observed radial velocity signature of most planetary systems with one planet or a few well-separated planets can be accurately modeled with approximately five model parameters per planet. For such systems, the star's motion can be modeled as a linear superposition of Keplerian orbits, allowing for very rapid model evaluation. In these cases, even simple Markov chain Monte Carlo algorithms provide a practical method for inferring posterior distributions for planet masses and orbits [4].

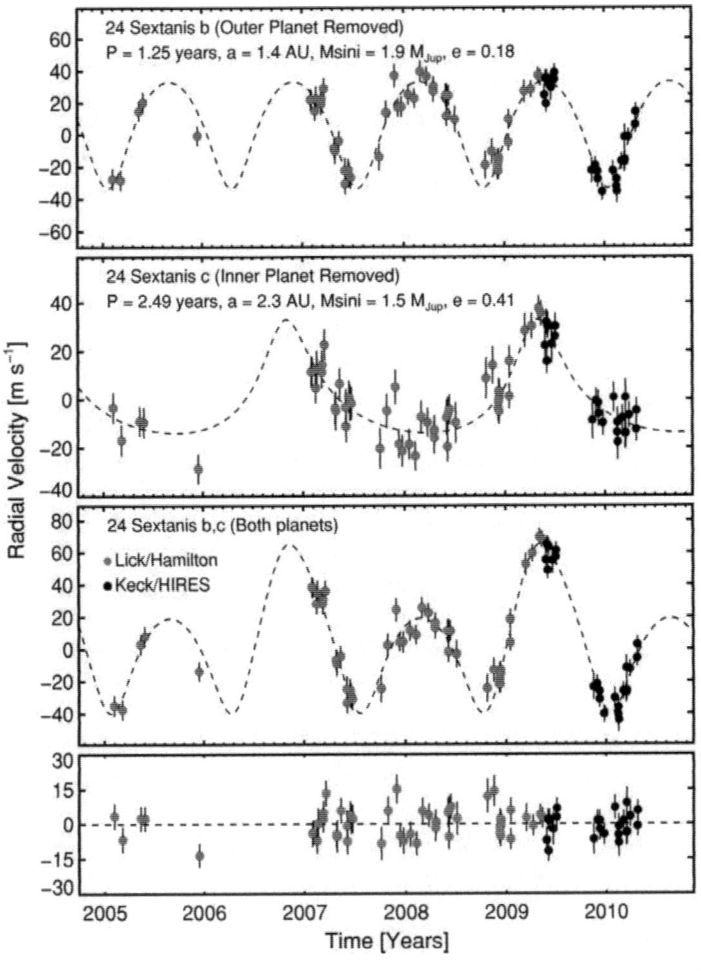

Figure 2 Time series of measured radial velocities of the star 24 Sextanis [5]. The third panel from the top shows the original measurements, along with a two-panel model (dashed curve). The two panels above show the same measurements, after subtracting a model for either the outer (top) or inner (second from top) planet. The bottom panel shows the residuals to the two planet model. Reproduced by permission of the AAS.

When multiple planets orbital at similar distances from their host star, their mutual gravitational interactions can cause changes in their orbits. Such effects are particularly important when two or more planets have nearly commensurate orbital periods, such as the near 1:2 mean-motion resonance for the planets orbiting 24 Sextanis (see Figure 2). For such systems, accurately modeling the star's motion requires performing n-body integrations that are much more computationally demanding. In these cases, more efficient algorithms for Markov chain Monte Carlo are particularly valuable. Two ensemble samplers, differential evolution MCMC (DEMCMC) and affine invariant sampler, have found utility among astronomers [6,7]. They offer a reduced number of model evaluations and increased suitability for parallelization. With these algorithms, astronomers have been able to analyze even the most challenging of such extant data sets. For example, Nelson et al. (2016) combined DEMCMC with the Swarm-NG package for integrating an ensemble of planetary systems on a graphical processing unit (GPU) [7]. They applied the resulting code to analyze several of the most challenging planetary systems and Doppler data sets. For example, 1418 Doppler measurements collected from four observatories over more than two decades of observations (plus transit and inteferometric constraints) were included in an analysis of five planets orbiting 55 Cancri, including a super-Earth, Jupiter-analog and two planets near a 1:3 mean motion resonance [8]. In another example, they characterized the three dimensional architecture of a system of four planets orbiting GJ 876, including a super-Earth-mass planet and three planets near a 1:2:4 mean-motion resonance [9]. These analyzes required weeks of computation on GPUs to perform the posterior sampling over parameter spaces with ~30 to ~40 dimensions. In the case of GJ 876, the authors quantified the evidence for the fourth planet via Bayesian model comparison. After computing posterior samples conditioned on the number of planet, they constructed an importance sampling density for each model. When combined with the ratio estimator [3], this yielded robust estimates of Bayes factors and evidence for the four-planet model.

2.3. *The Challenge of Stellar Activity*

So far, we have assumed that astronomical observations directly measure the velocity of the host star. In practice, a variety of astrophysical effects (e.g., starspots, plage, pulsations and convection) can affect the star's spectrum, perhaps causing perturbations to the star's apparent radial velocity. We refer to these effects collectively as "stellar activity". To test whether stellar activity might be masquerading as a planet, astronomers often measure one or more astronomically-motivated activity indicators from each spectrum. Once one finds evidence for a statistically significant periodic variation of the measured radial velocity, it is wise to check whether there is a similar periodicity in any of the observed activity indicators. If so, then a cautious astronomer would refrain from announcing the discovery of a planet, out of concern that stellar activity might be responsible for the apparent radial velocity signal. Of course, sometimes indications of stellar activity are identified as part of follow-up analyses after the initial claims of planet detection [10,11,12]. In other cases, further analysis can provide evidence that a periodicity previously assumed due to activity is more likely a planet [11]. In some high-profile cases, different analyses have led to different interpretations about the most likely cause for an apparent Doppler signal [10,13,14,15]. These differing interpretations underscore the important of increased statistical rigor as the field of exoplanets pushes to identify increasingly Earth-like planets.

In the early days of Doppler planet searches, the measurement uncertainty for a star's radial velocity was "large" enough that stellar activity was a relatively minor concern. Once stellar activity could no longer be completely ignored, astronomers began treating stellar noise as "jitter", an additional source of uncorrelated Gaussian noise [4]. While this approach has served astronomers extremely well, the quest for Earth-mass planets, and particularly Earth-mass planets in the "habitable zone" of their host star, has motivated the development of increasingly precise instruments. Several current spectrographs are so precise that stellar activity can be the dominant source of "noise".

2.4. Improving Doppler Surveys through Improved Modeling of Stellar Activity

In order to realize the potential of current and next-generation spectrographs, astronomers must change their paradigm from treating stellar activity as noise to treating it as a signal that is included in their model [e.g., 15]. Unfortunately, modeling how astrophysics at the surface of a star affects the apparent radial velocity is extremely computationally intensive. While first-principles simulations can provide valuable physical insights, they will not be a practical tool for analyzing radial velocity observations for decades to come. Therefore, it is necessary to develop physically-motivated, semi-empirical models for stellar activity. For example, one could consider the starlight to be a linear superposition of light from many patches on a star's surface, with each patch emitting the same spectrum, but shifted in wavelength (due to stellar rotation) and modulated in intensity (e.g., due to spots, plage, convection). While this represents a dramatic simplification, it still leaves a high-dimensional parameter space, since it requires specifying the intensity (or a proxy such as effective temperature) over the entire stellar surface and how it varies with time. Since the stellar rotation causes velocities order of magnitude larger than the gravitational perturbation of planets, even small star spots can cause an appreciable effect. Given the high contrast and sharp edges of star spots, they are not well-described by the obvious smooth functions (e.g., low-order spherical harmonics). Instead, one typically uses a high-resolution grid over the stellar surface (often approximated as a rotating sphere). Since distant main-sequence stars are not spatially resolved, solving the inverse problem - inferring the spot pattern from astronomical data - is impractical. Therefore, we must resort to marginalizing over uncertainty in the specific details stellar activity. The problem is further complicated by the temporal evolution of stellar activity. Star spots and plage can persist for timescales comparable to a stellar rotation period, but change in intensity and location. Thus, we must go beyond models consisting of a map on a rotating sphere. Developing models which are both astronomically accurate and sufficiently computationally efficient to enable marginalization over such a high-dimensional parameter space represents one major challenge in maximizing the scientific potential of next-generation Doppler planet surveys. Astronomers are making significant progress in developing empirically-motivated forward models [e.g., 16]. By performing such simulations for large ensembles of astrophysicaly plausible realizations of stellar activity, it may be possible to approximately marginalize over uncertainties in stellar activity. It will soon be possible to test such models using observations of the sun [17]. Since elements of the models were developed using data for the Sun, it will also be important to test such models on other stars. Verifying the accuracy of such models for stars other than the sun will be extremely demanding, astronomers must collect an unusually large number of extremely precise measurements.

2.5. Improving Doppler Surveys through Improved Analysis of Spectroscopic Observations

If the Doppler method is to discover Earth-mass planets in the habitable zone of sun-like stars, then astronomers must improve our ability to separate stellar activity from the Doppler signature of low-mass planets. So far, astronomers have used only a few indicators of stellar activity, each based on a very narrow range of wavelengths in the stellar spectrum. Further, these activity indicators were often developed for ease of measurement when computational resources were much more limited. Fortunately, there is reason for optimism. Remember, that each original observation contains megapixels of data. Rather than immediately reducing this down to measurements of the radial velocity and a few activity indicators, astronomers could analyze the full spectrum to extract additional information about the magnitude and nature of stellar activity. Just as combining measurements of the shift of many spectral lines allows for a much more precise measurement of the Doppler shift than is possible from any single spectral line, analyzing the full spectrum can provide information about changes in the shape of spectral lines that contain information about stellar activity. Characterizing this relationship will be challenging. Machine learning algorithms can be applied to the large number of spectroscopic observations obtained as part of planet search program, as well as to simulated datasets based on semi-empirical models [17]. Some common statistical algorithms (e.g., principal components analysis, independent components analysis) have been applied to stellar spectra [e.g., 18] and suggest that future applications to planet search data may allow for a dramatic dimensional reduction from megapixels of data to a relatively small number of components that encode information about the extent, location and nature of stellar activity. Generalizing these machine learning methods to take advantage of physical intuition may provide additional benefits for separating stellar activity and the Doppler signature of low-mass planets.

3. Additional Roles for High-Dimensional Data Analysis in Exoplanet Science

3.1. Characterizing Exoplanets via Transit Timing Variations

Recently, the method of transit timing variations has proven extremely powerful for measuring masses and densities of low mass planets. While the astronomical data differs significantly, the analysis methods developed for analysing Doppler observations of strongly-interacting planetary systems have proved invaluable for interpreting such transit timing observations [19]. The computational demands are considerable. The parameter spaces typically have a few dozen dimensions and each model evaluation is computationally more expensive than present analyses of Doppler observations. The resulting posterior distributions often have very strong and sometimes non-linear correlations, further complicating the sampling and interpretation of results. While current methodology is effective for exploring a posterior mode, further improvement in sampling algorithms will be important for accurately characterizing the wings of the posterior distributions for such systems (e.g., 99.9% credible regions) and for characterizing transit timing variations in systems where the perturbing planet is not observed to transit the host star.

3.2. Characterizing the Exoplanet Population

So far, we have focused on the characterization of an individual planetary system. As the number of known exoplanets has grown, interest has shifted from individual planets to questions about the population of planets in general. A variety of detection limits and observational biases significantly complicate the interpretation of the overall population. During a 2013 SAMSI workshop on Modern Statistical and Computational Methods for Analysis of Kepler Data, several astronomers began applying hierarchical Bayesian modeling to characterize the exoplanet population [20]. Even when the number of parameters describing the properties of the population is relatively small, performing inference over the population often requires dozens, hundreds or thousands of model parameters. Thus, these analyses can be computationally demanding. It can be challenging to develop models with a minimal number of parameters, but sufficient flexibility to model observations accurately. Visualizing the results to assess convergence, robustness and draw conclusions is also challenging. While we have made great progress, we have a long way to go. I look forward to learning from statisticians and practitioners in other fields and to sharing insights with the exoplanet community.

4. Acknowledgements

E.B.F. thanks the organizers of the 2015 meeting on High-Dimensional Data-Driven Science in Kyoto, Japan for inviting this manuscript and for enabling his participation in the meeting. E.B.F. acknowledges discussions and correspondence with Jim Berger, Jessi Cisewki, Merlise Clyde, Allen Davis, Xavier Dumusque, Debra Fischer, Daniel Jontof-Hutter, Tom Loredo, Paul Robertson, Sophia Araceli Sanchez-Maes, Robert Morehead, Benjamin Nelson, Darin Ragozzine, Megan Shabram, Sharon Wang, Angie Wolfgang, Robert Wolpert, Jason Wright and an anonymous referee. E.B.F. acknowledges support from the Pennsylvania State University's

Department of Astronomy & Astrophysics, Eberly College of Science, Institute for CyberScience, Center for Astrostatistics and Center for Exoplanets and Habitable Worlds. The Center for Exoplanets and Habitable Worlds is supported by the Pennsylvania State University, the Eberly College of Science, and the Pennsylvania Space Grant Consortium. This material was based upon work partially supported by the National Science Foundation under Grant DMS-1127914 to the Statistical and Applied Mathematical Sciences Institute (SAMSI). Any opinions, findings, and conclusions or recommendations expressed in this material are those of the author(s) and do not necessarily reflect the views of the National Science Foundation. The results reported herein benefitted from collaborations and/or information exchange within NASA's Nexus for Exoplanet System Science (NExSS) research coordination network sponsored by NASA's Science Mission Directorate.

References

[1] Butler R P *et al* 1996 *Proc. Astronomical Society of the Pacific* **108** 500
[2] Kurucz R L *et al* 1984 *National Solar Observatory Atlas* No. 1 (Sunspot, NM: NOAO)
[3] Ford E B and Gregory P C 2007 Statistical Challenges in Modern Astronomy IV *Astronomy Society of the Pacific Conference Series* **371** 189
[4] Ford E B 2005 *Astronomical J.* **129** 1706
Gregory P C 2005 *Astrophysical J.* **631** 1198
Ford E B 2006 *Astrophysical J.* **642** 505
[5] Johnson J A *et al* 2010 *Astronomical J.* **141** 16
[6] Ter Braak C J F 2006 *Stat. Comput.* **16** 239
Foreman-Mackey D *et al* 2013 *Proc. Astronomical Society of the Pacific* **125** 306
[7] Nelson B *et al* 2014 *Astrophysical J. Sup.* **210** 11
[8] Nelson B E *et al* 2014 *Monthly Notices of the Royal Astronomical Soc.* **441** 442
[9] Nelson B E *et al* 2016 *Monthly Notices of the Royal Astronomical Soc.* **455** 2484
[10] Robertson P *et al* 2014 *Science* **345** 440
[11] Robertson P and Mahadevan S 2014 *Astrophysical J. Letters* **793** L24
[12] Robertson P *et al* 2015 *Astrophysical J.* **801** 79
Robertson P *et al* 2015 *Astrophysical J. Letters* **805** L22
[13] Vogt S S *et al* 2010 *Astrophysical J.* **723** 954
Gregory P C 2011 *Monthly Notices of the Royal Astronomical Soc.* **415** 2523
Baluev R V 2013 *Monthly Notices of the Royal Astronomical Soc.* **429** 2052
Hatzes A P 2013 *Astronomische Nachrichten* **334** 616
Anglada-Escudé G and Tuomi M 2015 *Science* **347** 1080
Robertson P *et al* 2015 *Science* **345** 440
[14] Dumusque X *et al* 2012 Nature **491** 207
Hatzes A P 2013 *Astrophysical J.* **770** 133
[15] Rajpaul V *et al* 2015 *Monthly Notices of the Royal Astronomical Soc.* **452** 2269
Rajpaul V *et al* 2015 *Monthly Notices of the Royal Astronomical Soc. Letters* **456** L6
[16] Dumusque X *et al* 2014 *Astrophysical J.* **796** 132
[17] Dumusque X *et al* 2015 *Astrophysical J. Letters* **814** L21
[18] Martínez González M J *et al* 2008 *Astronomy & Astrophysics* **486** 637
[19] Carter J A *et al* 2012 *Science* **337** 556
Jontof-Hutter D *et al* 2015 *Nature* **522** 321
Jontof-Hutter D *et al* 2015 *Preprint* arXiv:1512.02003
[20] Foreman-Mackey D *et al* 2014 *Astrophysical J.* **795** 64
Rogers L A 2015 *Astrophysical J.* **801** 41
Wolfgang A *et al* 2015 *Preprint* arXiv:1504.07557
Shabram M *et al* 2015 *Astrophysical J.* in press (*Preprint* arXiv:1511.02861)

Journal of Physics: Conference Series **699** (2016) 012008 doi:10.1088/1742-6596/699/1/012008

Sparse Modeling for Astronomical Data Analysis

Shiro Ikeda[1], Hirokazu Odaka[2] and Makoto Uemura[3]

[1] The Institute of Statistical Mathematics, 10-3, Midoricho, Tachikawa 190-8562, Japan
[2] Kavli Institute for Particle Astrophysics and Cosmology (KIPAC), Menlo Park, CA 94025, U.S.A.
[3] Hiroshima Astrophysical Science Center, Hiroshima University, Kagamiyama 1-3-1, Higashi-Hiroshima 739-8526, Japan

E-mail: `shiro@ism.ac.jp`

Abstract. For astronomical data analysis, there have been proposed multiple methods based on sparse modeling. We have proposed a method for Compton camera imaging. The proposed approach is a sparse modeling method, but the derived algorithm is different from LASSO. We explain the problem and how we derived the method.

1. Introduction

In astronomy, different imaging methods are used depending on the wavelength. Compton camera imaging is one of the imaging method to visualize gamma-ray sources from 100keV to several MeV. It is an important technology and high-resolution Compton cameras will be installed in the next-generation X-ray observatory ASTRO-H [1, 2], which is scheduled for launch soon.

Imaging of Compton camera is not straightforward because only a small portion of photons are absorbed after Compton scattering and the direction of arrival of each photon is not known directly. Since photon detection is a stochastic process, the image reconstruction problem can be formulated as a statistical estimation problem [3, 4, 5]. We follow the framework developed for COMPTEL [6], and employ the bin-mode estimation (BME) method.

Since the gamma-ray source is sparsely distributed in the image, it is natural to apply sparse modeling method. However, LASSO type estimator does not work in the original form. We had to develop a new method for the Compton camera imaging. In this paper, we explain the method we have developed [7, 8] from a Bayesian viewpoint.

We have developed an estimation method for Compton camera imaging [7, 8].

2. Compton camera imaging

2.1. Compton camera system

Compton camera system is schematically shown in figure 1. Most of the photons will travel through the camera without any interaction, but some of the photons will be scattered by a scatterer and absorbed by an absorber.

When a photon from $\boldsymbol{u} = (\alpha, \delta)$ (in equatorial coordinates) is detected, it is scattered at $\boldsymbol{x} = (x_1, x_2)$ of the scatterer and absorbed at $\boldsymbol{y} = (y_1, y_2)$ of the absorber. Let E_1 and E_2 be the energies of the recoil electron and the scattered photon, respectively. At each event, $\boldsymbol{x}$, $\boldsymbol{y}$, E_1 and E_2 are measured.

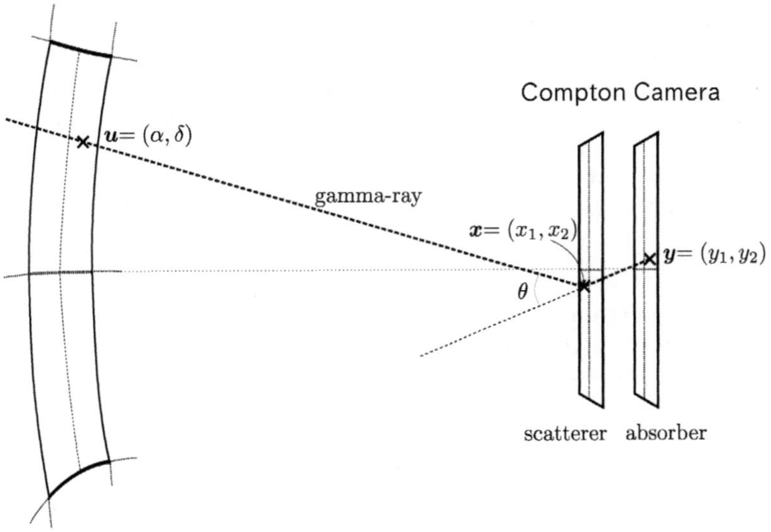

Figure 1. Compton camera.

This is the so called Compton scattering and the relation between the scattering angle θ and the energies is known. It is

$$\cos\theta = 1 - m_e c^2 \left(\frac{1}{E_2} - \frac{1}{E_1 + E_2} \right), \tag{1}$$

where m_e and c are the mass of an electron and the speed of light, respectively. Thus, from a single event, we know the photon was from a point on a cone (figure 1). The goal of the Compton camera imaging is to reconstruct the gamma-ray intensity map $\lambda(\boldsymbol{u})$ on the celestial sphere from collected information. This is not straightforward since a single scattering event does not provide direct information of the direction of arrival.

From the measurement of a single scattering event, we receive the scattering angle θ and the relative position of the event on the scatterer and absorber, i.e. $(\boldsymbol{x} - \boldsymbol{y})$ (because the distance from the astronomical objects are large). Let us define $\boldsymbol{v} = ((\boldsymbol{x} - \boldsymbol{y}), \cos\theta)$ as the information collected from a single event. In the following of the paper, the vector $\boldsymbol{u}$ is discretized into pixels, and $\boldsymbol{v}$ are discretized into bins, therefore they are finite discrete variables.

We would like to formulate the estimation problem with a probabilistic framework. Let us start by defining distributions. Let $s(\boldsymbol{u})$ be the probability that a photon from pixel $\boldsymbol{u}$ is absorbed. We define $\rho(\boldsymbol{u})$ as the probability that an absorbed photon arrives from $\boldsymbol{u}$, more precisely,

$$\rho(\boldsymbol{u}) = \frac{\lambda(\boldsymbol{u}) s(\boldsymbol{u})}{\sum_{\boldsymbol{u}} \lambda(\boldsymbol{u}) s(\boldsymbol{u})}. \tag{2}$$

Next, we define $p(\boldsymbol{v} \mid \boldsymbol{u})$ as the probability that an absorbed photon from $\boldsymbol{u}$ is detected at $\boldsymbol{v}$, and $q(\boldsymbol{v})$ as the probability that an absorbed photon is detected at $\boldsymbol{v}$. The relation between these probability distributions are summarized in the following equation

$$q(\boldsymbol{v}) = \sum_{\boldsymbol{u}} p(\boldsymbol{v} \mid \boldsymbol{u})\, \rho(\boldsymbol{u}). \tag{3}$$

All of $\rho(\boldsymbol{u})$, $p(\boldsymbol{v} \,|\, \boldsymbol{u})$, and $q(\boldsymbol{v})$ are multinomial distributions conditional to the events that a photon is absorbed by the camera. After observing many photons, the empirical distribution of $q(\boldsymbol{v})$ is observed. If $p(\boldsymbol{v} \,|\, \boldsymbol{u})$ and $s(\boldsymbol{u})$ is known, it is possible to estimate $\rho(\boldsymbol{u})$ and $\lambda(\boldsymbol{u})$.

The problem is how to prepare $p(\boldsymbol{v} \,|\, \boldsymbol{u})$ and $s(\boldsymbol{u})$. Here we utilize a numerical method. A Compton camera system was simulated and a lot of photons were randomly drawn numerically using a software [9]. The results were accumulated to compute $p(\boldsymbol{v} \,|\, \boldsymbol{u})$ and $s(\boldsymbol{u})$. This method is general and easily implemented.

2.2. Estimation

Equation (3) shows the relation between $q(\boldsymbol{v})$ and $\rho(\boldsymbol{u})$. Since $\boldsymbol{u}$ and $\boldsymbol{v}$ are discrete, $q(\boldsymbol{v})$ and $\rho(\boldsymbol{u})$ are positive vectors and $p(\boldsymbol{v} \,|\, \boldsymbol{u})$ is a positive matrix. We assume $p(\boldsymbol{v} \,|\, \boldsymbol{u})$ is known and $q(\boldsymbol{v})$ is observed empirically. Our goal is to estimate $\rho(\boldsymbol{u})$ under this condition. We further know $\rho(\boldsymbol{u})$ is sparse.

Although this is an estimation of a sparse vector, we cannot use LASSO type estimator [10]. We explain the reason from a Bayesian viewpoint.

The LASSO estimation can be viewed as the MAP (maximum a posteriori) estimation. The cost function is equivalent to the negative of the log posterior distribution. The posterior distribution consists of two terms: the log likelihood and the log of sparse prior. Let us consider the posterior distribution of the present problem.

Likelihood term Suppose N photons are detected independently and t th photon is absorbed at $\boldsymbol{v}_t$, $(t = 1, \cdots, N)$. The log likelihood function is defined as follows

$$L(\rho) = \sum_{t=1}^{N} \log q(\boldsymbol{v}_t) = \sum_{t=1}^{N} \log \sum_{\boldsymbol{u}} p(\boldsymbol{v}_t \,|\, \boldsymbol{u})\ \rho(\boldsymbol{u}). \tag{4}$$

This is different from the squared loss term of LASSO.

Prior term The widely used prior term of LASSO is proportional to $\|\cdot\|_{\ell_1}$. However the vector $\rho(\boldsymbol{u})$ is a multinomial distribution, therefore $\|\rho(\boldsymbol{u})\|_{\ell_1} = \sum_{\boldsymbol{u}} \rho(\boldsymbol{u}) = 1$ and the L1 norm of the vector is constant. Instead of using L1 norm, we employ $\alpha \sum_{\boldsymbol{u}} \log(1/\rho(\boldsymbol{u}))$. Since the term becomes large for a small $\rho(\boldsymbol{u})$, it encourages the solution to be sparse. The induced prior distribution is the Dirichlet distribution, which is the conjugate prior of the multinomial distribution.

Now, the log of the posterior distribution is

$$\sum_{t=1}^{N} \log \sum_{\boldsymbol{u}} p(\boldsymbol{v}_t \,|\, \boldsymbol{u})\ \rho(\boldsymbol{u}) - \alpha \sum_{\boldsymbol{u}} \log \rho(\boldsymbol{u}).$$

cost function to minimizes is the negative of the los posterior, which becomes

$$C = -\sum_{t=1}^{N} \log \sum_{\boldsymbol{u}} p(\boldsymbol{v}_t \,|\, \boldsymbol{u})\ \rho(\boldsymbol{u}) + \alpha \sum_{\boldsymbol{u}} \log \rho(\boldsymbol{u}).$$

Algorithm The algorithm for computing the optimal $\rho(\boldsymbol{u})$ can be easily implemented with a modified EM (Expectation-Maximization) algorithm [11]. It looks like the Iterative Shrinkage

Thresholding algorithm for LASSO.

$$\text{(E-step)} \qquad q^{(l)}(\boldsymbol{v}) = \sum_{\boldsymbol{u}} p(\boldsymbol{v}\,|\,\boldsymbol{u})\, \rho^{(l)}(\boldsymbol{u}), \tag{5}$$

$$\text{(M-step)} \qquad \rho^{(l+1)}(\boldsymbol{u}) \propto \max\left[0,\ \Big(\sum_{t=1}^{N} \frac{p(\boldsymbol{v}_t\,|\,\boldsymbol{u})}{q^{(l)}(\boldsymbol{v}_t)}\, \rho^{(l)}(\boldsymbol{u}) - \alpha\Big)\right]. \tag{6}$$

where l starts from 0 and increases by 1 at each iteration. By choosing α between 0 and 1, the optimal vector becomes sparse. When the difference between $q^{(l+1)}(\boldsymbol{v})$ and $q^{(l)}(\boldsymbol{v})$ becomes sufficiently small, the update is terminated. For more details of the algorithm, see [8].

3. Conclusion

Since the spatial distribution of astronomical objects is sparse in many cases, the idea of sparse modeling is often effective for astronomical data analysis. We have utilized the idea for Compton camera imaging and explained the method in this paper.

The popular LASSO type estimation employs L1-norm regularization term. However it does not work in the current situation since the target vector for estimation is a multinomial distribution whose L1-norm is constant. We employed the logarithm to obtain the sparse solution. In the Bayesian framework, our choice corresponds to setting the prior as the Dirichlet distribution, which is conjugate to the multinomial distribution.

Acknowledgments

The present study was financially supported by JSPS KAKENHI Grant Number 25120007 and 25120008.

References

[1] Takahashi T *et al* 2010 The ASTRO-H mission *Proc. SPIE* **7732** 77320Z

[2] Tajima H and *et al* 2010 Soft gamma-ray detector for the ASTRO-H mission *Proc. SPIE* **7732** 773216

[3] Tomitani T and Hirasawa M 2002 Image reconstruction from limited angle Compton camera data *Phys. Med. Biol.* **47** (12) 2129–45

[4] Parra L C 2000 Reconstruction of cone-beam projections from Compton scattered data *IEEE trans. Nucl. Sci.* **47** (4) 1543–50

[5] Xu D and He Z 2006 Filtered back-projection in 4π Compton imaging with a single 3D position sensitive CdZnTe detector *IEEE trans. Nucl. Sci.* **53** (5) 2787–96

[6] Bloemen H *et al* 1994 COMPTEL imaging of the Galactic disk and the separation of diffuse emission and point sources *Astrophys. J. Suppl. S.* **92** (2) 412–23

[7] Ikeda S, Odaka H, Uemura M, Takahashi T, Watanabe S and Takeda S 2013 Compton camera imaging *Proc. 2013 7th Int. Conf. on Sensing Technology (Wellington)* 674–7

[8] Ikeda S, Odaka H, Uemura M, Takahashi T, Watanabe S and Takeda S 2014 Bin mode estimation methods for Compton camera imaging *Nucl. Instrum. Methods Phys. Res.* A **760** 46–56

[9] Odaka H *et al* Development of an integrated response generator for Si/CdTe semiconductor Compton cameras *Nucl. Instrum. Methods Phys. Res.* A **624** (2) 303–9

[10] Tibshirani R 1996 Regression shrinkage and selection via the lasso *J. R. Stat. Soc. Series* B *Stat. Methodol.* 267–88

[11] Dempster A P, Laird N M and Rubin D B 1977 Maximum likelihood from incomplete data via the EM algorithm *J. R. Stat. Soc. Series* B *Stat. Methodol.* **39** 1–38

Data-driven approach to Type Ia supernovae: variable selection on the peak luminosity and clustering in visual analytics

Makoto Uemura[1], Koji S Kawabata[1], Shiro Ikeda[2], Keiichi Maeda[3], Hsiang-Yun Wu[4], Kazuho Watanabe[5], Shigeo Takahashi[6] and Issei Fujishiro[4]

[1] Hiroshima Astrophysical Science Center, Hiroshima University, Kagamiyama 1-3-1, Higashi-Hiroshima 739-8526, Japan
[2] The Institute of Statistical Mathematics and CREST, JST, Tachikawa 190-8562, Japan
[3] Department of Astronomy, Kyoto University, Kitashirakawa-Oiwake-cho Sakyo-ku, Kyoto 606-8502, Japan
[4] Department of Information and Computer Science, Keio University 3-14-1 Hiyoshi, Kohoku-ku, Yokohama 223-8522, Japan
[5] Toyohashi University of Technology, 1-1 Hibarigaoka Tempaku-cho Toyohashi 441-8580, Japan
[6] Department of Computer Science and Engineering, University of Aizu, Tsuruga, Ikki-machi, Aizu-Wakamatsu 965-8580, Japan

E-mail: uemuram@hiroshima-u.ac.jp

Abstract. Type Ia supernovae (SNIa) have an almost uniform peak luminosity, so that they are used as "standard candle" to estimate distances to galaxies in cosmology. In this article, we introduce our two recent works on SNIa based on data-driven approach. The diversity in the peak luminosity of SNIa can be reduced by corrections in several variables. The color and decay rate have been used as the explanatory variables of the peak luminosity in past studies. However, it is proposed that their spectral data could give a better model of the peak luminosity. We use cross-validation in order to control the generalization error and a LASSO-type estimator in order to choose the set of variables. Using 78 samples and 276 candidates of variables, we confirm that the peak luminosity depends on the color and decay rate. Our analysis does not support adding any other variables in order to have a better generalization error. On the other hand, this analysis is based on the assumption that SNIa originate in a single population, while it is not trivial. Indeed, several sub-types possibly having different nature have been proposed. We used a visual analytics tool for the asymmetric biclustering method to find both a good set of variables and samples at the same time. Using 14 variables and 132 samples, we found that SNIa can be divided into two categories by the expansion velocity of ejecta. Those examples demonstrate that the data-driven approach is useful for high-dimensional large-volume data which becomes common in modern astronomy.

1. Introduction

Type Ia supernovae (SNIa) are thermonuclear explosions that occur when an accreting white dwarf reaches its critical mass, and starts a runaway reaction. The critical mass, so-called the "Chandrasekhar mass", is about 1.38 times the mass of the sun ($M_\odot$). The presence of this critical mass makes the luminosity of SNIa uniform. Then, the apparent magnitude of SNIa measured on the earth is a function of the distance to the object. Hence, SNIa have been used as a "standard candle" which enables us to measure the distance of galaxies in cosmology. The accelerating expansion of the Universe was discovered based on the distance estimation using SNIa [1, 2, 3]. In recent years, the amount of the SNIa data has rapidly been increasing owing to large survey projects [4, 5]. When the data set was small, it can be handled based on the experience of domain experts. However, the numbers of both samples and variables are large nowadays, so that the data-driven approach is expected to be useful. Here, we introduce two issues of SNIa, that is, the variable selection of the peak magnitude and the classification of SNIa.

The observed peak magnitude of SNIa (M) has a small, but significant diversity. It is due to the interstellar extinction, and hence calibrated by their reddened color. In addition, it is well known that M depends on the decay rate [6]. The corrected magnitude, M_0, is expressed as:

$$M = M_0 + \beta_1 c + \beta_2 x, \tag{1}$$

where c and x denote the color and decay rate, and $\beta_{1,2}$ are their coefficients. Both c and x are values obtained by photometric observations with broad-band filters. The photometric data is more readily obtained compared with spectroscopic ones because the fluxes are integrated over the transmission curve of the broad-band filter in photometric observations. Recent survey projects provide large and uniform spectroscopic data. Using those spectra, a new set of explanatory variables of M has been searched [7, 8, 9]. However, the number of the candidates of explanatory variables, that is, all data points in spectra, is larger than the number of sample in those studies. We need to select the best set of variables for a given data set of M, although the standard least-square method cannot solve this problem.

The other example is about the classification of SNIa. Several sub-types have been proposed for SNIa based on observed features. The classification of SNIa is crucial for cosmology, because some sub-types possibly have different peak magnitudes or colors. For example, Wang, et al. 2009 classified SNIa using the expansion velocity which is measured from the absorption line of Si II 6355Å [10]. They proposed that SNIa having high expansion velocities form a distinct sub-type which has an intrinsically red color, and appropriate color corrections improve the distance estimation. Now, we can revisit past classifications using large data of samples and variables. A problem is that it is difficult to see the structure in the data in the case that the dimension of the data is high. We need to find both a good set of variables and sub-types at the same time.

In this article, we review our recent works related to the above two problems. Uemura, et al. 2015 use a variable selection approach for modeling the peak magnitude of SNIa [11]. We use cross-validation to control the generalization error and a LASSO-type estimator to choose the set of variables (section 3). In section 4, we present a visual analytics approach for clustering SNIa. This visual analytics tool is based on asymmetric biclustering method, and enable us to visually check the result of clustering to find the potential structures in the data.

2. Data

All data of SNIa was obtained from the supernova database operated by the UC Berkeley group[1] [5]. They present both photometric (the peak magnitude, color, decay rate, and distance) and spectroscopic data. We show an example of spectra in figure 1. The spectra of SNIa are characterized by broad absorption-lines, for example, of Si, Ca, and S. The right panel shows the spectra around Si II 5982 and 6355Å. We also show the definitions of some astronomical values related to the lines.

[1] http://heracles.astro.berkeley.edu/sndb/

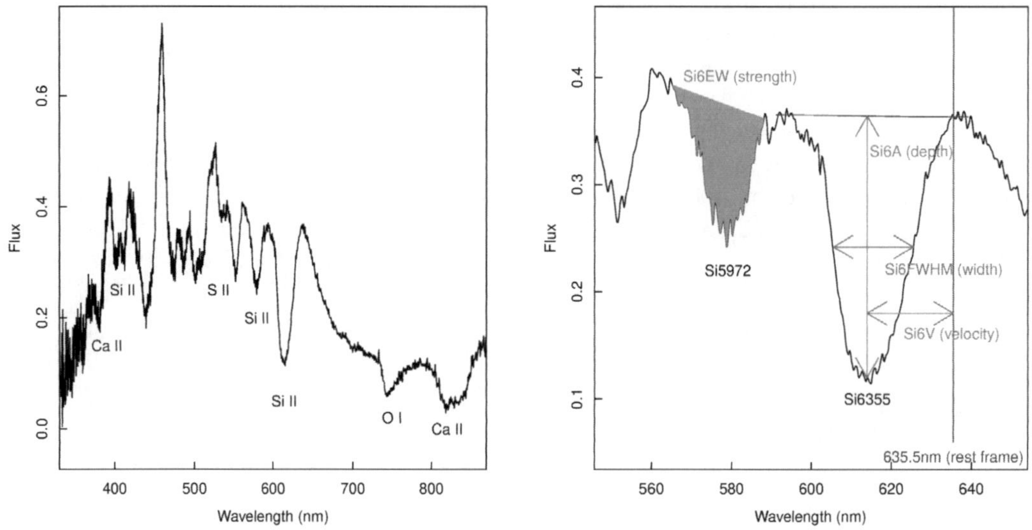

Figure 1. Left panel: Typical spectra of Type Ia supernovae (SNIa) between 350 and 850 nm. Right panel: Close-up view of the spectra around two silicon lines, Si II 5982 and 6355Å. Some definitions of astronomical values are also indicated. The data is from the database of the Berkeley group [5].

In section 3, we use the spectra, as well as the photometric data as the candidates of explanatory variables of the peak magnitude. Among the database, we selected samples having spectral data between 3500 and 8500Åtaken within five days from the maximum. This selection reduced the number of samples to 78. Each spectrum contains 138 data points. In our sample, we excluded Type Iax supernovae, which have recently been recognized as a peculiar faint class of SNIa [12]. In section 4, we use the absorption line parameters which are measured from the spectral data. In conjunction with the photometric data, such as the distance, decay rate, color, and peak magnitude, we use them as the candidates of variables for classification of SNIa.

3. Variable selection for modeling the peak magnitude of SNIa

In this section, we review our recent work about the variable selection of the peak absolute magnitude (M) of SNIa [11]. We consider a linear model of M:

$$\boldsymbol{M} = X\boldsymbol{\beta} + \boldsymbol{e}, \tag{2}$$

where $\boldsymbol{M}$ is a vector containing N samples of the peak absolute magnitude, M. X is a $N \times L$ matrix of the explanatory variables, including a constant term, and $\boldsymbol{\beta}$ is co-efficients. We consider a Gaussian noise, $\boldsymbol{e}$. The number of samples, N, is larger than that of the candidate variables, L if the spectral data are included as the explanatory variables [7, 8, 9]. However, our interest focuses on a solution having a few variables which control the peak magnitude of SNIa. In other words, most of elements of $\boldsymbol{\beta}$ should be zero. In order to obtain such a solution, we use a LASSO-type estimator:

$$\hat{\boldsymbol{\beta}} = \arg\min_{\boldsymbol{\beta}} \left\{ \|\boldsymbol{M} - X\boldsymbol{\beta}\|_2^2 + \lambda \|\boldsymbol{\beta}\|_1 \right\}, \tag{3}$$

where λ is a tunable constant. Using LASSO, we can obtain a sparse solution, $\hat{\boldsymbol{\beta}}$, and choose the best set of explanatory variables. We use cross-validation in order to control the generalization error. A 10-fold cross-validation was applied to determine λ in the following results.

Journal of Physics: Conference Series **699** (2016) 012009 doi:10.1088/1742-6596/699/1/012009

Table 1. Coefficients of variables [11].

	Coefficients		
Variables	Model 1	Model 2	Model 3
c	0.376	—	—
$f_{\mathrm{tot}}(6373)$	0.100	0	0
x	−0.050	−0.014	—
$f_{\mathrm{cnt}}(6084)$	−0.034	0	0
$f_{\mathrm{cnt}}(6289)$	−0.045	0	0
$f_{\mathrm{cnt}}(6631)$	−0.061	0	0
$R(3780/4580)$	−0.050	0	0
$f_{\mathrm{tot}}(3752)$	0.063	0	0

The candidates of the explanatory variables are the color (c), decay rate (x), six flux ratios that were previously proposed ($R(\lambda_1/\lambda_2)$), and spectral data. About the spectral data, we used two kinds of normalized spectra instead of the arbitrary flux ratios that were used in the previous studies. First, the variables of most interest are the flux ratios of the line areas to the continuum level. It can be obtained by the spectra normalized by the continuum level ($\boldsymbol{f}_{\mathbf{cnt}}(\lambda)$). Second, the local colors in the continuum which may have information independent of the broadband colors (c) are also variables of interest. They can be obtained by the spectra normalized by the total flux between 3500 and 8500Å($\boldsymbol{f}_{\mathbf{tot}}(\lambda)$). Each spectra was re-binned into 134 wavelength bins between 3500 and 8500Å, as in the past studies [9]. The number of all candidate variables is $L = 276$. The number of sample is $N = 78$. Each variable was normalized to have zero mean and unit variance by a linear scaling.

We start the model with all candidate variables (Model 1). Table 1 shows the selected variables having non-zero coefficients. As proposed in many past studies, the color, c, and decay rate, x have non-zero coefficients. $f_{\mathrm{tot}}(6373)$ denotes the flux at 6373Å of the total flux normalized spectra, having a relatively large coefficient. This variable represent a local continuum color at 6373Å, and could be interesting if it has information independent of the broad-band color, c. The other non-zero elements are related to the absorption lines: $f_{\mathrm{cnt}}(6084)$ and $f_{\mathrm{cnt}}(6289)$,the fluxes at 6084 and 6289Å of the continuum normalized spectra, are related to the absorption line, Si II 6355Å. $R(3780/4580)$, a flux ratio between 3780 and 4580Å, and $f_{\mathrm{tot}}(3752)$ are related to Ca II. $f_{\mathrm{cnt}}(6631)$ can be considered as a noise because it is a continuum flux of continuum normalized spectra.

The local color, $f_{\mathrm{tot}}(6373)$, is potentially interesting, while its non-zero coefficients could be due to a high correlation between c and $f_{\mathrm{tot}}(6373)$. To evaluate it, we used a new model in which M is corrected for the effect of c (Model 2). As a result, no variable was selected, except for the decay rate, x. The result was confirmed another model in which M is corrected for the effect of c and x (Model 3). No variable has non-zero coefficient in this model. This result suggests that $f_{\mathrm{tot}}(6373)$ and the other spectroscopic variables were selected in the first model simply because of their high correlations with c or x. Our analysis confirmed the past understanding about SNIa, that is, the peak absolute magnitude depends on the color and decay rate, and does not support adding any other variables in order to have a better generalization error.

In recent studies, arbitrary flux ratios were taken into account as the candidates of the explanatory variables of the peak magnitude [7, 8, 9]. Each spectra contains 138 data points, and the number of the flux combination is $138 \times 137 = 18906$, which is over two orders of magnitude larger than the number of samples. The results of those studies are not completely consistent. Our approach has two advantages against those past studies. First, we reduced the

Table 2. Variables used in the ABC analysis

Variable	Definition
Si6EW	Equivalent width of Si6355.
Si6DEW	Corrected Si6EW for the observation epoch.
Si6V	Velocity of Si6355.
Si6A	Depth of Si6355.
Si6FWHM	Full-width of half maximum of Si6355.
Si5EW	Equivalent width of Si5972.
Si5DEW	Corrected Si5EW for the observation epoch.
z	Redshift, corresponding to the distance to the object.
x	Decay rate.
c	Color.
mB	Apparent magnitude.
MB	Absolute magnitude calculated from z and mB.
MBc	Color (c) corrected absolute magnitude.
MBcx	Color (c) and decay-rate (x) corrected absolute magnitude.

number of candidate variables by using normalized spectra, instead of the arbitrary flux ratios. The reduction of the candidate variables leads to the reduction of false positive signals. Second, this data-driven approach enable us to determine the number of explanatory variables from the data itself. In past studies, only one or two variables were evaluated. In our study, we conclude that no spectroscopic variable can provide a better model for M. Of course, our conclusion is based on our samples. A better model could be found by using further larger data sets in future.

4. Classification of SNIa via visual analytics for asymmetric biclustering

In the last section, we developed a model of the peak magnitude of SNIa. This modeling was based on the assumption that our samples of SNIa have the common peak magnitude, while it is not trivial as introduced in section 1. A search for possible sub-types of SNIa and the axes for their classification is crucial not only for the cosmology, but also for the understanding their explosion mechanism. In this section, we present our analysis of SNIa using a visual analytics tool for this classification problem.

We use the Asymmetric Bi-Clustering (abbreviated as ABC hereafter) tool for our analysis [13]. This tool has recently been developed for an asymmetric biclustering-based subspace search of multivariate data. Highly correlated dimensions are automatically grouped to form feature subspaces in an interactive and progressive manner. The asymmetric biclustering combines spherical k-means for grouping highly correlated dimensions, together with ordinary k-means for identifying subsets of data samples. Lower dimensional representations of data in feature subspaces are successfully visualized by parallel coordinate plot (PCP).

Using the ABC tool, we analyzed 132 SNIa samples from the Berkeley database and 14 variables, which are summarized in table 2. The definitions of the equivalent width (EW), velocity (V), depth (A), and full-width of half maximum (FWHM) of lines are depicted in figure 1. We show an example set of the analysis process step by step. Figure 2 shows the initial state. The 132 samples are plotted in 14 parallel coordinates. Figure 3 shows the results of biclustering for 6 data and 6 axis clusters. In the upper panel, axes having high correlation are clustered, and the sample clusters are expressed by colors. The block matrix diagram shown in the lower panel indicates block errors of each axis and sample cluster. We can see that a small, blue-colored cluster has a large block error in the axis cluster of x and MBcx, as shown in

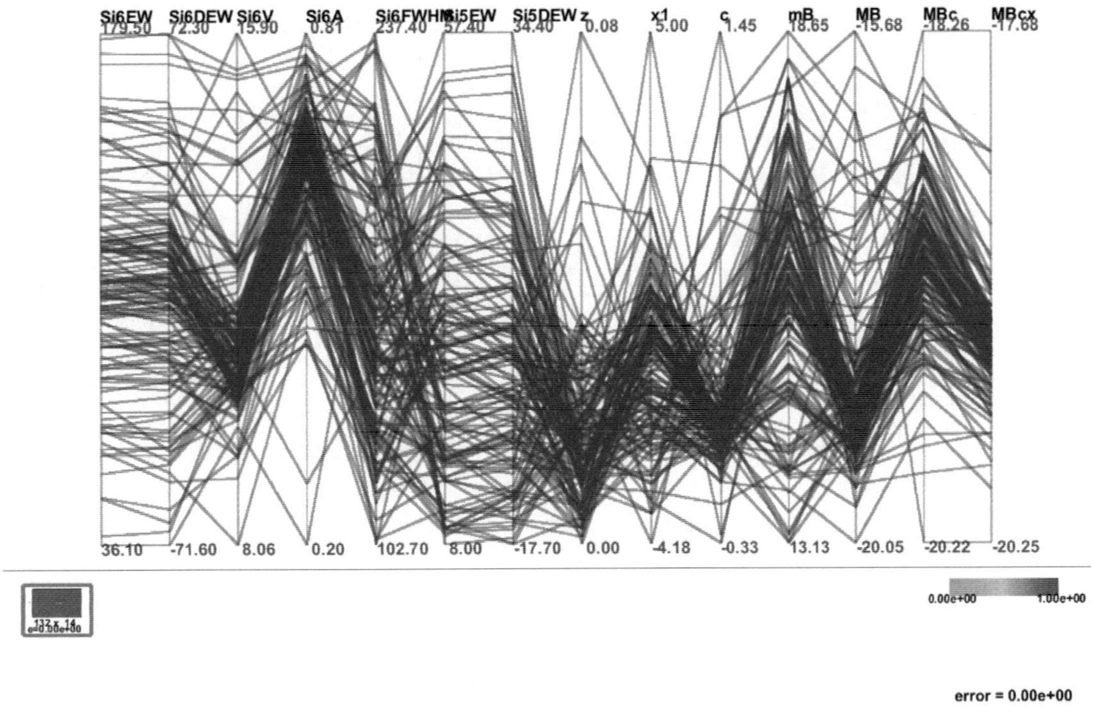

Figure 2. Initial state of the target SNIa dataset of 132 samples in 14 variables.

the lower panel. These are outliers which appears due to their peculiar variation pattern. We excluded them in the ABC tool, and then obtained a new result shown in figure 4. We found the axis cluster {MBc, MBcx} has large block errors in several sample clusters. Both of these variables are the absolute magnitudes, that is, the intrinsic luminosity of SNIa. It is well known that MBc correlates with x [6]. MBcx is the corrected value of MBc for this effect. Clustering those two variables can be understood because, in our data, the trend in MBc against x is small compared with the dispersion in MBc. Judging from the large block errors, we eliminated this axis cluster.

After further eliminating axes or axis clusters which have large block errors, we obtain three axis and three sample clusters, as shown in figure 5. We consider this state as the final result because the total error, which is indicated at the right-bottom area in figures 3–5, reaches its minimum at this state. We note that the order of the process can change run by run because it uses random values to set the initial cluster number for each sample. We have confirmed that most trials converge to the result shown in figure 5. In figure 5, we show the results in contracted PCP with both polylines (upper panels) and strip rendering (lower panels). We can see three axis and three sample clusters. The axis clusters are about the expansion velocity of Si II 6355Å (the left axis), the strength of Si II 6355 (middle), and 5972Å (right). We found that the objects having high velocity, forming a cyan-colored cluster, have strong Si II 6355 and weak Si II 5972, while the low velocity clusters, indicated by blue and magenta, exhibits weak, but positive correlation between Si II 6355 and 5972Å.

The two axes, Si II 6355 and 5972Å have been used for the classification of SNIa [14]. This classification scheme was proposed by a clustering analysis of spectra, and divided SNIa into four sub-types: "Normal", "Broad line", "Shallow silicon", and "Cool" types. The expansion velocity of Si II 6355Å has been used for the definition of the high velocity group of SNIa, which

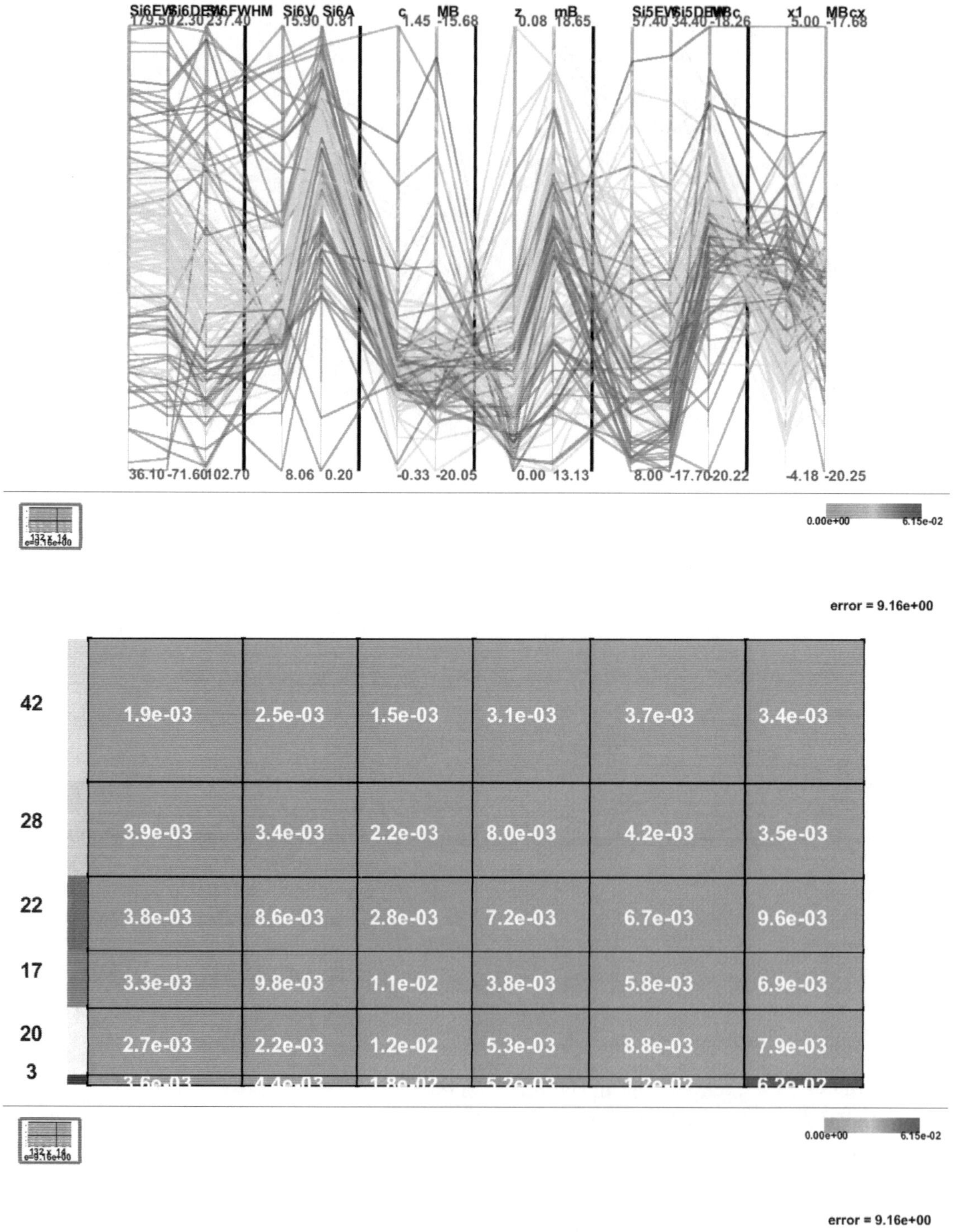

Figure 3. Biclustering 14 axis and 132 samples to 6 axis and 6 sample clusters. Upper panel: Clustered PCP. Lower panel: Block matrix diagram.

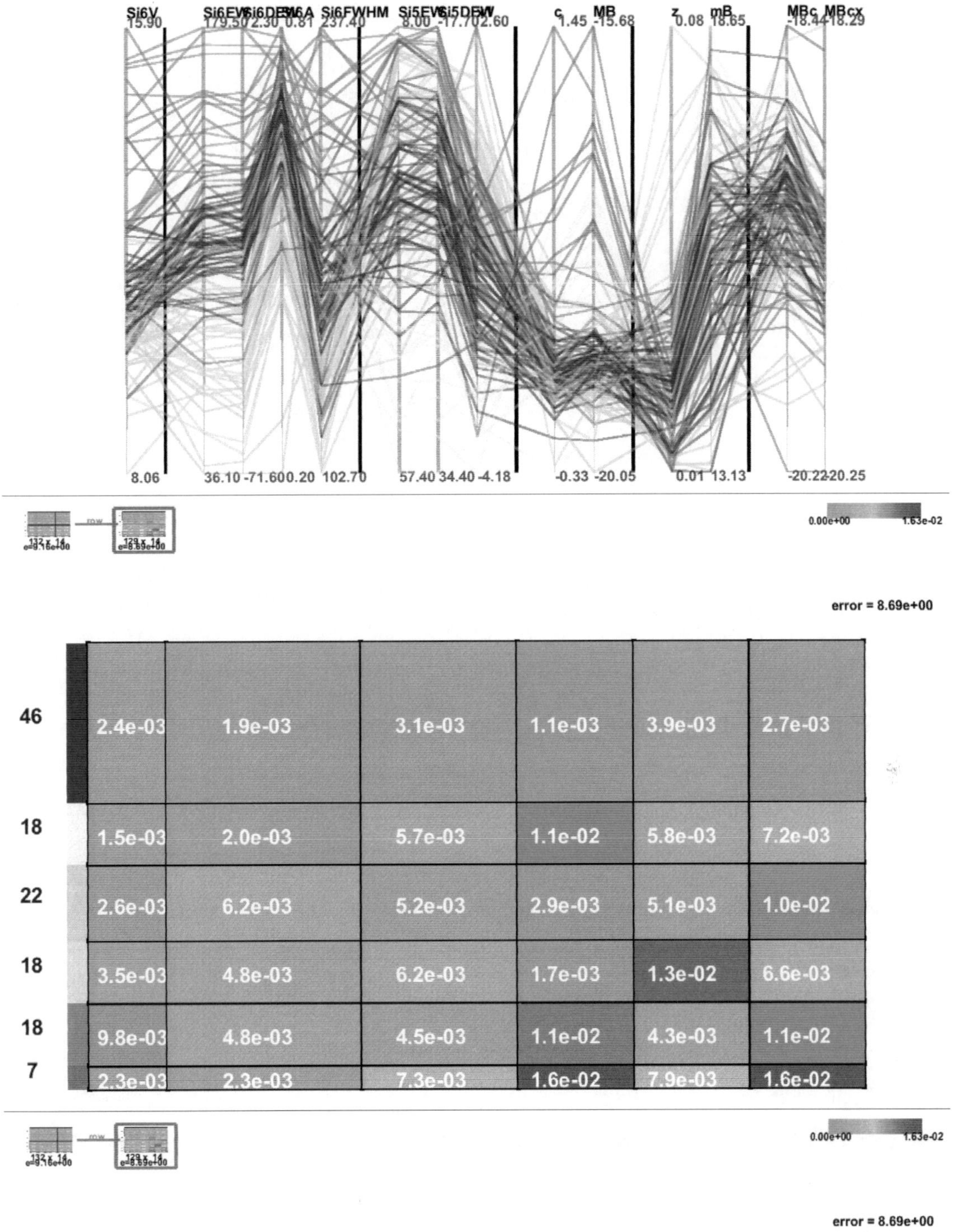

Figure 4. Biclustering 14 axis and 129 samples to 6 axis and 6 sample clusters. Upper panel: Clustered PCP. Lower panel: Block matrix diagram.

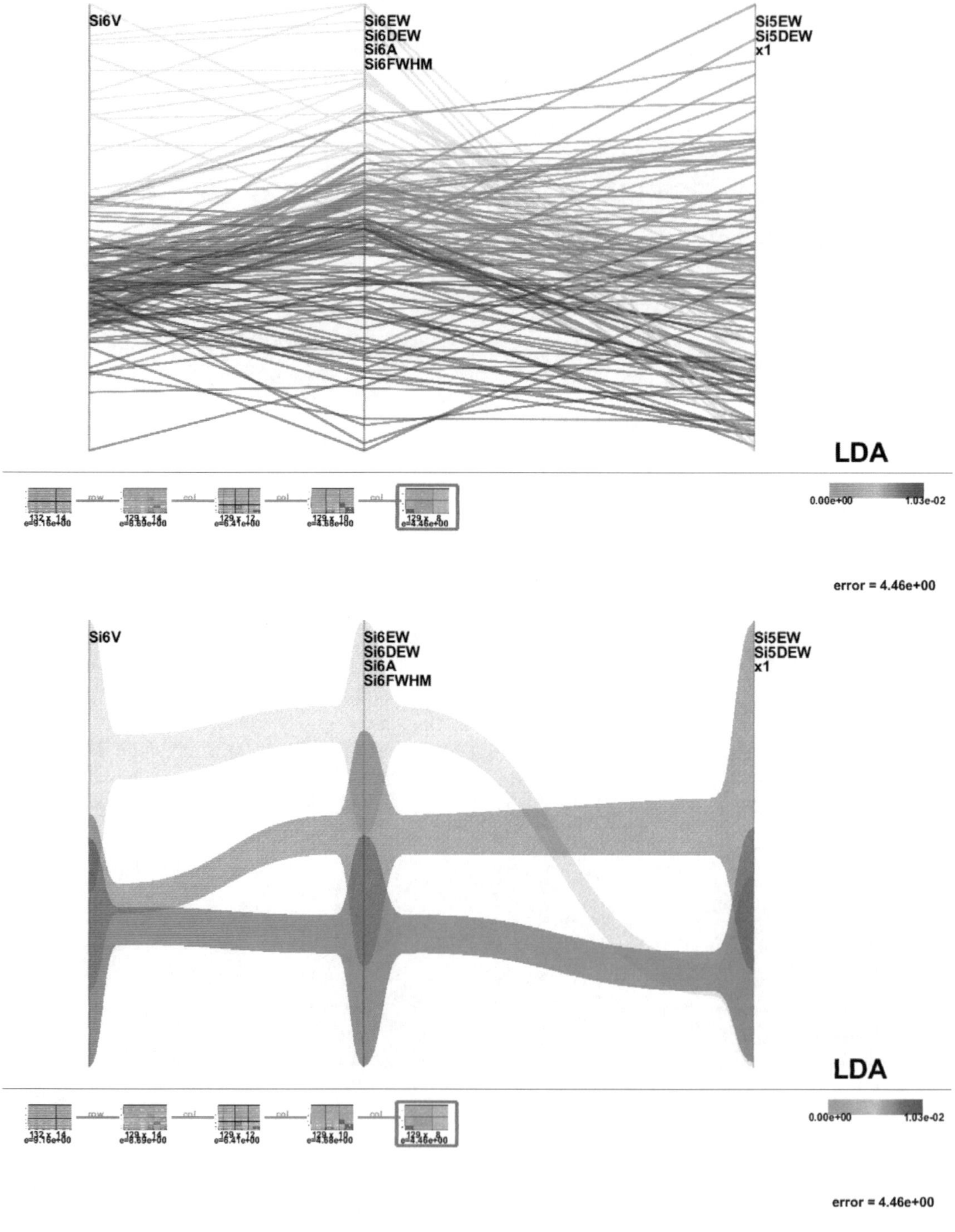

Figure 5. Biclustering 8 axis and 129 samples to 3 axis and 3 sample clusters. Upper panel: Contracted PCP depicted in polylines. Lower panel: Contracted PCP depicted in strip rendering.

is proposed to have a different color behavior [10]. Our analysis also suggests the presence of a high velocity cluster. Furthermore, its behavior of the line strengths corresponds to the "Broad line" type in the Si II 6355—5972Å classification scheme. The axis of the peak magnitude was eliminated because of large errors in our analysis process. This indicates that the diversity of the peak magnitude is not significant between each sample classes in our analysis. Thus, our result obtained by data-driven approach is successfully consistent with those previously proposed classifications based on the experience of domain experts.

Our study demonstrates that the asymmetric biclustering tool is useful to find a meaningful classification scheme hidden in high-dimensional data. A potential problem of this tool is the ambiguity of the number of clusters. As mentioned above, we selected it based on the total error, while it may be determined by more proper model selection criteria. The revised version of this tool could overcome this difficulty [15].

5. Summary

In this article, we introduced our study on SNIa using data-driven approach. The LASSO-type estimator enables us to determine the number of variables and select their best set. Our analysis confirmed the classical picture, that is, the model with the color and decay rate (section 3). The ABC tool is useful to find hidden structures in multivariate data. Using this tool, we revisit the classification of SNIa. The result is consistent with those proposed based on the experience of domain experts (section 4). Owing to recent survey projects, we can access large uniform datasets of SNIa. However, old-style analysis might overlook intriguing features in such high-dimensional large dataset. The two examples shown in this article demonstrate that the data-driven approach is useful for modern astronomy.

Acknowledgments

We would like to thank Drs. J. M. Silverman and A. V. Filippenko for providing the Berkeley supernova database. The present study was financially supported by JSPS KAKENHI Grant Number 25120007, 25120008, and 26800100.

References

[1] Schmidt B P, Suntzeff N B, Phillips M M, Schommer R A *et al* 1998 *Astrophysical Journal* **507** (1) 46
[2] Perlmutter S, Turner M S and White M 1999 *Physical Review Letters* **83** (4) 670
[3] Riess A G, Strolger L-G, Tonry J, Casertano S *et al* 2004 *Astrophysical Journal* **607** (2) 665
[4] Law N M, Kulkarni S R, Dekany R G, Ofek E O *et al* 2009 *Publ. of Astron. Soc. of Pacific* **121** (886) 1395
[5] Silverman J M, Foley R J, Filippenko, A V, Ganeshalingam M *et al* 2012 *Monthly Notices of the Royal Astronomical Society* **425** 1789
[6] Phillips M M 1993 *Astrophysical Journal Letters* **413** L105
[7] Bailey S, Aldering G, Antilogus P, Aragon C, Baltay C, Bongard S, Buton C, Childress M *et al* 2009 *Astronomy & Astrophysics* **500** L17
[8] Blondin S, Mandel K S, and Kirshner R P 2011 *Astronomy & Astrophysics* **526** A81
[9] Silverman J M, Ganeshalingam M, Li W and Filippenko A V 2012 *Monthly Notices of the Royal Astronomical Society* **425** 1889
[10] Wang X, Filippenko A V, Ganeshalingam M, Li W *et al* 2009 *Astrophysical Journal Letters* **699** L139
[11] Uemura M, Kawabata K S, Ikeda S and Maeda K 2015 *Publ. of the Astron. Soc. of Japan* **67** 55
[12] Foley R J, Challis P J, Chornock R, Ganeshalingam M, Li W, Marion G H, Morrell N I, Pignata G *et al* 2013 *Astrophysical Journal* **767** (1) 57
[13] Watanabe K, Wu H Y, Niibe Y, Takahashi S and Fujishiro I 2015 *Proceedings of IEEE Pacific Visualization Symposium 2015*
[14] Branch D, Dang L C, Hall N, Ketchum W *et al* 2006 *Publ. of Astron. Soc. of Pacific* **118** (842) 560
[15] Watanabe K, Wu H Y, Takahashi S and Fujishiro I 2016 Asymmetric biclustering with constrained von Mises-Fisher models *J. Phys. Conf. Ser.* (this volume) in press

Journal of Physics: Conference Series **699** (2016) 012010 doi:10.1088/1742-6596/699/1/012010

Sparsity and inverse problems in astrophysics

Jean-Luc Starck

Laboratoire AIM, UMR CEA-CNRS-Paris 7,
Irfu, Service d'Astrophysique,
CEA Saclay, F-91191 Gif-Sur-Yvette CEDEX, France

Abstract. Sparsity constraints are now very popular to regularize inverse problems. We review how sparsity can be used to solve inverse problems such as inpainting, deconvolution or blind source separation.

1. Introduction

Many data processing problems in astrophysics can be formalized as a linear inverse problem,

$$Y = HX \bullet N \; , \tag{1}$$

where Y are a set of noisy measurements, N is the unknown noise/error term, and the $\bullet$ operator translates the way the noise contaminates the data (e.g. "+" for additive noise). This noise/error can be either a stochastic measurement noise induced by the sensor (e.g. additive Gaussian noise, Poisson noise, or mixed Gaussian and Poisson), or a deterministic perturbation due, for example, to an imperfect forward model. X is the solution of our problem, and H is a linear operator. Finding X is an inverse problem.

Typical cases of such inverse problems are:

- *Deconvolution*: If the optical system is linear and shift-invariant, the relation between the data Y and the unknown X in the same coordinate frame is a convolution by a blurring kernel. H is the point spread function, PSF, of the optical system. We want to determine X knowing Y and H. This deconvolution problem has led to a large amount of work, the main difficulties being the existence of: (i) a cut-off frequency of the point spread function, and (ii) the additive noise.
- *Blind Deconvolution*: It is similar to the deconvolution problem, except the point spread function, i.e. matrix H, is unknown. Hence. we want to determine X and H knowing Y.
- *Myopic deconvolution*: It is an intermediate problem between deconvolution and blind deconvolution, the PSF H is partially known.
- *Superresolution*: H is the convolution by a blurring kernel followed by a sub-sampling. The goal is to recover spatial frequency information outside the spatial bandwidth of the data formation system.
- *Inpainting*: Missing data is a recurrent problem in astronomical data processing. H is a pixelwise multiplication by a binary mask; 1 for the pixels to be kept and 0 otherwise.

- *Radio-Interferometry Image Reconstruction:* this is a specific case of the deconvolution problem. Indeed the observed data, called visibilities, correspond to a subset of the Fourier component of the unknown X. The PSF, called *Dirty Beam*, is not compact in the direct space, and contains many zero in the Fourier domain.
- *Blind Source Separation:* In the blind source separation (BSS) setting, the instantaneous linear mixture model assumes that we are given m observations $Y = [y_1^T, \cdots, y_m^T]^T$ where each $(y_i)_{i=1,\cdots,m}$ is a row-vector of size t; each measurement is the linear mixture of n source processes . As the measurements are m different mixtures, source separation techniques aim at recovering the original sources $X = [x_1^T, \cdots, x_n^T]^T$ by taking advantage of some information contained in the way the signals are mixed in the observed data. The linear mixture model is rewritten in matrix form, $Y = HX + N$, where Y is the $m \times t$ measurement matrix (i.e. observed data), X is the $n \times t$ source matrix and H is the $m \times n$ mixing matrix. H defines the contribution of each source to each measurement. An $m \times t$ matrix N is added to account for instrumental noise or model imperfections. Similarly to the blind deconvolution, the H matrix is not known.
- *Image reconstruction:* an image is formed from projections.

When the inverse problem has not a unique and stable solution, it is an *ill-posed problem*, and a regularization is necessary to reduce the space of candidate solutions, and we need to incorporate some prior knowledge on the typical structure of the original signal X.

This prior information accounts typically for the "smoothness" class of the solution and can range from the uniform smoothness assumption to more complex knowledge of the geometrical structures of the solution. A very popular regularization in astronomy is the well-known Bayesian maximum entropy method (MEM), which is based on the principle that we want to select the simplest solution which fits the data. Sparsity is an alternative to describe the *simplest solution*, and has emerged as very powerful approach for regularization [1].

Sparse recovery involves four main ingredients which are the keys for its success. These ingredients are:

- **Data representation though the use of a dictionary:** this will be discussed in next section.
- **The noise modeling:** having the right dictionary is only a part of the story and is not enough. In most inverse problems, one needs to disentangle the useful signal/information from the noise. This requires to take properly into account the noise behavior in the observed data. Different kinds of noise models such as Gaussian, Poisson, mixture of Gaussian and Poisson, correlated noise, etc., can be used to derive the distribution of the coefficients in the chosen dictionary and therefore to detect, with a given confidence level, significant coefficients that should be used to reconstruct the signal of interest.
- **The variational problem and minimization algorithm:** piecing together the first ingredient and the second one (i.e. regularization and data fidelity), solving the inverse problem at hand as solving a variational (optimization) problem. To solve the obtained optimization problem, very general class of optimization schemes can be used, namely proximal splitting algorithms, that are fast, easy to implement and enjoy provable convergence. A very efficient proximal minimization method is presented in Section 5.3.
- **The choice of parameter(s):** variational formulation of inverse problems involve parameters that must be chosen. The scientist does not have to choose these parameters based on a tedious trial-and-error, by running many times the algorithm with a range of parameters. Furthermore, one definitively does not want to make scientific conclusions based on an ad-hoc parameter. We show in Section 5.4 how these parameters can be derived from the noise modeling.

A brief tour of sparsity is given in Section 2, and the different kind of disctionaries are discussed in Section 3. Specific inverse problems will be addressed, such as denoising and deconvolution in Sections 4 and 5, inpainting in Section 6, and blind source separation in Section 7.

2. Sparsity

A signal $X = (X[1], \ldots, X[N])$ considered as a vector in $\mathbb{R}^N$, is said to be sparse if most of its entries are equal to zero. If k number of the N samples are equal to zero, where $k \ll N$, then the signal is said to be k-sparse. In the case where only a few of the entries have large values and the rest are zero or close to zero the signal is said to be weakly sparse (or compressible). With a slight abuse of terminology, in the sequel, we will call compressible signals sparse. Generally signals are not sparse in direct space, but can be sparsified by transforming them to another domain. Think for instance of a purely sinusoidal signal which is 1-sparse in the Fourier domain, while it is clearly dense in the original one. In the so-called sparsity synthesis model, a signal can be represented as the linear expansion

$$X = \Phi\alpha = \sum_{i=1}^{T} \phi_i \alpha[i] \quad , \tag{2}$$

where $\alpha[i]$ are the synthesis coefficients of X, $\Phi = (\phi_1, \ldots, \phi_T)$ is the dictionary, and ϕ_i are called the atoms (elementary waveforms) of the dictionary Φ. In the language of linear algebra, the dictionary Φ is a $N \times T$ matrix whose columns are the atoms normalized, supposed here to be normalized to a unit ℓ_2-norm, i.e. $\forall i \in [1, T], \|\phi_i\|_2^2 = \sum_{n=1}^{N} |\phi_i[n]|^2 = 1$ [1]. A function can be decomposed in many dictionaries, but the best dictionary is the one with the sparsest (most economical) representation of the signal. In practice, it is convenient to use dictionaries with fast implicit transform (such as Fourier transform, wavelet transform, etc.) which allow us to directly obtain the coefficients and reconstruct the signal from these coefficients using fast algorithms running in linear or almost linear time (unlike matrix-vector multiplications). The Fourier, wavelet and discrete cosine transforms provide certainly the most well known dictionaries.

Sparsity as a very flexible model to represent data

Most natural signals however are not exactly sparse but rather concentrated near a small set. Such signals are termed *compressible* or *weakly sparse* in the sense that the sorted magnitudes $|\alpha_{(i)}|$, i.e. $|\alpha_{(1)}| > |\alpha_{(2)}|, ..., > |\alpha_{(t)}|$, of the sequence of coefficients α decay quickly according to a power law, i.e. $|\alpha_{(i)}| \leq Ci^{-1/r}$, $i = 1, \ldots, t$, where C is a constant. The larger r is, the faster the amplitudes of coefficients decay, and the more compressible the signal is. In turn, the non-linear ℓ_2 approximation error of α (and x) from its M largest entries in magnitude also decrease quickly. One can think, for instance, of the wavelet coefficients of a smooth signal away from isotropic singularities, or the curvelet coefficients of a piecewise regular image away from smooth contours. A comprehensive account on sparsity of signals and images can be found in [2].

3. Dictionaries

Obviously, the best dictionary is the one which leads to the sparsest representation. Hence we could imagine having a huge overcomplete dictionary (i.e. $T \gg N$), but we would be faced with prohibitive computation time cost for calculating the α coefficients. Therefore there is a trade-off between the complexity of our analysis step (i.e. the size of the dictionary) and the

[1] The l_p-norm of a vector X, $p \geq 1$, is defined as $\|X\|_p = \left(\sum_i |X[i]|^p\right)^{1/p}$, with the usual adaptation $\|X\|_\infty = \max_i X[i]$.

computation time. Some specific dictionaries have the advantage of having fast operators and are very good candidates for analyzing the data. The Fourier dictionary is certainly the most famous, but many others have been proposed in the literature such as wavelets [3], ridgelets [4], curvelets [5, 6] or shearlets [7], to name only a few.

Toward Morphological Diversity
The morpholocal diversity concept was introduced in [8] in order to model a signal as a a finite linear mixture, each component of the mixture being sparse in a given dictionary. The idea is that a single transformation may not always represent an image well, especially if the image contains structures with different spatial morphologies. For instance, if an image is composed of edges and a locally oscillating texture, we can consider edges to be sparse in the curvelet domain while the oscillating texture is better sparsified in the local DCT domain. It has been shown by several authors that choosing a dictionary as a combination of several sub-dictionaries, each sub-dictionary having a fast transformation/reconstruction, allows us to enjoy the advantages of all sub-dictionaries, still having fast and efficient algorithms.

Adaptive representations
Different approaches have also been proposed in order to build a dictionary directly from the data. This is the case in learned dictionaries [9], for instance using e.g. the KSVD algorithm [10], the grouplet decomposition [11] or the GMCA method for multichannel/hyperspectral data [12].

4. Sparse regularization for inverse problems

In the following, for a vector z we denote $\|z\|_p^p = \sum_i |z_i|^p$ for $p \geq 0$. In particular, for $p \geq 1$, this is the p-th power of the ℓ_p norm, and for $p = 0$, we get the ℓ_0 pseudo-norm which counts the number of non-zero entries in z. The ℓ_0 regularized problem amounts to minimizing

$$\tilde{\alpha} \in \mathrm{argmin}_\alpha \|Y - H\Phi\alpha\|_2^2 + \lambda\|\alpha\|_0, \tag{3}$$

where λ is a regularization parameter. A solution $\tilde{X}$ is reconstructed as $\tilde{X} = \Phi\tilde{\alpha}$. Clearly, the goal of Eq. 3 is to minimize the number of non-zero coefficients describing the sought after signal while ensuring that the forward model is faithful to the observations.

Solving Eq. 3 is however known to be NP-hard. The ℓ_1 norm has been proposed as a tight convex relaxation of Eq. 3 leading to the minimization problem

$$\tilde{\alpha} \in \mathrm{argmin}_\alpha \|y - H\Phi\alpha\|_2^2 + \lambda\|\alpha\|_1, \tag{4}$$

where λ is again a regularization parameter different from that of Eq. 3. There has been a tremendous amount of work where researchers spanning a wide range of disciplines have studied the structural properties of minimizers of Eq. 4 and its equivalence with Eq. 3. Eq. 4 is computationally appealing and can be efficiently solved, and it has also been proved that under appropriate circumstances, Eq. 4 produces exactly the same solutions as Eq. 3, see e.g., [13].

5. Astronomical Image Restoration

5.1. Denoising

The denoising problem corresponds to the case where H is the identity (i.e. $Y = X + N$). The classical modus operandi is then to first apply the transform operator to the noisy data Y, then apply a nonlinear estimation rule to the coefficients (each coefficient individually or as a group of coefficients), and finally compute the inverse transform to get an estimate $\tilde{X}$. In brief, A classic approach is to derive the threshold t from the noise modeling. Many thresholding or shrinkage rules have been proposed. Among them, hard and soft thresholding are certainly the most well known.

5.2. Hard and Soft Thresholding

Once coefficients $\alpha = \Phi^t Y$ have been calculated, *Hard thresholding* [14, 15] consists of setting to zero all coefficients whose magnitude is less than a threshold t:

$$\tilde{\alpha}_k = \mathrm{HardThresh}_t(\alpha_k) = \begin{cases} \alpha_k & \text{if } |\alpha_k| \geq t, \\ 0 & \text{otherwise.} \end{cases} \tag{5}$$

Hard thresholding is a keep-or-kill procedure.

Soft thresholding [16, 17] is defined as the kill-or-shrink rule:

$$\tilde{\alpha}_k = \begin{cases} \mathrm{sign}(\alpha_k)(|\alpha_k| - t) & \text{if } |\alpha_k| \geq t, \\ 0 & \text{otherwise.} \end{cases} \tag{6}$$

The coefficients above the threshold are shrunk toward the origin. This can be written in the compact form

$$\tilde{\alpha}_k = \mathrm{SoftThresh}_t(\alpha_k) = \mathrm{sign}(\alpha_k)(|\alpha_k| - t)_+ \ , \tag{7}$$

where $(\cdot)_+ = \max(\cdot, 0)$.

Thresholding as a Minimization Problem Suppose that Φ is orthonormal. It can be proved that equations Eq. 6 and 7 are the unique closed form solution to the following minimization problems for a hard or soft thresholding with a threshold t:

$$\tilde{\alpha} = \underset{\alpha}{\mathrm{argmin}} \ \|Y - \Phi\alpha\|^2 + t^2 \|\alpha\|_0 \quad \textbf{Hard thresholding} \tag{8}$$

$$\tilde{\alpha} = \underset{\alpha}{\mathrm{argmin}} \ \tfrac{1}{2} \|Y - \Phi\alpha\|^2 + t \|\alpha\|_1 \quad \textbf{Soft thresholding} \tag{9}$$

Therefore, hard and soft thresholding have been used as building blocks to derive fast and efficient iterative thresholding techniques to minimize more complex inverse problems with sparsity constraints, such as inpainting and deconvolution.

For the orthogonal wavelet transform, hard thresholding is known to result in a larger variance estimate, while soft thresholding with the same threshold level creates undesirable bias because even large coefficients lying out of noise are shrunk. In practice, hard thresholding is generally preferred to soft thresholding.

5.3. Deconvolution

In a deconvolution problem, when the sensor is linear, H is the block Toeplitz matrix. A first iterative thresholding deconvolution method was proposed in [18] which consists in the fowollowing iterative scheme:

$$X^{(n+1)} = X^{(n)} + H^T \left(\mathbf{WDen}_{\Omega^{(n)}} \left(Y - HX^{(n)} \right) \right) \tag{10}$$

where **WDen** is an operator which performs a wavelet thresholding, i.e. applies the wavelet transform of the residual $R^{(n)}$ (i.e. $R^{(n)} = Y - HX^{(n)}$), threshold some wavelet coefficients, and applies the inverse wavelet transform. Only coefficients that belong to the so called *multiresolution support* $\Omega^{(n)}$ [18] are kept, while the others are set to zero. At each iteration, the multiresolution support $\Omega^{(n)}$ is updated by selecting new coefficients in wavelet transform of the residual which have an absolute value larger than a given threshold. The threshold is automatically derived assuming a given noise distribution such as Gaussian or Poisson noise.

More recently, it was shown [19, 20, 21] that a solution of (4) can be obtained through the following iteration:

$$\alpha^{(n+1)} = \mathrm{SoftThresh}_\lambda\left(\alpha^{(n)} + \Phi^T H^T\left(Y - H\Phi\alpha^{(n)}\right)\right) , \tag{11}$$

with $\|H\| = 1$. In the framework of monotone operator splitting theory, it is called Forward-Backward (FB) splitting algorithm and is classical gradient projection method for constrained convex optimization. It was shown that for frame dictionaries, it converges to the solution [21]. In the last ten years, many new optimization methods have been proposed to perform a sparse regularization in a very more efficient way. For instance, the primal-dual splitting algorithm of [22] can handle properly both sparse and positivity constraints, and leads to the following scheme:

(i) **Initialization:** Choose $\tau > 0$ and $\sigma > 0$ such that $1 - \tau\sigma \|\Phi\|^2 > \|H\|^2/2$; A sequence of relaxation parameters $\mu_n \in]0,1]$; (x_0, u_0).

(ii) **Main iteration**
- $p_{n+1} = \left(x_n - \tau\Phi u_n + \tau H^T(y - Hx_n)\right)_+$;
- $q_{n+1} = (\mathbf{I} - \mathrm{SoftThresh}_\lambda)\left(u_n + \sigma\Phi(2p_{n+1} - x_n)\right)$;
- $(x_{n+1}, u_{n+1}) = \mu_n(p_{n+1}, q_{n+1}) + (1-\mu_n)(x_n, u_n)$.

The parameter μ_n is a relaxation one that can be used to accelerate the algorithm. In the unrelaxed case, i.e. $\mu_n = 1$ for all n, we obtain $(x_{n+1}, u_{n+1}) = (p_{n+1}, q_{n+1})$.

5.4. Regularization parameters

Once the dictionary and the minimization method are chosen, a final problem remains: choosing parameters that are needed to control the algorithm. Most minimization methods, using l_0 and l_1, have a single thresholding step, where coefficients in the dictionary have to be soft or hard-thresholded using a threshold value, which is a value λ, common to projection in Φ. This parameter controls the trade-off between the fidelity to the observed visibility and the sparsity of the reconstructed solution. In real applications however, the sought-after coefficients are likely to have very different scalings, and it would be wiser to adapt the regularization parameter to each coefficient. This corresponds to solving

$$\min_\alpha \frac{1}{2}\|y - H\Phi\alpha\|^2 + \sum_{i=1}^{T} \lambda_i |\alpha[i]| . \tag{12}$$

where T is the size of the vector α (i.e. the number of coefficients).

5.4.1. Noise driven strategy Recall that in the iterative schemes discussed earlier, the noise impacts the solution through the residual $R^{(n)} = Y - H\Phi\alpha^{(n)}$ and its backprojected coefficients $\alpha^{(R)} = \Phi^T H^T R^{(n)}$. If we see the regularization as a denoising step, then we can consider that λ_i should be related to the noise in $\alpha^{(R)}$, $\lambda_i = k\sigma_i$, where k is generally chosen between 3 and 5, and σ_i is the noise standard deviation on the coefficient $\alpha[i]$. If the noise is zero-mean additive with known covariance matrix Σ, then σ_i is the square root of the ith entry on the diagonal of the covariance matrix $\Phi^T H^T \Sigma H \Phi$. Alternatively, if we know how to simulate a set of N_r realistic noise realizations, we can calculate the coefficients $\alpha^{(r)} = \Phi^T H^T Z_r$ for each realizations Z_r, and we have:

$$\sigma_i = \sqrt{\frac{1}{N_r}\sum_{r=1}^{N_r} \alpha^{(r)}[i]^2} .$$

Journal of Physics: Conference Series **699** (2016) 012010 doi:10.1088/1742-6596/699/1/012010

5.4.2. Residual-driven strategy for subband transforms In [23], it is proposed to fix the threshold only from the noise distribution at each wavelet scale j. A similar approach can be used for multiscale geometric transforms such as the curvelets.

Suppose that H is a convolution operator, Φ corresponds to a band-pass transform with J bands, and the noise N is stationary an Gaussian. Then $\alpha_j^{(N)} = (\Phi^T H^T N)_j$ in each band j is also Gaussian and stationary. We can therefore only consider one regularization parameter per band rather than one per coefficients, and Eq. 12 becomes

$$\min_{\alpha} \frac{1}{2} \|y - H\Phi\alpha\|^2 + \sum_{j=1}^{J} \lambda_j \|\alpha_j\|_1 , \tag{13}$$

where α_j are the coefficients in the band j and J is the number of bands. We could then use the $\alpha_j^{(R^{(n)})}$ to estimate standard deviation σ_j. As $R^{(n)}$ may contain also structure and not only noise, a more robust approach is to use the MAD (Median of Absolute Deviation) estimator. Then we have

$$\sigma_j = \mathrm{MAD}(\alpha_j^{(R^{(n)})})/0.6745 = \mathrm{median}(\left|\alpha_j^{(R^{(n)})} - \mathrm{median}(\alpha_j^{(R^{(n)})})\right|)/0.6745 ,$$

6. Inpainting

Missing data are a standard problem in astronomy. They can be due to bad pixels, or image area we consider as problematic due to calibration or observational problems. These masked areas lead to many difficulties for post-processing, especially to estimate statistical information such as the power spectrum or the bispectrum. The inpainting technique consists in filling the gaps. The classical image inpainting problem can be defined as follows. Let X be the ideal complete signal, Y the observed incomplete data and M the binary mask (i.e. $M_i = 1$ if we have information at pixel i, $M_i = 0$ otherwise). In short, we have: $Y = MX$. Inpainting consists in recovering X knowing Y and M. We thus want to minimize:

$$\min_X \|\Phi^T X\|_0 \quad \text{subject to} \quad Y = MX . \tag{14}$$

It was shown in [24] that this optimization problem can be efficiently solved through an iterative thresholding algorithm [24, 25] :

$$X^{(n+1)} = \Delta_{\Phi,\lambda_n}(X^{(n)} + Y - MX^{(n)}) . \tag{15}$$

where the nonlinear operator $\Delta_{\Phi,\lambda}(Z)$ consists in i) decomposing the signal Z in the dictionary Φ to derive the coefficients $\alpha = \Phi^T Z$, ii) thresholding the coefficients: $\tilde{\alpha} = \rho(\alpha, \lambda)$, where the thresholding operator ρ can either be a hard thresholding or a soft thresholding, and iii) reconstructing $\tilde{Z}$ from the thresholded coefficients $\tilde{\alpha}$.

The threshold parameter λ_n decreases with the iteration number and it plays a role similar to the cooling parameter of the simulated annealing techniques, i.e. it allows the solution to escape from local minima. More details on optimization in inpainting with sparsity can be found in [25, 2]. The case where the dictionary is a union of subdictionaries $\Phi = \{\Phi_1, ..., \Phi_K\}$ where each Φ_i has a fast operator has also been investigated in [24, 25].

Sparse inpainting has been used for different applications in astrophysics:

- high order statistics on weak lensing and Cosmic Microwave Background (CMB) data: for second order statistics such as power spectrum estimation, robust methods exist and it is generally not necessary to use more sophisticated methods. For higher order statistics (bispectrum, etc), dealing with the mask of missing data is a challenging task. It was shown that sparse inpainting can very well fill the gap, and keep a very high accuracy on the statistical estimators for both weak lensing [26] and CMB data [27]

- asteroseismology: Sparse inpainting has been included in the official CoRoT pipeline to correct the missing data in both asteroseismic and exoplanet channel [28, 29].
- Large scale studies on spherical maps for ISW (integrated Sachs-Wolfe effect) signal reconstruction [30] and CMB anomalies analysis [31].

7. Blind Source Separation

In the blind source separation (BSS) setting, the instantaneous linear mixture model assumes that we are given m observations $\{Y_1, \cdots, Y_m\}$ where each $\{Y_i\}_{i=1,\cdots,m}$ is a row-vector of size t; each measurement is the linear mixture of n source processes. As the measurements are m different mixtures, source separation techniques aim at recovering the original sources $\mathbf{X} = [X_1^T, \cdots, X_n^T]^T$ by taking advantage of some information contained in the way the signals are mixed in the observed data. The linear mixture model is rewritten in matrix form,

$$Y = HX + N, \tag{16}$$

where Y is the $m \times t$ measurement matrix (i.e. observed data), X is the $n \times t$ source matrix and H is the $m \times n$ mixing matrix. H defines the contribution of each source to each measurement. An $m \times t$ matrix $\mathbf{N}$ is added to account for instrumental noise or model imperfections.

It has been shown that sparsity is a very robust regularization to solve BSS for both underdetermined (i.e. we have less observations than unknown, $m < n$) [32] and overdetermined mixtures ($m \geq n$) [33, 12]. The GMCA (Generalized Morphological Component Analysis) algorithm [33] finds the sparsest solution through the following iterative scheme:

$$\begin{aligned} X^{(k+1)} &= \Delta_{\Phi,\lambda_k}(H^{+^{(k)}} Y) \ , \\ H^{(k+1)} &= Y X^{(k+1)^T} (X^{(k+1)} X^{(k+1)^T})^{-1} \ , \end{aligned} \tag{17}$$

where $H^{+^{(k)}}$ is the pseudo-inverse of the estimated mixing matrix $H^{(k)}$ at iteration k, λ_k is a decreasing threshold and Δ_{Φ,λ_k} is the nonlinear operator which consists in decomposing each source X_i on the dictionary Φ, threshold its coefficients and reconstruct it. Finally, for hyperspectral data, it was advocated to impose sparsity constraints on the columns of the mixing matrix to enhance source recovery [12]. GMCA has shown to be very impressive to remove Galactic and extragalactic foregrounds in order to detect the Epoch of Reionization 21-cm signal [34].

8. Conclusion

Signal processing in the 20th century was mainly considering band limited signals with an acquisition system based on the Shannon-Nyquist sampling theorem, and inverse problems were solved using least square estimators. The last ten years have seen emerging a new paradigm, based on sparse signals and the compressed sensing sampling theorem, where sparse regularization is used to solve inverse problems. We have presented how sparse representations can be used in different applications such denoising, deconvolution, inpainting or blind source separation. More details can be found in [1].

Acknowledgments

This work was funded by the PHySIS project (contract no. 640174) and the DEDALE project (contract no. 665044), within the H2020 Framework Program of the European Commission.

References

[1] Starck J -L, Murtagh F and Fadili M J 2015 *Sparse Image and Signal Processing: Wavelets and Related Geometric Multiscale Analysis* Cambridge University Press
[2] Starck J -L, Murtagh F and Fadili M J 2010 *Sparse Image and Signal Processing* Cambridge University Press
[3] Mallat S 1998 *A Wavelet Tour of Signal Processing* Academic Press
[4] Candès E J and Donoho D L 1999 Ridgelets: the key to high dimensional intermittency? *Philosophical Transactions of the Royal Society of London* A **357** 2495–2509
[5] Starck J -L, Candès E and Donoho D L 2002 The curvelet transform for image denoising *IEEE Transactions on Image Processing* **11**(6) 131–141
[6] Candès E J, Demanet L, Donoho D L and Ying L 2006 Fast discrete curvelet transforms *Multiscale Modeling and Simulation* **5**(3) 861–899
[7] Kutyniok G and Lim Wang-Q 2011 Compactly supported shearlets are optimally sparse *Journal of Approximation Theory* **163**(11) 1564–1589
[8] Starck J -L, Elad M and Donoho D L 2004 Redundant multiscale transforms and their application for morphological component analysis *Advances in Imaging and Electron Physics* 132
[9] Olshausen B A and Field D J 1996 Emergence of simple-cell receptive-field properties by learning a sparse code for natural images *Nature* **381**(6583) 607–609
[10] Aharon M, Elad M and Bruckstein A 2006 K-SVD: An algorithm for designing overcomplete dictionaries for sparse representation *IEEE Transactions on Signal Processing* **54**(11) 4311–4322
[11] Mallat S 2009 Geometrical grouplets *Applied and Computational Harmonic Analysis* **26**(2) 161–180
[12] Bobin J, Starck J -L, Moudden Y and Fadili M J 2008 Blind source separation: The sparsity revolution *Advances in Imaging and Electron Physics* **152** 221–306
[13] Donoho D L 2006 For most large underdetermined systems of linear equations, the minimal ℓ_1 solution is also the sparsest solution *Communications on Pure and Applied Mathematics* **59**(7) 907–934
[14] Starck J -L and Bijaoui A 1994 Filtering and deconvolution by the wavelet transform *Signal Processing* **35** 195–211
[15] Donoho D L, Johnstone I M, Kerkyacharian G and Picard D 1995 Wavelet shrinkage: asymptopia? *Journal of the Royal Statistical Society Series* B **57** 301–369
[16] Weaver J B, Yansun X, Healy D M and Cromwell L D 1991 Filtering noise from images with wavelet transforms *Magnetic Resonance in Medicine* **21**(2) 288–295
[17] Donoho D L 1995 De-noising by soft-thresholding *IEEE Transactions on Information Theory,* **41**(3) 613–627
[18] Starck J -L, Bijaoui A and Murtagh F 1995 Multiresolution support applied to image filtering and deconvolution *CVGIP: Graphical Models and Image Processing* **57** 420–431
[19] Figueiredo M A and Nowak R 2003 An EM algorithm for wavelet-based image restoration *IEEE Transactions on Image Processing* **12**(8) 906–916
[20] Daubechies I, Defrise M and De Mol C 2004 An iterative thresholding algorithm for linear inverse problems with a sparsity constraint *Communications on Pure and Applied Mathematics* **57** 1413–1541
[21] Combettes P L and Wajs V R 2005 Signal recovery by proximal forward-backward splitting *Multiscale Modeling and Simulation* **4**(4) 1168–1200
[22] Vũ B 2013 A splitting algorithm for dual monotone inclusions involving cocoercive operators *Advances in Computational Mathematics* **38**(3) 667–681
[23] Lanusse F, Paykari P, Starck J -L, Sureau F, Bobin J and Rassat A 2014 PRISM: Recovery of the primordial spectrum from Planck data *Astron. Astrophys.* **571** L1
[24] Elad M, Starck J -L, Donoho D L and Querre P 2005 Simultaneous cartoon and texture image inpainting using morphological component analysis (MCA) *Applied and Computational Harmonic Analysis* **19** 340–358
[25] Fadili M J, Starck J -L and Murtagh F 2009 Inpainting and zooming using sparse representations *The Computer Journal* **52** 64–79
[26] Pires S, Starck J -L, Amara A, Réfrégier A and Teyssier R 2009 Cosmological model discrimination with weak lensing *Astron. Astrophys.* **505** 969–979
[27] Perotto L, Bobin J, Plaszczynski S, Starck J -L and Lavabre A 2010 Reconstruction of the cosmic microwave background lensing for Planck *Astron. Astrophys.* **519** A4
[28] García R A, Mathur S, Pires S, Régulo C, Bellamy B, Pallé P L, Ballot J, Barceló Forteza S, Beck P G, Bedding T R, Ceillier T, Roca Cortés T, Salabert D and Stello D 2014 Impact on asteroseismic analyses of regular gaps in Kepler data *Astron. Astrophys.* **568** A10
[29] Pires S, Mathur S, García R A, Ballot J, Stello D and Sato K 2015 Gap interpolation by inpainting methods: Application to ground and space-based asteroseismic data *Astron. Astrophys.* **574** A18
[30] Dupé F -X, Rassat A, Starck J -L and Fadili M J 2011 Measuring the integrated Sachs-Wolfe effect *Astron. Astrophys.* **534** A51
[31] Starck J -L, Fadili M J and Rassat A 2013 Low-ℓ CMB analysis and inpainting *Astron. Astrophys.* **550** A15

[32] Gribonval R, Rauhut H, Schnass K and Vandergheynst P 2008 Atoms of all channels, unite! Average case analysis of multi-channel sparse recovery using greedy algorithms *Journal of Fourier Analysis and Applications* **14**(5) 655–687
[33] Bobin J, Starck J -L, Fadili M J and Moudden Y 2007 Sparsity and morphological diversity in blind source separation *IEEE Transactions on Image Processing* **16**(11) 2662–2674
[34] Chapman E, Zaroubi S and Abdalla F 2014 Foreground Removal vs. Foreground Avoidance: Contamination of the EoR Window ArXiv e-prints arXiv:1408.4695

Extracting nonlinear spatiotemporal dynamics in active dendrites using data-driven statistical approach

Toshiaki Omori[1] and Koji Hukushima[2,3]

[1] Department of Electrical and Electronic Engineering, Graduate School of Engineering, Kobe University, 1-1 Rokkodai-cho, Nada-ku, Kobe, Hyogo 657-8501, Japan
[2] Department of Basic Science, Graduate School of Arts and Sciences, The University of Tokyo, 3-8-1 Komaba, Meguro-ku, Tokyo 153-8902, Japan
[3] Center for Materials Research by Information Integration, National Institute for Materials Science, 1-2-1 Sengen, Tsukuba, Ibaraki 305-0047, Japan

E-mail: omori@eedept.kobe-u.ac.jp

Abstract. We propose a data-driven statistical method for extracting nonlinear spatiotemporal membrane dynamics of active dendrites. We employ a framework of probabilistic information processing to extract the nonlinear spatiotemporal dynamics obeying the reaction-diffusion equation from partially observable data. By employing sequential Monte-Carlo method and other statistical methods, membrane dynamics and their underlying electrical properties are simultaneously estimated in the proposed method. Using the proposed method, we show that nonlinear spatiotemporal dynamics in active dendrites can be extracted from partially observable data.

1. Introduction

Recent findings in experimental neuroscience suggest that active dendrite plays important roles in neural information processing such as directional selectivity [1], integration of sensory and motor inputs [2], and so on [3, 4, 5, 6, 7, 8, 9, 10]. For example, recent experimental results showed that active dendritic computation in retinal ganglion cells plays a key role in directional selectivity for visual stimuli [1, 2]. However, the mechanism of active dendritic computation is still unclear, since the spatiotemporal dynamics of active dendrite remains unknown.

Dynamical behaviors in active (nonlinear) dendrites depend on nonlinear ion channels and neuronal morphology and can be described by a reaction-diffusion equation called a nonlinear cable equation, while classical theoretical studies of dendrites have assumed passive (linear) dendrites [11, 12, 13]. Great advances in measurement technology enable us to deal with spatiotemporal data from neural systems including active dendrites as imaging data. However, the observable information in the measurements is quite limited, when compared with the complexity of the entire system of active dendrites.

Some estimation techniques using the state-space modeling approach to extract the spatiotemporal dynamics of the dendrites have been proposed. In some of previous methods, only membrane potentials are estimated while assuming the parameters underlying spatiotemporal dynamics are known [14], and most of the previous methods only estimate linear dynamics in multi-compartment models or nonlinear dynamics in single-compartment models [15], even though to reveal the active dendritic computation, it is important to establish methods for extracting nonlinear dynamics for spatiotemporal membrane evolution in multi-compartment models.

In this study, we propose a data-driven statistical method to estimate nonlinear spatiotemporal membrane dynamics of active dendrites in order to extract nonlinear dynamics of dendritic membrane. We employ a data-driven statistical approach to extract the nonlinear spatiotemporal dynamics obeying the nonlinear reaction-diffusion equation from partially observable data. First, we formulate a generalized state-space model of active dendrite that is based on multi-compartment model, and then we derive sequential estimation algorithm for the generalized state-space model. Membrane dynamics and the electrical properties underlying them are simultaneously estimated by using sequential Monte-Carlo method and other statistical methods. Using the proposed method, we show that nonlinear spatiotemporal dynamics in active dendrites can be extracted from partially observable data.

This paper is organized as follows. In section 2, we formulate spatiotemporal dynamics in active dendrites from the nonlinear cable equation and derive a nonlinear state-space model for active dendrites. In section 3, we propose a statistical method for simultaneously extracting membrane responses and membrane properties from partially observable data. In section 4, we show the effectiveness of the proposed method by using multi-compartment models, including the one for hippocampal pyramidal neurons. Concluding remarks are given in section 5. A preliminary version of this work appeared in [16].

2. Theory

In this section, we propose a statistical method for extracting spatiotemporal nonlinear dynamics of active dendrites. As shown in Fig. 1, we consider simultaneous estimation of membrane responses and electrical properties underlying dendritic dynamics from partially observable noisy data. To this end, we formulate a generalized state-space model of dendritic dynamics. We first derive a system model that describes spatiotemporal nonlinear dynamics of dendrite. Next we formulate an observation model that reflects the partially observable situation, as in imaging experiments. Finally we derive a data-driven statistical method for estimating membrane responses including membrane potential and channel variables and underlying parameters such as membrane properties from partially observable data.

2.1. System Model

Here we derive a probabilistic model of membrane potential and channel variables in active dendrites. A nonlinear reaction-diffusion equation for active dendrites is employed to the system model.

2.1.1. Dynamics of Membrane Potential A membrane potential $V(x,t)$ at position x and time t in active dendrite is assumed to obey the following reaction-diffusion equation called nonlinear cable equation [13]:

$$C\frac{\partial V}{\partial t} = -\sum_{\mathrm{X}} g_{\mathrm{X}}(m_{\mathrm{X}})^{M_{\mathrm{X}}}(h_{\mathrm{X}})^{N_{\mathrm{X}}}(V - E_{\mathrm{X}}) - D\frac{\partial^2 V}{\partial x^2} + I_{\mathrm{ext}}(x,t) + \xi^{(V)}(x,t), \quad (1)$$

where m_{X} and h_{X} show the activation and inactivation for ion channel variables, respectively. Each compartment is assumed to have some kinds of membrane currents $\sum_{\mathrm{X}} g_{\mathrm{X}} m_{\mathrm{X}}^{M_{\mathrm{X}}} h_{\mathrm{X}}^{N_{\mathrm{X}}}(V_x - E_{\mathrm{X}})$, axial currents $D\frac{\partial^2 V}{\partial x^2}$, external input currents $I_{\mathrm{ext}}(x,t)$ and noise current $\xi^{(V)}(x,t)$. The maximal membrane conductances, reversal potentials and membrane capacitance are expressed by $g_{\mathrm{X},x}$, E_{X} and C, respectively.

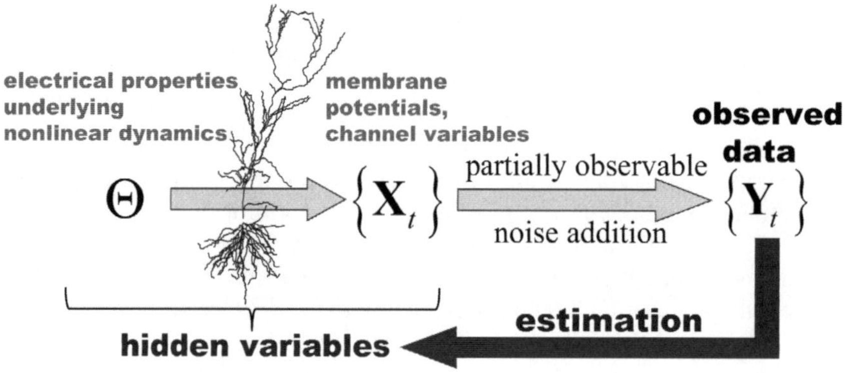

Figure 1. Framework for extracting spatiotemporal dynamics of active dendrites. By formulating statistical models of forward process, we estimate both membrane responses $\boldsymbol{X}_t$ and membrane properties Θ from partially observable noisy data $\boldsymbol{Y}_t$.

By discretizing Eq. (1) with respect to space and time, we derive the following equation:

$$\begin{aligned} v_{x,t+1} &= v_{x,t} - \Delta \sum_{\mathrm{X}} g_{\mathrm{X},x}(m_{\mathrm{X},x,t})^{M_{\mathrm{X}}}(h_{\mathrm{X},x,t})^{N_{\mathrm{X}}}(v_{x,t} - E_{\mathrm{X}}) - \Delta g_{\mathrm{axial}} \sum_{y}(v_{x,t} - v_{y,t}) \\ &+ \Delta I_{\mathrm{ext},x,t} + \sqrt{\Delta}\xi_{x,t}^{(v)}, \end{aligned} \tag{2}$$

where time width is set to Δ and we put $C = 1$ without loss of generality. Each position in the spatially discretized system is called *a compartment*, and compartments are connected according to neuronal spatial structure as shown in Fig. 2(a).

Based on the statistics of noise, the probabilistic density function of membrane potential is described by a probabilistic model $p(v_{x,t+1}|\{v_{x,t}\}, m_{\mathrm{X},x,t}, h_{\mathrm{X},x,t})$ that depends on the state of neighboring compartments at preceding time $\{v_{x,t}\}$. If the noise is white Gaussian, the probabilistic density function can be described by

$$p(v_{x,t+1}|\{v_{x,t}\}, m_{\mathrm{X},x,t}, h_{\mathrm{X},x,t}) = \mathcal{N}(v_{x,t+1}|\mu_{x,t+1}, \sigma_v^2), \tag{3}$$

where the average is expressed by $\mu_{x,t+1} = v_{x,t} - \Delta \sum_{\mathrm{X}} g_{\mathrm{X},x}(m_{\mathrm{X},x,t})^{M_{\mathrm{X}}}(h_{\mathrm{X},x,t})^{N_{\mathrm{X}}}(v_{x,t} - E_{\mathrm{X}}) - \Delta g_{\mathrm{axial}} \sum_y (v_{x,t} - v_{y,t}) + \Delta I_{\mathrm{ext},x,t}$, and the variance is expressed by σ_v^2.

2.1.2. Dynamics of Channel Variables In conductance-based models such as the Hodgkin-Huxley model, channel variables $m_{\mathrm{X},x,i}$ and $h_{\mathrm{X},x,i}$ obey the following first-order kinetics:

$$\frac{dm_{\mathrm{X},x}}{dt} = \alpha_{m_{\mathrm{X}}}(V_x)(1 - m_{\mathrm{X},x}) - \beta_{m_{\mathrm{X}}}(V_x)m_{\mathrm{X},x} + \xi_x^{(m_{\mathrm{X}})}(t), \tag{4}$$

$$\frac{dh_{\mathrm{X},x}}{dt} = \alpha_{h_{\mathrm{X}}}(V_x)(1 - h_{\mathrm{X},x}) - \beta_{h_{\mathrm{X}}}(V_x)h_{\mathrm{X},x} + \xi_x^{(h_{\mathrm{X}})}(t), \tag{5}$$

where $\alpha_{m_{\mathrm{X}}}(V_x)$, $\alpha_{h_{\mathrm{X}}}(V_x)$, $\beta_{m_{\mathrm{X}}}(V_x)$, and $\beta_{h_{\mathrm{X}}}(V_x)$ are functions of membrane potential. By discretizing Eq. (4) and (5) with respect to time, we obtain the probabilistic model of the channel variables: $p(m_{\mathrm{X},x,t+1}|m_{\mathrm{X},x,t}, v_{x,t})$ and $p(h_{\mathrm{X},x,t+1}|h_{\mathrm{X},x,t}, v_{x,t})$. If the noise is white Gaussian, the probabilistic density function can be described by

$$p(m_{\mathrm{X},x,t+1}|m_{\mathrm{X},x,t}, v_{x,t}) = \mathcal{N}(m_{\mathrm{X},x,t+1}|\mu_{m_{\mathrm{X}},x,t+1}, \sigma_{\mathrm{X}_m}^2), \tag{6}$$

$$p(h_{\mathrm{X},x,t+1}|h_{\mathrm{X},x,t}, v_{x,t}) = \mathcal{N}(h_{\mathrm{X},x,t+1}|\mu_{h_{\mathrm{X}},x,t+1}, \sigma_{\mathrm{X}_h}^2). \tag{7}$$

Based on probabilistic models (Eqs. (3), (6), and (7)), the system models for all the hidden state vectors $\boldsymbol{X}_t = \{v_{x,t}, m_{\mathrm{X},x,t}, h_{\mathrm{X},x,t}\}$ are summarized as $p(\boldsymbol{X}_{t+1}|\boldsymbol{X}_t)$.

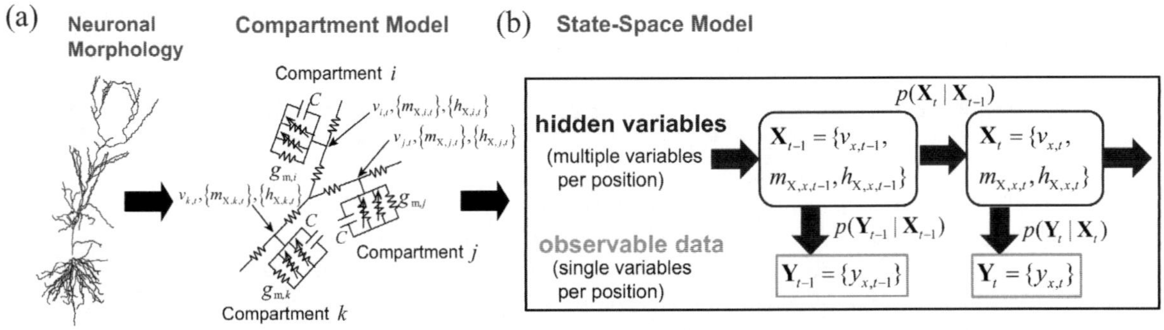

Figure 2. State-space modeling of active dendrite. (a) multi-compartment model of dendrite. Each compartment has passive and active ion channels and is assumed to be connected to neighboring compartments according to neuronal morphology. (b) state-space model of dendrite. Nonlinear spatiotemporal dynamics obeying reaction-diffusion equation called cable equation is reflected in the system model $p(\boldsymbol{X}_{t+1}|\boldsymbol{X}_t)$, whereas the partial and noisy observation process is reflected in the observation model $p(\boldsymbol{Y}_t|\boldsymbol{X}_t)$.

2.2. Observation Model

Each compartment has a multi-dimensional state such as membrane potential $v_{x,t}$ and channel variables $m_{\mathrm{X},x,t}$ and $h_{\mathrm{X},x,t}$ for each type of ion channel X. We assume the situation that only membrane potential can be observed; the observed variable $y_{x,t}$ is assumed to be expressed as follows:

$$y_{x,t} = g(v_{x,t}) + \xi_{x,t}^{(y)}, \tag{8}$$

where $g(\cdot)$ is observation function of true membrane potential $v_{x,t}$, and $\xi_{x,t}^{(y)}$ is an observation noise. According to the statistics of the noise, we can derive probabilistic version of observation model $p(y_{x,t}|v_{x,t})$. If the observation noise is white Gaussian, the probabilistic density function is described by

$$p(y_{x,t}|v_{x,t}) = \mathcal{N}(y_{x,t}|g(v_{x,t}), \sigma_y^2). \tag{9}$$

The observation model for the entire multi-compartment model is expressed as $p(\boldsymbol{Y}_t|\boldsymbol{X}_t)$.

3. Estimation of Hidden Variables

Here we describe the method for estimating latent variables $\{\boldsymbol{X}_t\}$ from observable data $\{\boldsymbol{Y}_t\}$. Hidden variables at time $t+1$, $\boldsymbol{X}_{t+1}$, are estimated using the observable data up to the previous time $\boldsymbol{Y}_{1:t}$ based on the predictive distribution as follows:

$$p(\boldsymbol{X}_{t+1}|\boldsymbol{Y}_{1:t}) = \int p(\boldsymbol{X}_{t+1}|\boldsymbol{X}_t)p(\boldsymbol{X}_t|\boldsymbol{Y}_{1:t})d\boldsymbol{X}_t, \tag{10}$$

where $p(\boldsymbol{X}_t|\boldsymbol{Y}_{1:t})$ shows a filtering distribution given the observable data up to the same time $\boldsymbol{Y}_{1:t}$ as follows:

$$p(\boldsymbol{X}_t|\boldsymbol{Y}_{1:t}) = \frac{p(\boldsymbol{Y}_t|\boldsymbol{X}_t)p(\boldsymbol{X}_t|\boldsymbol{Y}_{1:t-1})}{\int p(\boldsymbol{Y}_t|\boldsymbol{X}_t)p(\boldsymbol{X}_t|\boldsymbol{Y}_{1:t-1})d\boldsymbol{X}_t}. \tag{11}$$

Note that the predictive and filtering distributions can be obtained iteratively by using Eqs. (10) and (11). Using the filtering and predictive distributions, a smoothing distribution $p(\boldsymbol{X}_t|\boldsymbol{Y}_{1:T})$ is obtained as follows:

$$p(\boldsymbol{X}_t|\boldsymbol{Y}_{1:T}) = p(\boldsymbol{X}_t|\boldsymbol{Y}_{1:t}) \int \frac{p(\boldsymbol{X}_{t+1}|\boldsymbol{X}_t)p(\boldsymbol{X}_{t+1}|\boldsymbol{Y}_{1:T})}{p(\boldsymbol{X}_{t+1}|\boldsymbol{Y}_{1:t})} d\boldsymbol{X}_{t+1}. \tag{12}$$

Note that these three distributions have system model $p(\boldsymbol{X}_{t+1}|\boldsymbol{X}_t)$ and observation model $p(\boldsymbol{Y}_t|\boldsymbol{X}_t)$. By maximizing the filtering distribution $p(\boldsymbol{X}_t|\boldsymbol{Y}_{1:t})$ or the smoothing distribution $p(\boldsymbol{X}_t|\boldsymbol{Y}_{1:T})$, we obtain the estimated values for hidden variables $\boldsymbol{X}_t$.

Since we have assumed the nonlinearity in the generalized state-space model, the integrations in filtering, predictive and smoothing distributions become analytically intractable. In the present study, we employ a sequential Monte-Carlo method [17, 18, 19] to tackle this difficulty and perform the filtering and prediction iteratively.

In addition to hidden variables of $\boldsymbol{X}_t$, a set of parameters Θ including maximal membrane conductances $\{g_{\mathrm{X},x}\}$ is unknown. The EM algorithm [20, 21, 22] is employed in order to estimate those parameters underlying the nonlinear spatiotemporal dynamics. In the E-step, expectation of log-likelihood function is calculated

$$Q(\Theta|\Theta_k) = \langle \log p(\{\boldsymbol{X}_t\}, \{\boldsymbol{Y}_t\}|\Theta)\rangle_{p(\{\boldsymbol{X}_t\}|\{\boldsymbol{Y}_t\},\Theta_k)}, \tag{13}$$

where Θ_k shows a set of parameters estimated at step k of the EM algorithm. In the M-step, we obtain the set of parameters Θ that maximizes $Q(\Theta|\Theta_k)$ as Θ_{k+1},

$$\Theta_{k+1} = \arg\max_{\Theta} Q(\Theta|\Theta_k). \tag{14}$$

By performing the E-step and the M-step iteratively, we obtain estimated values of parameters by the converged value of Θ_k.

4. Results

In this section, we evaluate the effectiveness of the proposed method by using simulated data of multi-compartment models with active channels. We assume that each compartment has passive and active channels, and only noisy membrane potentials are partially observable. Using the proposed method, we estimate not only membrane potentials but also other hidden variables such as activation and inactivation variables for active channels for every compartments. We also estimate underlying parameters, including maximal membrane conductances of active channels, which govern the nonlinear spatiotemporal dynamics of dendritic membrane.

Here we consider a compartment model with serially connected compartments and extract true membrane potentials $v_{x,t}$ and channel variables of sodium current m_{Na}, h_{Na} and potassium current m_{K}, h_{K}. The estimated time evolution of these hidden variables is shown in Fig. 3(a). We find that true membrane potential at each compartment x can be estimated accurately. Furthermore, other non-observable hidden variables such as those for sodium channels can be estimated as shown in the bottom figures in Fig. 3(a). Hidden variables for potassium channels can be estimated as well (data not shown). These results show that the proposed method enables us to extract hidden variables underlying nonlinear spatiotemporal dynamics of dendritic membrane. Electrical properties such as maximal membrane conductances should be estimated since they govern nonlinear spatiotemporal dynamics of dendritic membrane potentials and channel variables. The estimated maximal membrane properties of sodium and potassium channels are shown in Fig. 3(b). In Fig. 3(b), we find that estimated maximal membrane conductances converge to true value (dashed line). In these results, we see that the proposed method simultaneously estimates only electrical properties but also membrane potentials and channel variables.

To evaluate the proposed method in a more realistic situation, we extract nonlinear spatiotemporal dynamics in a hippocampal CA1 pyramidal neuron model by the proposed method. We utilize simulated data of action potentials backpropagating from soma to apical dendrites as shown Fig. 4(a). We assume that only noisy membrane potentials in parts of dendritic compartments can be observed. Using this limited data, we simultaneously

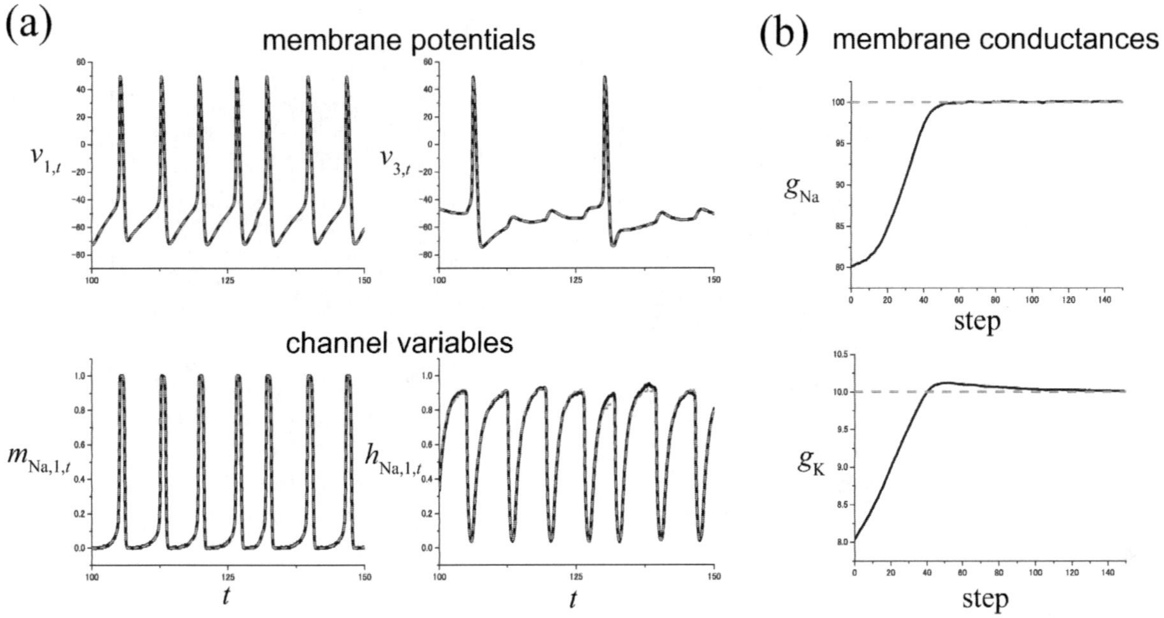

Figure 3. Estimation of nonlinear spatiotemporal dynamics in multi-compartment model. (a) True and estimated time courses of membrane potentials $v_{x,t}$ and channel variables $m_{X,x,t}$ and $h_{X,x,t}$. Dashed and solid lines show true and estimated time courses, respectively. (b) Estimated maximal conductances of ion channels g_{Na} and g_K. As iteration of estimation proceeds, the estimated values (solid lines) converge to true values of the maximal conductances (dashed lines).

estimate membrane response and membrane properties. The estimated distribution of membrane potentials is shown in Fig. 4(b). We find that the estimated distribution is similar to the true one, and that underlying parameters can be estimated successfully (data not shown), and that backpropagating action potentials can be reproduced by the proposed method. These results show that the proposed method can extract spatiotemporal nonlinear dynamics of active dendrite from partially observable noisy data.

5. Concluding Remarks

In this study, we have proposed a data-driven statistical method for estimating spatiotemporal membrane dynamics of active dendrites. The generalized state-space model of active dendrite based on the multi-compartment model of dendrites has been derived. A novel spatiotemporal dynamics extraction technique has been realized by using the sequential Monte-Carlo method and the EM algorithm. Using the proposed method, we have shown that inner state of neurons, such as membrane potential and ion channel variables and their underlying parameters, are simultaneously estimated in the multi-compartment model of the hippocampal CA1 pyramidal neuron. These results show that nonlinear spatiotemporal dynamics of active dendrites can be extracted from partially observable data by means of the proposed method.

Acknowledgments

This study was partially supported by a Grant-in-Aid for Scientific Research for Innovative Areas "Initiative for High-Dimensional Data-Driven Science through Deepening of Sparse Modeling" (No. 25120010 for T.O. and K.H.), a Grant-in-Aid for Young Scientist (B) (No. 25120005 for T.O.), a Grant-in-Aid for Challenging Exploratory Research (No. 25610102 for K. H.),

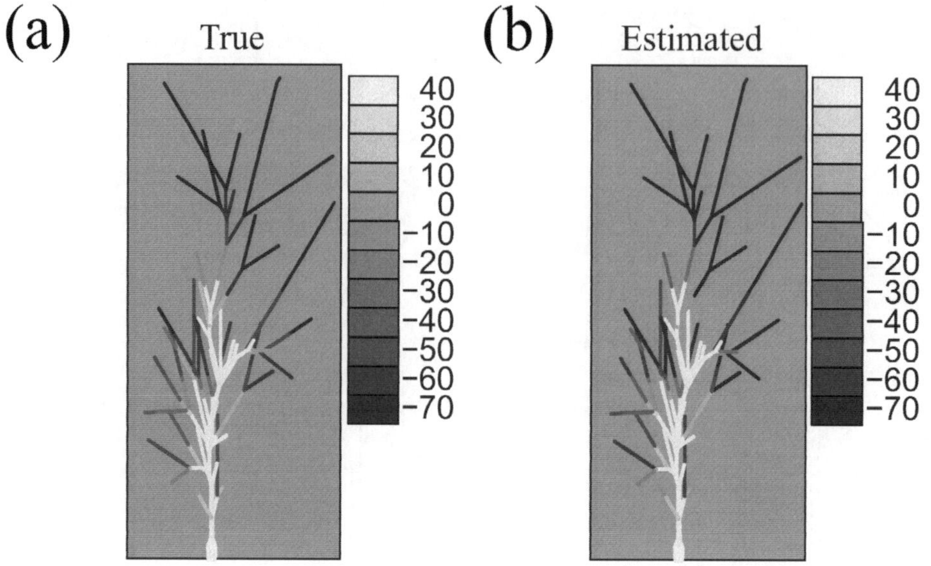

Figure 4. Estimation of spatiotemporal dynamics in multi-compartment model with neural morphology of hippocampal CA1 pyramidal neuron. True and estimated distributions of membrane potentials are shown in figures (a) and (b), respectively. Backpropagating action potentials along apical dendrite are simulated and a part of compartments is assumed to be observable for estimation. Maximal values of membrane potentials at each position in active dendrite $\max_t v_{x,t}$ are shown.

and a Fund for the Promotion of Joint International Research (Fostering Joint International Research, No. 15KK0010 for T. O.) from the Ministry of Education, Culture, Sports, Science, and Technology of Japan.

References

[1] Sivyer B and Williams S R 2013 *Nature Neurosci.* **16** 1848
[2] Xu N, Harnett M T, Williams S R, Huber D, O'Connor D H, Svoboda K and Magee J C 2012 *Nature* **492** 247
[3] Häusser M 2000 *Science* **290** 739
[4] Shepherd G M 2004 *The Synaptic Organization of the Brain* (Oxford University Press)
[5] Omori T, Aonishi T, Miyakawa H, Inoue M and Okada M 2006 *Brain Res.* **1125** 199
[6] Spruston N 2008 *Nat. Rev. Neurosci.* **9** 206
[7] Omori T, Aonishi T, Miyakawa H, Inoue M and Okada M 2009 *Neurosci. Res.* **64** 83
[8] Silver A S 2010 *Nat. Rev. Neurosci.* **11** 474
[9] Zhang D, Li Y, Rasch M J and Wu S 2013 *Front. Comput. Neurosci.* **7** 56
[10] Yates D 2013 *Nat. Rev. Neurosci.* **15** 815
[11] Rall W 1959 *Exp. Neurol.* **1** 491
[12] Rall W 1960 *Exp. Neurol.* **2** 503
[13] Rall W 1999 An historical perspective on modeling dendrites *Dendrites* ed Stuart G, Spruston N and Häusser M (Oxford University Press)
[14] Paninski L 2010 *J. Comput. Neurosci.* **28** 211
[15] Omori T, Aonishi T and Okada M 2013 *Advances in Cognitive Neurodynamics* **3** 649
[16] Omori T 2014 *Neural Infor. Proc.* **1** 27
[17] Kitagawa G 1987 *J. American Stat. Assoc.* **82** 1032
[18] Doucet A, Godsill S and Andrieu C 2000 *Statistics and Computing* **10** 197
[19] Doucet A, de Freitas N and Gordon N 2001 *Sequential Monte Carlo Methods in Practice* (Springer-Verlag)
[20] Bishop C M 2006 *Pattern Recognition and Machine Learning* (Singapore: Springer)
[21] Dempster A P, Laird N M and Rubin D B 1977 *J. R. Stat. Soc.* B **39** 1

[22] McLachlan G J and Krishnan T 1997 *The EM Algorithm and its Extensions* (New York: Wiley)

Graph structure modeling for multi-neuronal spike data

Shotaro Akaho[1], Sho Higuchi[2], Taishi Iwasaki[3], Hideitsu Hino[4], Masami Tatsuno[5] and Noboru Murata[3]

[1] The National Institute of Advanced Industrial Science and Technology, 1-1-1 Umezono, Tsukuba, Ibaraki 305-8568 Japan
[2] NTT DATA Corporation, 3-9 Toyosu 3-chome, Koto-ku, Tokyo 135-8671 Japan
[3] Waseda University, 3-4-1 Okubo, Shinjuku-ku, Tokyo 169-8555 Japan
[4] University of Tsukuba, 1-1-1 Tennodai, Tsukuba, Ibaraki 305-8577 Japan
[5] University of Lethbridge, Lethbridge, AB, T1K3M4, Canada

E-mail: s.akaho@aist.go.jp

Abstract. We propose a method to extract connectivity between neurons for extracellularly recorded multiple spike trains. The method removes pseudo-correlation caused by propagation of information along an indirect pathway, and is also robust against the influence from unobserved neurons. The estimation algorithm consists of iterations of a simple matrix inversion, which is scalable to large data sets. The performance is examined by synthetic spike data.

1. Introduction

Due to recent advances in experimental technologies, we can record activities of a large number of neurons simultaneously. It has been suggested that cell assemblies can be formed dynamically in the course of motor and cognitive functions [1]. In order to understand how the cell assemblies contribute to brain functions, it is important to clarify the relationship between neurons. For example, Wilson and McNaughton estimated a functional connection in the hippocampal CA1 network using a cross-correlation function [2]. Recently, researchers have also started investigating higher-order interactions using information-theoretic approaches [3–5]. However, many of the existing methods have the following difficulties:

(i) Pseudo-correlation: Neuron activities can propagate to other neurons via indirect pathways. This causes pseudo correlation.

(ii) Direction of connection: Functional connectivity estimated by correlation-based methods is often symmetric. It is not possible to extract information about the direction of connections.

(iii) Openness of the system: In neurophysiological experiments, only a limited number of neurons can be recorded. The activity of measured neurons is also affected by other neurons that are not observed.

(iv) Positive and negative connectivity: The previously proposed method based on diffusion process [6] was developed to analyze a network of excitatory (positive) connections. However, there are also inhibitory (negative) connections in the brain.

Journal of Physics: Conference Series **699** (2016) 012012 doi:10.1088/1742-6596/699/1/012012

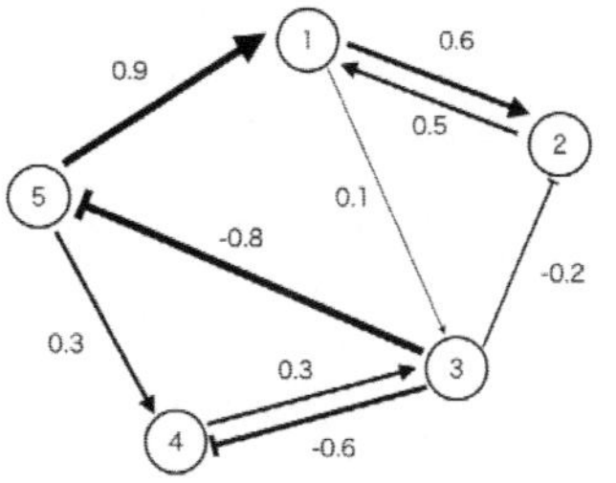

$$\mathbf{\Lambda} = \begin{pmatrix} 0 & 0.6 & 0.1 & 0 & 0 \\ 0.5 & 0 & 0 & 0 & 0 \\ 0 & -0.2 & 0 & -0.6 & -0.8 \\ 0 & 0 & 0.3 & 0 & 0 \\ 0.9 & 0 & 0 & 0.3 & 0 \end{pmatrix}$$

Figure 1. An example of a graph structure representing direct connectivity Λ

To overcome these difficulties, we propose a novel method to extract dependency from multivariate time series data.

2. Model

Suppose there are N neurons measured, membrane potential $v_i(t+\Delta)$ of the neuron i at time $t+\Delta$ is affected by the neuron j at time t by

$$v_i(t+\Delta) = b_i(t) + \theta_{ij} Y_j(t), \tag{1}$$

where Δ is a small period of time, $Y_j(t)$ is a binary random variable representing spike of the neuron j, and $b_i(t)$ is input from all other neurons including unmeasured ones except the neuron j. The coefficient θ_{ij} represents the connection from the neuron j to the neuron i, including not only a direct connection but also indirect connection through two-steps, three-steps and longer pathways through other neurons. Let λ_{ij} be the direct connection from the neuron j to the neuron i. We assume that indirect connection propagates in a multiplicative way. For instance, the strength of connection from the neuron j to the neuron i through the neuron k is equal to $\xi_k^{(1)} \lambda_{ik} \lambda_{kj}$, where $\xi_k^{(1)}$ denotes a decaying factor. Therefore θ_{ij} is decomposed as the sum of multi-step connections,

$$\theta_{ij} = \lambda_{ij} + \sum_k \xi_k^{(1)} \lambda_{ik} \lambda_{kj} + \sum_{k,l} \xi_{kl}^{(2)} \lambda_{ik} \lambda_{kl} \lambda_{lj} + \cdots, \tag{2}$$

where in this paper we assume there are no direct self-connection, hence $\lambda_{ii} = 0$. For simplicity, we further assume decaying factors $\xi_k^{(1)}$, $\xi_{kl}^{(2)}$, ... are all equal to one by constraining $\lambda_{ij} < 1$ to incorporate the decaying effect in λ_{ij}. As a result, we have a simple relation between the accumulated total connection θ_{ij} and the direct connection λ_{ij},

$$\Theta = \Lambda + \Lambda^2 + \cdots = (I - \Lambda)^{-1} - I, \tag{3}$$

where Θ and Λ are matrices whose ij elements are θ_{ij} and λ_{ij} respectively.

In this paper, we assume that we can construct a matrix X whose off-diagonal element is related to Θ as

$$x_{ij} = \alpha \theta_{ij} + \epsilon_{ij}, \qquad i \neq j \tag{4}$$

where x_{ij} is the ij element of X, α is a scalar parameter, and ϵ_{ij} is a random noise. We assume the diagonal elements of X are not obtained, and let $\mathrm{diag}X = O$, then we have

$$X + D \simeq \alpha \Theta = \alpha \left\{ (I - \Lambda)^{-1} - I \right\}, \tag{5}$$

where D is a diagonal matrix representing unknown diagonal elements.

As a consequence, the problem is to obtain the connectivity matrix Λ for given X by estimating parameters D, α and solving eq. (5). An example of Λ is shown in figure 1, which can be regarded as weighted edges of a graphical structure.

In the rest of the paper, first we propose an iterative algorithm to find α, D, Λ for a given X. Next, we give a way of constructing X from measured neural spike data, and show some simulation results.

3. Optimization algorithm

The objective function to be minimized is the squared loss defined from eq. (5),

$$J(\alpha, D, \Lambda) = \left\| X + D - \alpha\{(I - \Lambda)^{-1} - I\} \right\|_F^2 \tag{6}$$

where $\| \quad \|_F$ denotes the Frobenius norm.

We further introduce *sparsity* of the connectivity matrix Λ as a constraint. Otherwise, most neurons are connected each other, which makes it difficult to interpret the results. The sparsity constraint is also consistent with a neuroscientific finding that approximately 10% of excitatory neurons are connected [7]. Here we use L_0 norm constraint as sparsity, then the problem is to minimize the objective function $J(\alpha, D, \Lambda)$ under the constraint

$$\|\Lambda\|_0 \leq M. \tag{7}$$

In this paper, the hyperparameter M, which represents an upperbound of the number of nonzero components of Λ, is estimated by StARS criterion as described later.

3.1. Objective minimization

We apply alternating iterative procedures to minimize the objective function $J(\alpha, D, \Lambda)$, i.e., each parameter α, D or Λ is optimized by fixing the other parameters. For sparsity, we apply a greedy approach.

First, D is initialized by rI, where r is a constant so that $\det(X + D) \neq 0$. α is randomly initialized from uniform distribution on $(0, 1]$. Fixing D and α, we obtain $\hat{\Lambda}'$, which is an estimator of Λ, by solving eq. (5) and taking $\lambda_{ii} = 0$ into account,

$$\hat{\Lambda}' = \text{offdiag}\left[-\left(\frac{1}{\alpha}(X + D) + I \right)^{-1} \right], \tag{8}$$

where offdiag$[A]$ is an operator that sets A's diagonal elements to zero. Since $\hat{\Lambda}'$ is not a sparse matrix, we obtain a sparse estimate of Λ by cutting off the small values of $\hat{\Lambda}'$,

$$\hat{\lambda}_{ij} = \begin{cases} \hat{\lambda}'_{ij} & \text{if } |\hat{\lambda}'_{ij}| \geq \zeta \\ 0 & \text{otherwise} \end{cases} \tag{9}$$

where ζ is the M-th largest value among $\{|\hat{\lambda}'_{ij}|\}$.

In order to optimize α for a fixed $\hat{\lambda}_{ij}$, by letting $D = \text{diag}\left(\alpha(I - \hat{\Lambda})^{-1} - I\right)$ in $J(\alpha, D, \hat{\Lambda})$, we have a quadratic optimization function, and the solution is given by

$$\hat{\alpha} = \frac{\sum_{i \neq j} x_{ij} \left[(I - \hat{\Lambda})^{-1} - I \right]_{ij}}{\sum_{i \neq j} \left[(I - \hat{\Lambda})^{-1} - I \right]_{ij}^2}. \tag{10}$$

Then D is updated by

$$D = \text{diag}\left(\hat{\alpha} \left\{ (I - \hat{\Lambda})^{-1} - I \right\} \right), \tag{11}$$

and the procedures above are repeated for a predefined number of times.

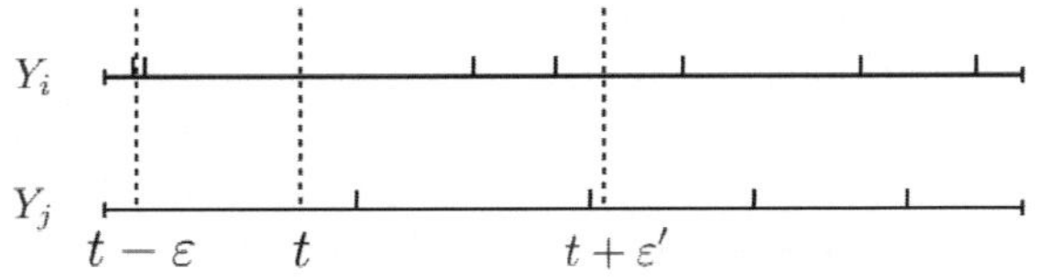

Figure 2. Two time-windows for measuring spikes

3.2. Hyperparameter optimization

In order to determine the number of nonzero elements M in Λ, AIC and BIC are often used. However, in this paper, we apply StARS (Stability Approach to Regularization Selection) [8] because neural data is noisy and thus the criterion is required to be robust against noise. StARS tends to give rise to less sparse graphs than AIC and BIC. Thus it is more suitable if we want to avoid relevant connections are omitted.

Since the original proposal of StARS is formulated for undirected graph, in the following, we describe the method extended to the case of directed graph.

Suppose we have multidimensional neural spike time series (N neurons $\times$ time length T). We randomly extract K subsequences $S_1, S_2, \ldots, S_K$, where each S_i has time length b (i.e., S_i is $N \times b$ matrix), and S_i and S_j can be overlapped.

We apply the algorithm described in the subsection above for a subsequence S_k with a fixed number of connections M, and we define the indicator variable,

$$\psi_{ij}^{M}(S_k) = \begin{cases} 1 & \text{if a connection from } j \text{ to } i \text{ exists,} \\ 0 & \text{otherwise.} \end{cases} \tag{12}$$

Collecting those values for all subsequences, we estimate the probability of existence of the connection,

$$\phi_{ij}^{b}(M) = \frac{1}{K} \sum_{k} \psi_{ij}^{M}(S_k). \tag{13}$$

We also calculate

$$\xi_{ij}^{b}(M) = 2\phi_{ij}^{b}(M)(1 - \phi_{ij}^{b}(M)) \tag{14}$$

which represents instability of the estimation. This is twice as large as the variance of Bernoulli probability of existence of the connection. Let us define average of $\xi_{ij}^{b}(M)$ by

$$D_b(M) = \frac{\sum_{i,j} \xi_{ij}^{b}(M)}{N(N-1)}. \tag{15}$$

We choose a maximal number of M so that the value of $D_b(M')$ is less than a certain threshold γ for all $M' \leq M$.

4. Construction of statistics

The algorithm described in the section above assumes that off-diagonal elements of X is given. In this section, we propose to estimate those values from observed neural spike data.

In eq. (1), we formulated a generative model of neuron connectivity. However, since neuron spike trains are point processes, we have to observe them within a certain width of time window. Here we consider two time windows $[t-\epsilon, t]$ and $[t, t+\epsilon']$ as shown in figure 2, where the former corresponds to the firing $Y_j(t)$ in eq. (1) and the latter corresponds to the time for propagation

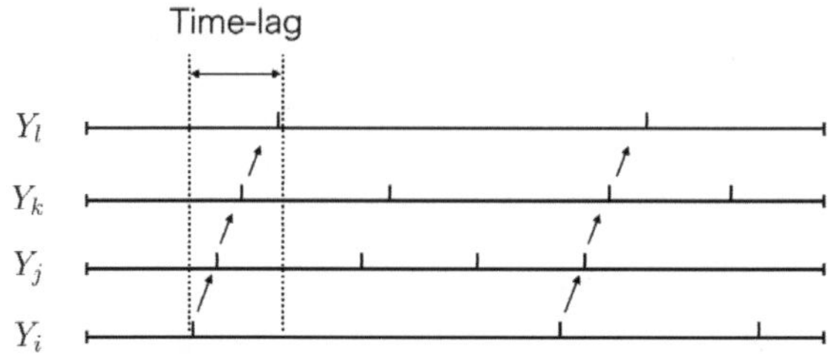

Figure 3. Time lag for multi-step propagation

to the i-th neuron $v_i(t+\Delta)$. For simplicity, we use the notation $Y_i(t) = 1$ for the event $\{Y_i(\tau) = 1, \exists\tau \in [t-\epsilon, t]\}$, and $Y_i(t+\Delta) = 1$ for the event $\{Y_i(\tau) = 1, \exists\tau \in [t, t+\epsilon']\}$.

The values of ϵ and ϵ' are determined based on the characteristics of neuron's behavior. ϵ should be taken a little longer than refractory period. On the other hand, for ϵ', since multi-step propagation through other neurons requires more time as shown in figure 3, it should be long enough to capture the time-lag of the propagation.

In eq. (1), the i-th neuron fires at time $t+\Delta$ randomly depending on the membrane potential $v_i(t+\Delta)$. Although this relation is typically modelled by a sigmoidal function, for simplicity, we assume it is a linear function, thus eq. (1) can be written as

$$\Pr[Y_i(t+\Delta) = 1 \mid Y_j(t)] \propto v_i(t) = b_i(t) + \theta_{ij} Y_j(t). \tag{16}$$

For $Y_j(t) = 0$ and 1, the expected values of eq. (16) are

$$\mathrm{E}[Y_i(t+\Delta) \mid Y_j(t) = 0] \propto \mathrm{E}[b_i(t)], \tag{17}$$

$$\mathrm{E}[Y_i(t+\Delta) \mid Y_j(t) = 1] \propto \mathrm{E}[b_i(t)] + \theta_{ij}. \tag{18}$$

By taking difference between eq. (17) and eq. (18), we have

$$\mathrm{E}[Y_i(t+\Delta) \mid Y_j(t) = 1] - \mathrm{E}[Y_i(t+\Delta) \mid Y_j(t) = 0] \propto \theta_{ij}. \tag{19}$$

The left hand side of eq. (19) can be estimated from observed time sequence, and can be used as an estimator of x_{ij} that is proportional to θ_{ij}. Remark that the influence $b_i(t)$ from other neurons including unmeasured neurons is cancelled out by taking difference.

5. Simulations

5.1. Synthetic data

We apply the proposed method to synthetic data generated by neural models proposed by Izhikevich [9]. The Izhikevich model is known to be able to simulate various types of realistic neuron behaviors even though the model can be formulated by small number of parameters.

Let $v_i(t)$[mV] ($i = 1, \ldots, N'$) be the membrane potential of the neuron i, $u_i(t)$ be the inactivator giving negative feedback, $I_i(t)$ be the post-synaptic current with random external input I_i^{ext}, W_{ij} be the connection weight from the neuron j to the neuron i, $Y_i(t)$ be the binary state (0 or 1) of neuron i. Izhikevich model is described by the following difference equations:

For $v_i(t) \geq 30$[mV],

$$
\begin{aligned}
Y_i(t) &= 1 && (20)\\
I_i(t) &= I_i^{\text{ext}}(t) + \sum_j W_{ij} Y_j(t) && (21)\\
v_i(t+dt) &= c_i && (22)\\
u_i(t+dt) &= u_i(t) + d_i && (23)\\
Y_i(t+dt) &= 0, && (24)
\end{aligned}
$$

and for $v_i(t) < 30$[mV],

$$
\begin{aligned}
v_i(t+dt) &= v_i(t) + 0.04v_i^2(t) + 5v_i(t) + 140 - u_i(t) + I_i(t) && (25)\\
u_i(t+dt) &= u_i(t) + a_i\{b_i v_i(t) - u_i(t)\}, && (26)
\end{aligned}
$$

where dt is a small time step(0.5[msec]), a_i, b_i, c_i, d_i are parameters specifying neuron behavior (excitatory or inhibitory).

In the numerical simulation, the total number of neurons is set as $N' = 50$, among which 40 neurons are excitatory and 10 neurons are inhibitory based on the biological evidence. In addition, the number of connections from one neuron is limited to at most 10. The strengths of connections are randomly generated from uniform distribution on $(0, 1]$ for excitatory connections and $[-1, 0)$ for inhibitory connections.

To investigate the performance for partial observations, we assume only $N = 32$ neurons (all excitatory) among 50 neurons are observable and used for estimation.

The resting membrane potential $v_i(t)$ is set to be between -70[mV] and -60[mV], and the neuron fires when $v_i(t)$ exceeds 30[mV] and it is reset to c_i. In this simulation, we do not take into account the signal delay and the synaptic plasticity.

For the time window parameters ϵ and ϵ', we fixed ϵ by 3.0[msec] and we examined 4 different length of ϵ', 3.0, 4.0, 5.0, 6.0[msec].

The number of iterations of the algorithm is set to 5.

5.2. Results

Since the resulting connection matrix represents a weighted graph structure, we evaluate the performance by two kinds of consistency: (1) topology of graph structure, and (2) strength of connection weights.

5.2.1. Topology of graph structure Suppose $\hat{\Lambda}$ and Λ^* be the estimation and the ground truth of connection weight matrices respectively. Here we define the TP (true positive) by the number of connections correctly estimated that satisfy both $\hat{\lambda}_{ij} \neq 0$ and $\lambda^*_{ij} \neq 0$, and the FN (false negative) by the number of connections that satisfy $\hat{\lambda}_{ij} = 0$ and $\lambda^*_{ij} \neq 0$. The sensitivity is calculated by Sensitivity = TP / (TP + FN), which represents the ratio of correctly extracted connections among all existing connections.

It has been demonstrated that the majority of cortical excitatory connections is weak [7] and therefore long term continuous recordings (e.g., 25 hours or longer) would be necessary to detect weak connections from extracellular spike data [10]. The sensitivity using a small amount of spiking data (eg., 1 hour or less) like this investigation would be unavoidably low because of these weak connections. However, in terms of appropriate scientific inference of connections, it would be more important to avoid large FP (false positive) that is the number of estimated connections that do not exist, because FP suggests an effect that is not actually there. In the following, we calculate FP rate, which is defined by FP / (FP + TN), where TN denotes true negative.

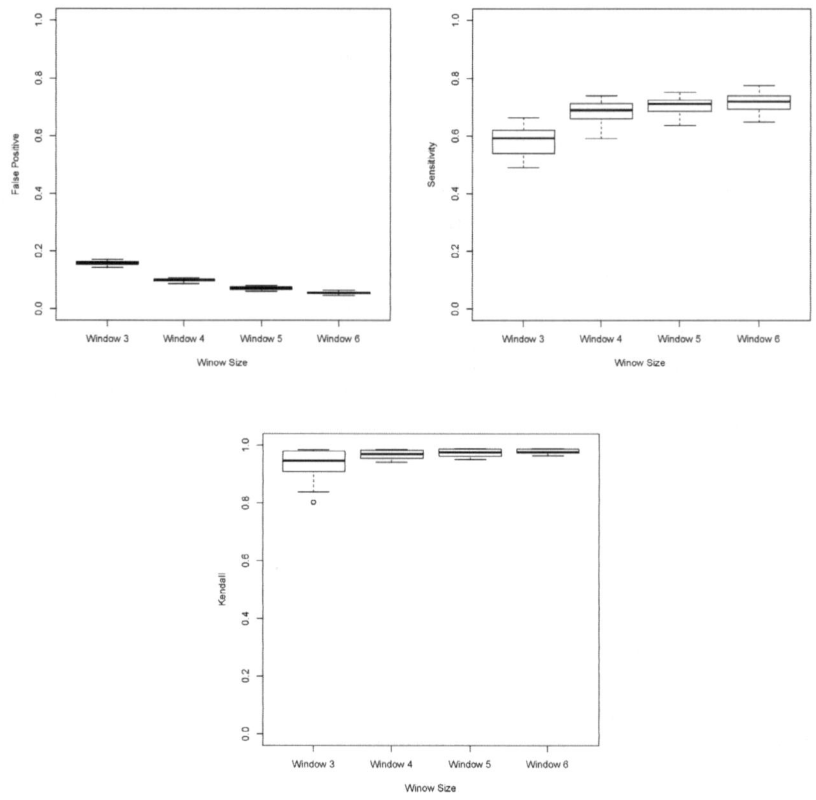

Figure 4. Performance of the proposed algorithm. Upper-left: False positive rate, upper-right: Sensitivity, and lower: Kendall's τ

5.2.2. Evaluation of connection weights In order to evaluate how the proposed method extract strengths of connections correctly, we use Kendall rank correlation coefficient (Kendall's τ) that measures the degree of coincidence of rank statistics between $\hat{\Lambda}$ and Λ^*. To focus on nonzero connections, first we exclude the pairs with no connections ($\lambda_{ij} = 0$). However, since the nonzero components are not always the same in the two matrices, we calculate Kendall's τ by choosing the connections included both in $\hat{\Lambda}$ and in Λ^*.

5.3. Evaluation

We performed the simulation for 20 different networks generated by different random seeds. Boxplots of FP rate, Sensitivity and Kendall's τ are plotted in figure 4. Horizontal axis represents the length of window size ϵ'. For $\epsilon' = 5$ and 6, FP rate is quite small, which means that the proposed method correctly estimates the absence of connections in the true network. We see relatively lower values of sensitivity mainly caused by the sparsity of the graph. Kendall's τ is quite high for $\epsilon' = 5$ and 6, so the strong connections are extracted correctly. In figure 5, examples of true and estimated network structures are shown.

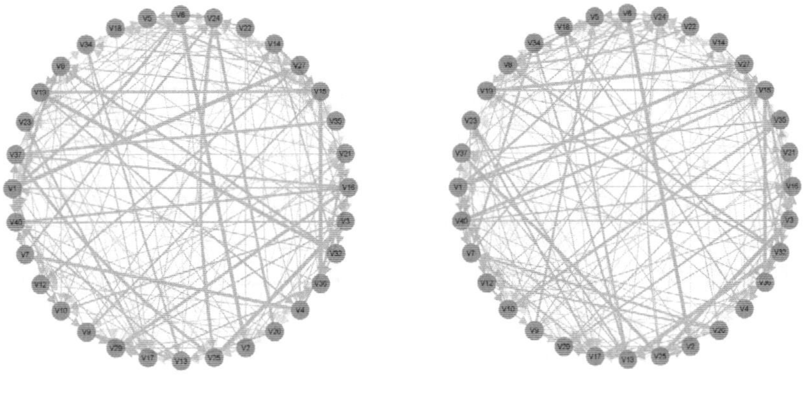

Figure 5. Examples of graph structure. Left: True, Right: Estimated. Red edges represent excitatory connections, and blue edges represent inhibitory connections.

6. Conclusion

In this paper, we proposed a method to extract directed connectivity between neurons based on a simple generative model. Using the synthetic data that were generated by the Izhikevich neuron model, we demonstrated that the proposed method successfully detected the direction and strength of both excitatory and inhibitory connections. In neuroscience, cross-correlation has been widely used to assess synaptic connections [10, 11]. Our method would complement the existing methods and can be easily applied to real data because it does not involve many tuning parameters. By designing appropriate statistics, the proposed method would be also applicable to other application areas.

Acknowledgments

The present study was financially supported by JSPS KAKENHI 25120009, 25120011, 26120504, 15H01678, 25870811.

References

[1] Hebb D O 1949 *The Organization of Behavior: A Neuropsychological Theory* (Wiley)
[2] Wilson M A and McNaughton B L 1994 *Science* **265** 676–8
[3] Ganmor E, Segev R and Schneidman E 2011 *PNAS* **108** 9679–84
[4] Ohiorhenuan I E, Mechler F, Purpura K P, Schmid A M and Hu Q 2010 *Nature* **466** 617–21
[5] Shimazaki H, Amari S, Brown E N and Grün S 2012 *PLoS Comp. Biol.* **8** e1002385
[6] Noda A, Hino H, Tatsuno M, Akaho S and Murata N 2014 *Neural Computation* **26** 1455–83
[7] Song S, Sjöström P J, Reigl M, Nelson S and Chklovskii D B 2005 *PLoS Biol.* **3** e68
[8] Roeder K, Liu H and Wasserman L 2010 *Advances in Neural Information Processing Systems* **23** 1432–40
[9] Izhikevich E M 2003 *IEEE Trans. on Neural Networks* **14** 1569–72
[10] Schwindel C D, Ali K, McNaughton B L and Tatsuno M 2014 *Journal of Neuroscience* **34** 5454–67
[11] Ostojic S, Brunel N and Hakim V 2009 *Journal of Neuroscience* **29** 10234–53

Scampi: a robust approximate message-passing framework for compressive imaging

Jean Barbier[†], Eric W. Tramel[‡], and Florent Krzakala[‡]

† Laboratoire de Théorie des Communications, Ecole Polytechnique Fédérale de Lausanne, Faculté Informatique et Communications, CH-1015, Suisse.
‡ Laboratoire de Physique Statistique CNRS UMR 8550, Université P. et M. Curie et Ecole Normale Supérieure, 24 rue Lhomond, 75005 Paris, France.

E-mail: `jean.barbier@epfl.ch`, `eric.tramel@ens.fr`, `florent.krzakala@ens.fr`

Abstract. Reconstruction of images from noisy linear measurements is a core problem in image processing, for which convex optimization methods based on total variation (TV) minimization have been the long-standing state-of-the-art. We present an alternative probabilistic reconstruction procedure based on approximate message-passing, Scampi, which operates in the compressive regime, where the inverse imaging problem is underdetermined. While the proposed method is related to the recently proposed GrAMPA algorithm of Borgerding, Schniter, and Rangan, we further develop the probabilistic approach to compressive imaging by introducing an expectation-maximization learning of model parameters, making the Scampi robust to model uncertainties. Additionally, our numerical experiments indicate that Scampi can provide reconstruction performance superior to both GrAMPA as well as convex approaches to TV reconstruction. Finally, through exhaustive best-case experiments, we show that in many cases the maximal performance of both Scampi and convex TV can be quite close, even though the approaches are *a prori* distinct. The theoretical reasons for this correspondence remain an open question. Nevertheless, the proposed algorithm remains more practical, as it requires far less parameter tuning to perform optimally.

1. Introduction and problem setting

Over the past decade, the study of compressed sensing (CS) [1–4] has blossomed into a large field of active research in signal processing. Of particular interest in this field is the development of sparse signal reconstruction algorithms from noisy, linear measurements. While the development of generic algorithms for this problem is well advanced, there still remains much work to be done on the development of reconstruction algorithms that exploit the statistics of particular signal classes. In this work, we are interested in the class of *natural images* which has its application within compressive imaging [5]. Natural images are considered as a class of 2D signals which posses a *piecewise-continuity*, exhibiting large regions of the signal support which are constant or slowly varying interspersed with singularities of rapid transitions, or *edges*. In this work we will see how this property of piecewise-continuity can be exploited to accurately reconstruct compressively sampled images, subject to varying degrees of measurement degradation.

For the present study, we consider the reconstruction of a vectorized image signal $\mathbf{s}^* \in \mathbb{R}^N$ composed of $N = L^2$ pixels from linear measurements $\mathbf{y} \in \mathbb{R}^M$ which represent the observations of the signal projected by the measurement matrix $\mathbf{F} \in \mathbb{R}^{M\times N}$. The measurements are corrupted by an additive white Gaussian noise (AWGN) which is zero mean and of variance Δ^*,

$\boldsymbol{\xi} \sim \mathcal{N}(\boldsymbol{\xi}; \mathbf{0}_{M\times 1}, \Delta^* \mathbf{I}_{M\times M})$ where $\mathbf{I}_{a\times a}$ is the identity matrix of size $a \times a$ and $\mathbf{0}_{a\times b}$ is a matrix of size $a \times b$ whose elements are all 0. With these definitions, the undersampled observation procedure can be written as

$$\mathbf{y} = \mathbf{F}\mathbf{s}^* + \boldsymbol{\xi} \quad \Leftrightarrow \quad y_\mu = \langle \mathbf{F}_\mu, \mathbf{s}^* \rangle + \xi_\mu, \tag{1}$$

where $\langle \cdot, \cdot \rangle$ represents an inner-product and $\mathbf{F}_\mu$ is the μth row of $\mathbf{F}$. The fundamental theory of CS asserts that with a properly chosen $\mathbf{F}$, usually random, one can accurately recover $\mathbf{s}^*$ in the under-determined setting of $M < N$, by introducing *a priori* knowledge of $\mathbf{s}^*$, namely, sparsity. A sparse signal is one which, either in the cardinal basis or some other transform basis, posses only a few non-zero components. To be more precise, in an *analysis model* [6], it is assumed that a linear transform $\Psi\mathbf{s}^*$ of the signal $\mathbf{s}^*$ to be reconstructed is sparse, while $\mathbf{s}^*$ is generally not. In this case, $\mathbf{s}^*$ is sometimes refered as *cosparse*. Complementary to this definition is the *synthesis model*, in which it is directly the signal $\mathbf{s}^*$ of interest that is assumed to be sparse, i.e. Ψ is the identity. The generic reconstruction task, in the case of AWGN corrupted observations, can thus be written as a Lasso [7] problem,

$$\hat{\mathbf{s}}_{\ell_1} = \underset{\mathbf{s}\in\mathbb{R}^N}{\text{argmin}} \; ||\mathbf{y} - \mathbf{F}\mathbf{s}||_2^2 + \lambda||\Psi\mathbf{s}||_1, \tag{2}$$

where λ is a regularization parameter whose optimal selection is generally unknown *a posteriori*, and the ℓ_p norm is defined as $||\mathbf{x}||_p \triangleq (\sum_i |x_i|^p)^{1/p}$. While one might entertain the use of generic sparse reconstruction techniques for images by recovering sparse coefficients in, say, the wavelet basis, the use of gradient variation minimization has remained state-of-the-art for image reconstruction in CS. Specifically, we are interested in the application of total variation (TV) minimization [3] for image reconstruction. In this approach, one biases the optimization towards solutions which exhibit a sparse 2D discrete gradient. In the case of anisotropic TV minimization for AWGN corrupted observations, the reconstruction problem can be written as

$$\hat{\mathbf{s}}_{\text{TV}} = \underset{\mathbf{s}\in\mathbb{R}^N}{\text{argmin}} \; ||\mathbf{y} - \mathbf{F}\mathbf{s}||_2^2 + \lambda \sum_{i=1}^{N} \left(|\nabla_{s_i}^{\text{H}}| + |\nabla_{s_i}^{\text{V}}| \right), \tag{3}$$

where $\nabla_{s_i}^{\text{H}}$ and $\nabla_{s_i}^{\text{V}}$ are the horizontal and vertical discrete gradients of $\mathbf{s}$, respectively, at pixel index i. This problem has been well studied from the perspective of convex optimization, with many recent developments focused on improving the rate-of-convergence and computational efficiency of algorithms to solve this problem [8]. One robust and efficient reconstruction algorithm which we utilize in this study is the TV-AL3 of [9], based on the augmented Lagrangian approach.

Rather than studying the TV minimization problem as a convex one, we will instead adopt a probabilistic framework that mimics the original TV problem, in the sense that it will try to output piecewise-continuous solutions of the system (1). The use of statistical approaches to TV minimization is not novel to this work. For instance, the 1D TV problem, and its associated phase transitions, was studied in [10] and a reconstruction method based on approximate message-passing (AMP) was first proposed. In a related vein, there also exist applications of AMP to image representation via non-local means [11] as well as AMP in conjunction with an amplitude-scale-invariant Bayes estimator [12]. Another recent work for 1D TV proposes a joint prior defined over nearest-neighbors components, characterizing the gradient with a spike-and-slab prior within AMP [13] (SS-AMP). Finally, one of the most exciting applications of the generalized AMP [14] to 2D signals has been the work of [15], which proposed the use of a cosparse analysis model. This approach, known as GrAMPA, is also general in the sense of non-AWGN *i.i.d.* observation channels and can be used with any sparsifying basis Ψ. This feature is inherited from the generalized AMP, which is capable to deal with any pointwise function of the compressed measurements $\mathbf{Fs}^*$, including the *i.i.d.* AWGN channel as a special case. In this case, the canonical AMP algorithm is recovered. GrAMPA was shown to be a state-of-the-art approach for image reconstruction in the benchmark tests of [13, 15]. Motivated by the results of both SS-AMP and GrAMPA, we present a new AMP-based algorithm for 2D CS image reconstruction particularly adapted to natural images.

Similar to GrAMPA, the proposed method will also work with the cosparse analysis model. Furthermore, we utilize the sparse non-informative parameter estimator (SNIPE) prior proposed in [15] for auxiliary gradient variables that will be introduced in the model. We term this approach "SNIPE-based cosparse analysis AMP for imaging" (Scampi). We will detail the novel modifications of our approach with respect to GrAMPA in the sequel, but they can be summarized as *i)* an expectation-maximization learning of the noise variance, which makes the algorithm more robust to channel uncertainty, and *ii)* the introduction of auxiliary variables that allow one to relax the constraints over the differences between neighboring pixels. In terms of statistical physics, our proposed approach can be seen as a "finite temperature" version of GrAMPA. This relaxation improves reconstruction accuracy for natural images, where algorithms must be more robust to violations of the piecewise-continuous assumption.

We also show that in the limit of oracle parameter selection, the performances of TV-AL3 and Scampi are comparable, and in the low-measurement regime, the differences are almost negligible. This limiting performance raises a number of interesting questions about the nature and maximal performance of TV-based CS reconstruction of natural images.

2. Proposed model and algorithm: Scampi

2.1. The cosparse TV analysis model with dual variables

Let us define the set of all edges between neighboring pixels in the image as $E \triangleq \{\nu = (i,j) : (i,j) \in \{1,\ldots,N\} \times \{1,\ldots,N\}, s_i \text{ neighbor of } s_j\}$. Pixel neighborhoods can be chosen arbitrarily, but to match the TV cost function, we will limit ourselves to neighborhoods corresponding to a 4-connected lattice, placing edges between horizontally and vertically co-located pixels. Next, we introduce a set of auxiliary, or *dual*, variables which exist on the edges which describe the differences between individual pixels,

$$\mathbf{d} \triangleq [(s_i - s_j) : (i,j) = \nu \in E]. \tag{4}$$

We now define the augmented signal $\mathbf{x} \triangleq [\mathbf{s}, \mathbf{d}]^\intercal$ as the concatenation of the image representation and the vector of dual variables. We can now write out an augmented system for (1) which introduces the dual variables

$$\begin{bmatrix} \mathbf{y}_{M\times 1} \\ \mathbf{0}_{|E|\times 1} \end{bmatrix} = \begin{bmatrix} \mathbf{F}_{M\times N} & \mathbf{0}_{M\times |E|} \\ \mathbf{D}_{|E|\times N} & -\mathbf{I}_{|E|\times|E|} \end{bmatrix} \times \begin{bmatrix} \mathbf{s}_{N\times 1} \\ \mathbf{d}_{|E|\times 1} \end{bmatrix} + \begin{bmatrix} \boldsymbol{\xi}_{M\times 1} \\ \boldsymbol{\zeta}_{|E|\times 1} \end{bmatrix}, \tag{5}$$

where $\mathbf{F}$ is the original measurement matrix and the dimensions of each vectors and matrices have been indicated in subscript for clarity. The entire system can be written and solved in the reduced form: $\mathring{\mathbf{y}} = \mathring{\mathbf{F}}\mathbf{x} + \mathring{\boldsymbol{\xi}}$.

$\mathbf{D}$ is a 2D finite difference operator, i.e. the concatenation of two matrices matrices which calculate both the horizontal and vertical gradients. As stated earlier, we consider for each pixel its four closest neighbors, thus $\mathbf{D} \triangleq [\mathbf{D}_\mathrm{r}, \mathbf{D}_\mathrm{d}]^\intercal$ is the concatenation of $\mathbf{D}_\mathrm{r}$ which is made of zeros everywhere except for a main diagonal of 1's, and a single -1 on each row, in a column corresponding to the *right* neighbor of the pixel at the index represented by the 1 on the main diagonal. The matrix $\mathbf{D}_\mathrm{d}$ is constructed in similar fashion, but for the *down* nieghbor. This choice of right/down differencing, versus left/up, is arbitrary. As long as $\mathbf{D}$ is constructed consistently, all pairwise interactions are encoded consistently. Finally, the vector $\boldsymbol{\zeta}$ is an error that can be thought of as the degree of correspondence between $\mathbf{d}$ and the true discrete 2D gradient of $\mathbf{s}$. We will discuss this point more throughly in the next section. As a whole, we refer to this augmented system as a *cosparse TV analysis model with dual variables.*

2.2. Probabilistic model for TV reconstruction

We now present a Bayesian framework for the TV problem. The above construction allows one to easily associate a factorized prior P_0 to $\mathbf{x}$, a key assumption for the AMP algorithm [16–21]. While some arbitrary *i.i.d.* distribution can be assigned to the elements of $\mathbf{s}$, we will assume a SNIPE prior $P_{\rm S}$ [15] for $\mathbf{d}$. This prior is defined as the limiting distribution of a spike-and-slab distribution with infinite variance,

$$P_{\rm S}(x;\omega) \triangleq \lim_{\sigma\to\infty} \frac{\rho(\sigma,\omega)}{\sigma}\mathcal{N}(x/\sigma;0,\sigma^2) + (1-\rho(\sigma,\omega))\delta(x), \tag{6}$$

for which, under a proper scaling of $\rho(\sigma,\omega)$, has ω as the only free parameter of the prior. This hyperparameter allows to tune the sparsity of the gradient of the image. By using $P_{\rm S}$ to bias the elements of $\mathbf{d}$ during inference, we place more probabilistic weight on images whose finite differences are sparse, according to ω. For further details on SNIPE, we direct the reader to [15].

If we consider that the elements of $\mathbf{s}$ obey a uniform distribution, and if we enforce that the elements of the concatenated signal $\mathbf{x}$ are ordered such that the first N elements correspond to pixel values and the last $|E|$ elements correspond to $\mathbf{d}$, then the factorized prior for $\mathbf{x}$ can be writen as

$$P_0(\mathbf{x}) \propto \prod_{i=1}^{N+|E|} P_0^i(x_i) = \prod_{i=1}^{N+|E|} \left[\mathbb{I}(i\le N)\mathcal{U}(x_i) + \mathbb{I}(i>N)P_{\rm S}(x_i;\omega)\right], \tag{7}$$

where $\mathcal{U}$ is a uniform distribution over some domain of $\mathbb{R}$ and $\mathbb{I}(\cdot)$ is the indicator function.

With the prior defined for $\mathbf{x}$, we would like to turn our attention to the posterior distribution of $\mathbf{x}$ given $\mathbf{y}$, however, first, we must take a moment to consider the likelihood induced by the distributions of the error terms $\boldsymbol{\xi}$ and $\boldsymbol{\zeta}$. We diverge from the GrAMPA approach, by the introduction of the "finite temperature" noted in the previous section. In order to mimic the GrAMPA approach thanks to the cosparse TV analysis model with dual variables (5), one would set $\boldsymbol{\zeta} = \mathbf{0}_{|E|\times 1}$, creating a model for which the elements of $\mathbf{d}$ are exactly the discrete gradient of $\mathbf{s}$. Indeed, this is an intuitive construction. However, imposing such a strict condition might impede the reconstruction. The fist key feature of our approach is that instead, we propose that the hard linear constraints $\{d_\nu - (s_i - s_j) = 0 : (i,j) = \nu \in E\}$ be "relaxed" by taking $\boldsymbol{\zeta} \sim \mathcal{N}(\boldsymbol{\zeta}; \mathbf{0}_{|E|\times 1}, T\,\mathbf{I}_{|E|\times|E|})$. Here, the amount of relaxation is determined by the temperature-like variance parameter, T. In the frozen zero-temperature limit, $T \to 0$, one recovers GrAMPA, up to the nuances of implementation (in the case of *i.i.d.* AWGN noise). However, for finite temperature, some slack is introduced into the problem, allowing the dual variables to fluctuate some distance away from the true gradient, a helpful feature to avoid poor local solutions.

Finally, we can now write the posterior distribution for our TV model (5) as

$$P_{\rm Scampi}(\mathbf{x} = [\mathbf{s},\mathbf{d}]|\mathring{\mathbf{y}} = [\mathbf{y},\mathbf{0}]) = \frac{1}{\mathcal{Z}} \prod_{\mu=1}^{M} \mathcal{N}\left(y_\mu; \langle \mathbf{F}_\mu, \mathbf{s}\rangle, \Delta\right) \prod_{\nu=1}^{|E|} \left[\mathcal{N}\left(d_\nu; \langle \mathbf{D}_\nu, \mathbf{s}\rangle, T\right) P_{\rm S}(d_\nu;\omega)\right], \tag{8}$$

where the normalization $\mathcal{Z} \triangleq Z(\mathbf{y},\mathbf{F},\mathbf{D},\Delta,T,\omega)$, also known as the partition function in physics, is a function of the problem instance, and is intractable to calculate, in practice. If known, the noise variance associated with the measurements in the posterior should match the true noise variance, $\Delta = \Delta^*$. If Δ^* is unknown, it may be estimated, as we will see in Sec. 2.4. Rather than attempting to calculate this posterior exactly, the message-passing

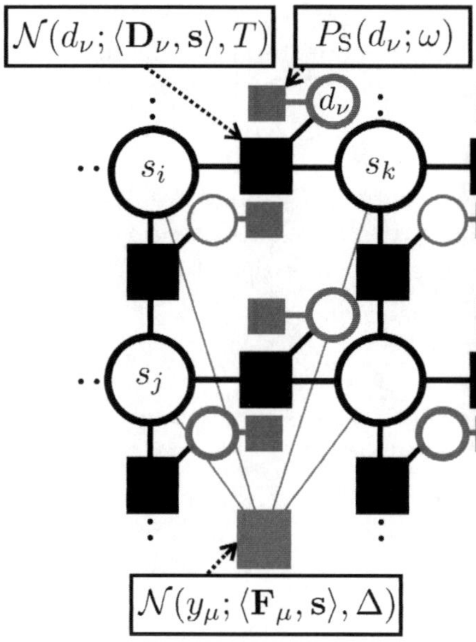

Figure 1. Factor graph associated to the probability measure (8) sampled by the Scampi algorithm for the image reconstruction. The large black circles are the pixels of $\mathbf{s}$, while the smaller red ones are the dual variables, $\mathbf{d}$. The squares represent the factors: the green ones connected to all the pixel variables are the "measurement" factors enforcing $\mathbf{s}$ to verify the linear measurements $\mathbf{y}$, up to Gaussian fluctuations of variance Δ. The red factors represent the factorizable prior we assume about the dual variables, which is the SNIPE prior P_S. Finally, the black factors are enforcing consistency between $\mathbf{s}$ and $\mathbf{d}$, as weighted by T. To avoid double counting, for each pixel, only two out of the four possible neighbor interactions are taken into account.

approach we discuss, here, attempt to *approximate* this posterior via information transfer between nodes on the factor graph representation of the posterior. Fig. 1 gives the factor graph representation of the proposed Scampi posterior. The factor graph associated with the posterior sampled by GrAMPA is the same, but lacks the red variables and factors. Also, the black factors would be of the form $P_S(\langle \mathbf{D}_\nu, \mathbf{s} \rangle; \omega)$ instead of the proposed Gaussian densities, i.e. $P_{\mathrm{GrAMPA}}(\mathbf{s}) \propto \prod_{\mu=1}^{M} \mathcal{N}(y_\mu; \langle \mathbf{F}_\mu, \mathbf{s} \rangle, \Delta) \prod_{\nu=1}^{|E|} P_S(\langle \mathbf{D}_\nu, \mathbf{s} \rangle; \omega)$.

Lastly, we would like to mention that one can also consider a 2D extension of SS-AMP without appealing to the augmented analysis model (5). In this case, the posterior becomes $P_{\mathrm{SS-AMP}}(\mathbf{s}) \propto \prod_{\mu=1}^{M} \mathcal{N}(y_\mu; \langle \mathbf{F}_\mu, \mathbf{s} \rangle, \Delta) \prod_{\nu=1}^{|E|} [\rho \mathcal{N}(\langle \mathbf{D}_\nu, \mathbf{s} \rangle; 0, T) + (1-\rho)\delta(\langle \mathbf{D}_\nu, \mathbf{s} \rangle)]$. However, empirical results indicate that this algorithm suffers from strong convergence issues. Additionally, the exploding complexity, and quantity, of the resulting message-passing equations create too many numerical issues for a practical implementation at scale.

2.3. AMP for the cosparse analysis model with dual variables: Scampi

The Scampi algorithm, detailed in Fig. 2, is *canonical AMP applied to the cosparse analysis model with dual variables* (5), with a few added features we detail in the sequel. In this sense, the algorithm (i.e. AMP) is already well known. The novelty resides in its use to the proposed cosparse analysis model with dual variables, in spite of its sparse nature. Scampi performs a minimum mean square error estimation of $\mathbf{x}$ by sampling from an iteratively computed approximation of the distribution (8). Similarly, GrAMPA is performing minimum mean square

$t \leftarrow 0$

$\tau \leftarrow 1 + \epsilon$

while $t < t_{max}$ **and** $\tau > \epsilon$ **do**

$$\tilde{\Theta}_\mu^{(t+1)} \leftarrow \sum_{i=1}^{N+|E|} \mathring{F}_{\mu i}^2 v_i^{(t)}$$

$$w_\mu^{(t+1)} \leftarrow \beta w_\mu^{(t)} + (1-\beta)\left(\sum_{i=1}^{N+|E|} \mathring{F}_{\mu i} a_i^{(t)} - \tilde{\Theta}_\mu^{(t+1)} \frac{\mathring{y}_\mu - w_\mu^{(t)}}{\mathring{\Delta}_\mu^{(t)} + \Theta_\mu^{(t)}}\right)$$

$$\Theta_\mu^{(t+1)} \leftarrow \beta \Theta_\mu^{(t)} + (1-\beta)\tilde{\Theta}_\mu^{(t+1)}$$

$$\Sigma_i^{(t+1)} \leftarrow \left(\sum_{\mu=1}^{M+|E|} \frac{\mathring{F}_{\mu i}^2}{\mathring{\Delta}_\mu^{(t)} + \Theta_\mu^{(t+1)}}\right)^{-\frac{1}{2}}$$

$$R_i^{(t+1)} \leftarrow a_i^{(t)} + \left(\Sigma_i^{(t+1)}\right)^2 \sum_{\mu=1}^{M+|E|} \mathring{F}_{\mu i} \frac{\mathring{y}_\mu - w_\mu^{(t+1)}}{\mathring{\Delta}_\mu^{(t)} + \Theta_\mu^{(t+1)}}$$

$$v_i^{(t+1)} \leftarrow f_{\mathrm{c}}\left(\left(\Sigma_i^{(t+1)}\right)^2, R_i^{(t+1)}; \omega\right)$$

$$a_i^{(t+1)} \leftarrow f_{\mathrm{a}}\left(\left(\Sigma_i^{(t+1)}\right)^2, R_i^{(t+1)}; \omega\right)$$

Update all $\mathring{\Delta}_\mu^{(t+1)}$ according to (15)

$\tau \leftarrow \frac{1}{N}||\mathbf{a}^{(t+1)} - \mathbf{a}^{(t)}||_2^2$

$t \leftarrow t+1$

end while

return $\mathbf{a}^{(t)}$

Figure 2. The Scampi algorithm with damping (controlled by $0 \leq \beta < 1$) for the co-sparse model with dual variables. The convergence tolerance is given by ϵ and t_{max} is the maximum number of iterations. A suitable initialization for the posterior mean and variance are, respectively, $a_i^{(0)} = \mathbb{E}_{P_0^i}(x_i)$ and $v_i^{(0)} = \mathrm{Var}_{P_0^i}(x_i)$, where P_0^i is the prior associated with component x_i. Additionally, one initializes $w_\mu^{(0)} = \mathring{y}_\mu$. Once the algorithm has converged, the minimum mean square error estimate of component x_i is given by $a_i^{(t)}$.

error estimation but for the posterior $P_{\mathrm{Gr\mathring{A}MPA}}(\mathbf{s})$. The full derivation of AMP in similar notations can be found in [20, 22]. Additionally, we define the augmented vector of per-factor noise variances as $\mathring{\Delta}_\mu \triangleq \Delta \mathbb{I}(\mu \leq M) + T\mathbb{I}(\mu > M)$.

Despite the fact that the augmented system matrix $\mathring{\mathbf{F}}$ is sparse, and that the derivation of AMP is based on "central limit"-like arguments justified for dense matrices, nothing prevents us from applying it effectively to such a problem, and as we will see, it achieves state-of-the-art performance. Nevertheless, we point out that other message-passing schemes which derivation is not (or less) based on the density of the matrix could be considered for possibly improved results. These include the belief propagation and relaxed belief propagation algorithms [20]. But the payoff with such methods is computational. Indeed using these algorithms can very quickly become intractable, such as in the present setting where a part of the matrix is dense (and its size grows quadratically with the image size). For this reason, we use the computationally efficient approximate message-passing algorithm.

The prior-dependent denoisers which iteratively produce the posterior estimates are given by

$$f_{\mathrm{a}}(\Sigma_i^2, R_i; \omega) \triangleq \int \mathrm{d}x\, x\, P_0^i(x)\, \mathcal{N}\left(x; R_i, \Sigma_i^2\right) = R_i\, \mathbb{I}(i \leq N) + f_{\mathrm{a}}^{\mathrm{S}}(\Sigma_i^2, R_i; \omega)\, \mathbb{I}(i > N), \tag{9}$$

$$f_{\mathrm{c}}(\Sigma_i^2, R_i; \omega) \triangleq \int \mathrm{d}x\, x^2\, P_0^i(x)\, \mathcal{N}\left(x; R_i, \Sigma_i^2\right) - f_{\mathrm{a}}(\Sigma_i^2, R_i; \omega)^2 = \Sigma_i^2\, \mathbb{I}(i \leq N) + f_{\mathrm{c}}^{\mathrm{S}}(\Sigma_i^2, R_i; \omega)\, \mathbb{I}(i > N) \tag{10}$$

where the prior terms are given by (7) and the SNIPE-specific denoisers [15] for the dual variables are given by

$$f_{\mathrm{a}}^{\mathrm{S}}(\Sigma^2, R; \omega) \triangleq \frac{R}{1 + e^{-\frac{R^2}{2\Sigma^2} + \omega}}, \quad f_{\mathrm{c}}^{\mathrm{S}}(\Sigma^2, R; \omega) \triangleq \frac{1}{1 + e^{-\frac{R^2}{2\Sigma^2} + \omega}} \left(\frac{R^2}{1 + e^{\frac{R^2}{2\Sigma^2} - \omega}} + \Sigma^2 \right). \tag{11}$$

2.4. Learning the noise variance via the Bethe free energy

We now present the second key feature of Scampi, an iterative estimation of both the noise variance and dual parameter relaxation, which improves reconstruction accuracy for natural images. However, we note that this approach in fact *degrades* reconstruction accuracy for strictly piecewise constant signals. Nevertheless, if the image of interest is *known* to be strictly piecewise constant, the temperature T can be fixed to 0 to improve performance. If not, the hyperparameters $\{\Delta, T\}$ can be learned via the expectation-maximization we now detail.

Given the definition of the Bethe free energy G derived in [23] for the posterior measure (8), we update $\{\Delta, T\}$ at each iteration according to their fixed-point values in G. We point that the derivation in [23] relies on the same assumptions required for the AMP algorithm to be formally justified, i.e. a *dense* measurement matrix with *i.i.d.* entries. In spite of the sparsity of the measurement matrix associated to the cosparse analysis model with dual variables, we assume that this free energy expression remains valid, which is consistent with the use of the AMP algorithm for the reconstruction. This choice is justified *a posteriori* by the excellent performances of the overall Scampi algorithm.

For our posterior, the Bethe free energy is written as

$$G = \frac{1}{2} \sum_{\mu} \left[\frac{(\mathring{y}_\mu - \sum_i \mathring{F}_{\mu i} a_i)^2}{\mathring{\Delta}_\mu} + \log\left(1 + \frac{\sum_i \mathring{F}_{\mu i}^2 v_i}{\mathring{\Delta}_\mu}\right) + \log\left(2\pi \mathring{\Delta}_\mu\right) \right] + \sum_i D_{\mathrm{KL}}(P_i || P_0^i), \tag{12}$$

where $D_{\mathrm{KL}}(P_i||P_0)$ is the Kullback-Leibler divergence between the prior P_0^i of the ith component and its estimated posterior $P_i(x_i|\Sigma_i^2, R_i) \propto P_0^i(x_i)\mathcal{N}(x_i; R_i, \Sigma_i^2)$, and the Gaussian fields $\{R_i, \Sigma_i^2\}$ and posterior mean and variance estimates $\{a_i, v_i\}$ of x_i are the fixed-point values of the Scampi algorithm in Fig. 2.

For the moment, let us consider that there is a unique noise parameter $\mathring{\Delta}$ for all of our factors. To minimize G, we start by isolating the terms which depend on $\mathring{\Delta}$,

$$G_{\mathring{\Delta}} = \frac{1}{2} \sum_{\mu=1}^{M+|E|} \left[\frac{(\mathring{y}_\mu - \sum_{i=1}^{N+|E|} \mathring{F}_{\mu i} a_i)^2}{\mathring{\Delta}} + \log\left(\mathring{\Delta} + \sum_{i=1}^{N+|E|} \mathring{F}_{\mu i}^2 v_i\right) \right]. \tag{13}$$

We then write a fixed-point condition which is dependent on $\mathring{\Delta}$, for all other variables fixed,

$$\frac{\partial G_{\mathring{\Delta}}}{\partial \mathring{\Delta}} = -\frac{1}{2} \sum_{\mu=1}^{M+|E|} \left[\frac{1}{\mathring{\Delta}^2} \left(\mathring{y}_\mu - \sum_{i=1}^{N+|E|} \mathring{F}_{\mu i} a_i \right)^2 - \left(\mathring{\Delta} + \sum_{i=1}^{N+|E|} \mathring{F}_{\mu i}^2 v_i \right)^{-1} \right] = 0. \tag{14}$$

It is possible to extract the optimal $\mathring{\Delta}$ by equating the two terms inside the sum, providing the set of solutions $\mathring{\Delta}_\mu \ \forall \ \mu \in \{1,\ldots,M+|E|\}$. Defining the auxiliary term $\chi_\mu \triangleq \mathring{y}_\mu - \sum_i \mathring{F}_{\mu i} a_i$, each $\mathring{\Delta}_\mu$ is obtained by solving the quadratic equation $\mathring{\Delta}_\mu^2 - \chi_\mu^2 \mathring{\Delta}_\mu - \chi_\mu^2 \sum_i \mathring{F}_{\mu i}^2 v_i = 0$. Since all variance terms are strictly positive, the feasible solution for this equation is simply $\mathring{\Delta}_\mu = g_\mu$ where $g_\mu \triangleq \frac{1}{2}\big(\chi_\mu^2 + \chi_\mu \sqrt{\chi_\mu^2 + 4\sum_i \mathring{F}_{\mu i}^2 v_i}\big)$. Averaging over $\{\mathring{\Delta}_\mu : \mu \in \{1,\ldots,M+|E|\}\}$, we obtain a single parameter $\mathring{\Delta}$. Recall, however, that we wish to consider two *different* noise variances $\{\Delta, T\}$. Since both $\mathbf{s}$ and $\mathbf{d}$ exist in different domains, it is not reasonable to expect that they both should share the same noise variance estimate. Instead, two averages are performed over the proper sets of variables in $\mathbf{x}$. Introducing the iterative time index, we obtain the fixed-point estimation

$$\Delta^{(t+1)} = \frac{1}{M}\sum_{\mu=1}^{M} g_\mu^{(t)}, \quad T^{(t+1)} = \frac{1}{|E|}\sum_{\mu=M+1}^{M+|E|} g_\mu^{(t)}, \tag{15}$$

where $g_\mu^{(t)}$ is a function of $\left\{\Delta^{(t)}, T^{(t)}, \mathbf{a}^{(t+1)}, \mathbf{v}^{(t+1)}, \mathring{\mathbf{y}}, \mathring{\mathbf{F}}\right\}$. In the final implementation of Scampi, we also introduce a damping on the update of $\{\Delta, T\}$.

Intuitively, this iterative estimation of the noise variances operates as a kind of annealing which plays a pivotal role in providing Scampi's superior reconstruction performance for natural images. Additionally, the estimation of $\{\Delta, T\}$ means that Scampi is more robust to uncertainty on these hyperparameters than GrAMPA, an important feature for practical use cases. We present results demonstrating this effect in Sec. 3. Our Scampi implementation can be downloaded on `https://github.com/jeanbarbier/scampi`.

2.5. Hadamard operators for large-scale compressive imaging

We now turn our attention to one important practical consideration for high-dimensional signals, such as natural images. As N increases, the construction, storage, and numerical use of random, dense projection matrices containing $O(N^2)$ quickly becomes computationally infeasible. To confront this issue, we use sub-sampled Hadamard operators, which have been empirically shown to provide performance very close to purely random matrices [24], but at a significant reduction in computational complexity, requiring $O(N \log N)$ operations for matrix multiplication.

Since each Hadamard mode, except the average mode, have exactly zero mean, the system (5) would be invariant by a constant shift in the signal components without the presence of the average mode in $\mathbf{F}$. To break this symmetry, we force the retention of the average mode, which fixes the mean of the signal, when randomly selecting Hadamard modes for the construction of $\mathbf{F}$. Once a realization of $\mathbf{F}$ is chosen (as well as $\mathbf{F}^\intercal$ and $\mathbf{F}^2$, $(\mathbf{F}^2)^\intercal$ where the squared operators are simply sums in the case of Hadamard matrices), it is trivial to construct $\mathring{\mathbf{F}}$ (and $\mathring{\mathbf{F}}^\intercal, \mathring{\mathbf{F}}^2, (\mathring{\mathbf{F}}^2)^\intercal$, as well). Furthermore, $\mathring{\mathbf{F}}$ does not generate memory issues since all sub-matrices, other than $\mathbf{F}$, are extremely sparse with complexity of $O(N)$ per matrix multiplication. Thus, the overall complexity is $O(N \log N)$ per Scampi iteration.

3. Experimental results

We now present the results of two experimental protocols for image reconstruction from AWGN corrupted CS measurements. In all cases, the tested images are rescaled to the range $s_i \in [0,1]$, though this property isn't leveraged by any of the tested algorithms. Reconstruction performance is quantified using the normalized signal to noise ratio (NSNR), measured in decibels, between the true image $\mathbf{s}^*$ and the reconstruction $\hat{\mathbf{s}}$ defined as $\mathrm{NSNR}(\hat{\mathbf{s}}, \mathbf{s}^*) \triangleq 10\log_{10}(||\mathbf{s}^* - \hat{\mathbf{s}}||_2^2/||\mathbf{s}^*||_2^2)$ dB. For all experiments, the level of the true experimental noise variance, Δ^*, is reported as the *input* SNR (ISNR) and calcualted as $\mathrm{ISNR} = 10\log_{10}(||\mathbf{y}||_2^2/(M\Delta^*))$ dB.

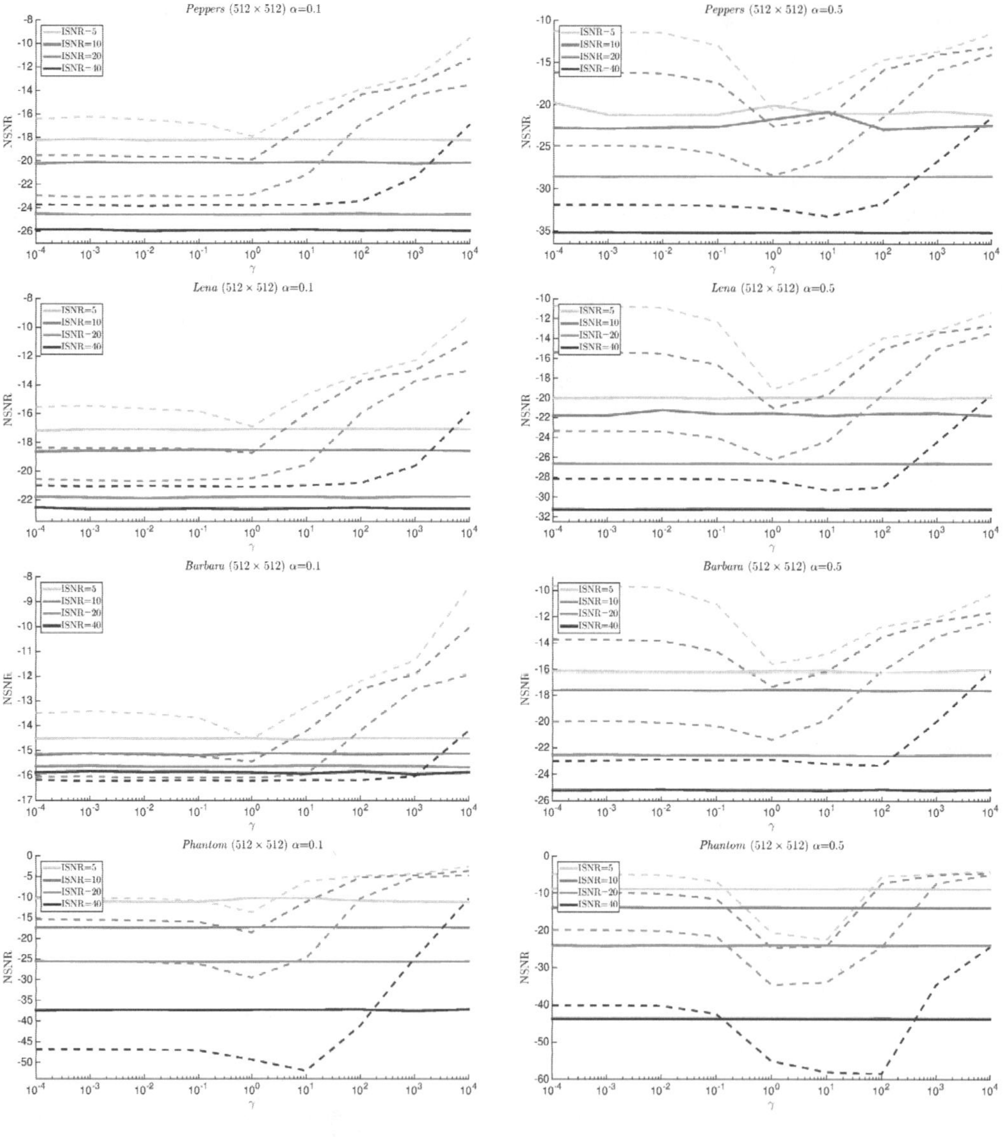

Figure 3. Reconstruction performance in NSNR (the lower the better) of Scampi with noise learning (solid curves) and GrAMPA (dashed curves), over the noise variance mismatch factor γ, for varying 512×512 test images at $\alpha = 0.1$ (left) and $\alpha = 0.5$ (right). Colors correspond to various levels of ISNR (solid and dashed curves of same color correspond to same dB). Each point is obtained through a single run for a random $(\mathbf{F}, \boldsymbol{\xi})$ instance common to both algorithms. At this problem dimensionality, finite size performance fluctuations are negligible.

We conduct our experiments for the common test images *Lena*, *Barbara*, *Peppers*, as well as the Shepp-Logan phantom (*Phantom*) at a resolution of 512×512 pixels. *Barbara* and *Lena* represent common examples of complex natural images, with *Lena* possessing a mixture of both smooth and high-variation regions, while *Barbara* contains very many edges due to the large amount of textured content in the image. *Peppers* represents a simpler case of a more piecewise-continuous natural image, and *Phantom* is, simply, a cartoon image of a few regions of *constant* pixel value.

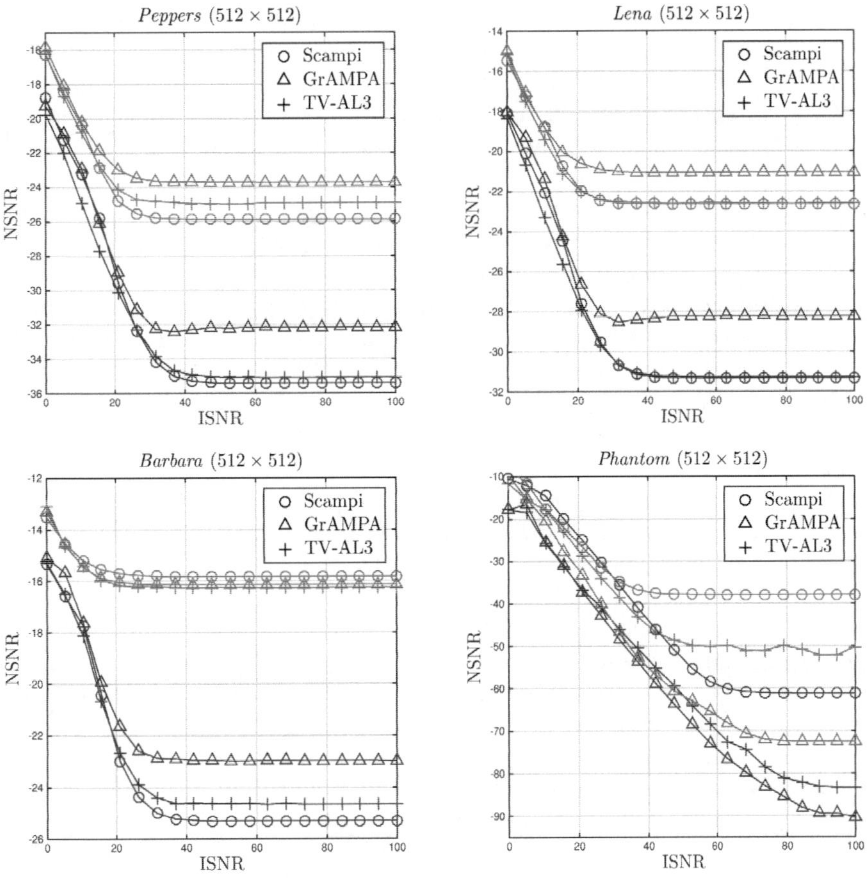

Figure 4. Comparison of optimal reconstruction accuracy as a function of the ISNR for Scampi, GrAMPA, and TV-AL3. All test points correspond to a search over 30 test-points for tuning the regularization parameter of TV-AL3 or the SNIPE ω parameter for GrAMPA and Scampi. Two measurement rates are tested for each method on each 512×512 image, $\alpha = 0.5$ (blue) and $\alpha = 0.1$ (red).

The first experiment, which results are presented in Fig. 3, compares the robustness of GrAMPA and Scampi, which includes the noise learning (15), to uncertainty of the measurement AWGN noise variance Δ^*. To simulate this uncertainty, both GrAMPA and Scampi are provided with an initial estimate of the noise variance, $\Delta = \gamma\Delta^*$, where $\gamma = 1$ represents perfect knowledge about the noisy channel. Each plot corresponds to the achieved reconstruction performance for a different image and two sub-sampling rates $\alpha = M/N \in \{0.1, 0.5\}$, as a function of the mismatch factor $\gamma \in [10^{-4}, 10^4]$. The dashed and solid curves correspond to the best GrAMPA and Scampi performances, respectively, over the range of the parameter $\omega \in \{0, 1, \ldots, 5\}$ for the SNIPE prior $P_S(\cdot; \omega)$. Generally, the best ω value for GrAMPA is the same (or close) to the one for Scampi. As seen in these charts, Scampi is able to provide better NSNR performance than GrAMPA, and without requiring any specific knowledge about Δ^*. GrAMPA, however, requires that Δ^* be known, in most cases, within an order of magnitude. Additionally, in the case of images, such as *Phantom*, for high ISNR, exact knowledge of Δ^* is in fact *detrimental* to the performance of the algorithm.

In the second experiment, see Fig. 4, we show the best-case performance of both GrAMPA and Scampi as a function of the ISNR for the specified test images at $\alpha = 0.1$ and $\alpha = 0.5$. We compare both approaches to TV-AL3 [9], as well, to give a reference point between these two probabilistic techniques and a state-of-the-art convex technique. For these experiments, best-case means that we intensively test over the free parameters of the algorithms, reporting only the best results. In the case of GrAMPA and Scampi, we test over ω, and for TV-AL3 we test over the weight of the TV regularization cost. In all cases, each point on the curves represents the best outcome over 30 reconstructions sweeping over each tuning parameter's domain. Additionally, we provide all algorithms with the true Δ^*. Our goal in reporting these best-case results is to compare the *maximal* performance of these techniques.

The findings we report are quite interesting and varied. The first topic we address is that for all tested natural images, TV-AL3 performs better than GrAMPA in terms of NSNR. Only for *Phantom*, whose gradient is very well modeled by the SNIPE prior, does GrAMPA outperform TV-AL3. However, one should note that the tuning of the free parameters of TV-AL3 is more problematic *a posteriori* than GrAMPA, as the domain of the optimal regularization term can be quite large, depending on the problem instance, while the optimal value of ω is generally around 0 or 1. But, these maximally best-case performance results imply that the convex treatment of TV is superior to the probabilistic model used by GrAMPA in the case of natural images.

However, the model used by Scampi, along with the dynamic estimation of both the measurement noise variance and constraint slack on the dual variables, provides best-case performance equal to or beyond TV-AL3's ones for most test cases, notably on *Peppers* and *Lena*, and at high measurement rates for an edge-heavy image such as *Barbara*. Finally, as we pointed out earlier, we see the expected subpar performance of Scampi for the cartoon *Phantom* image. However, we did not test the fixed $T = 0$ case for this image, which may yet improve the performance of Scampi for such images.

We also show on Fig. 5 some reconstruction results obtained with Scampi and GrAMPA for different settings, without optimization of ω which is set to 0 in all cases. As expected, the images obtained with Scampi are smoother and match better the three first natural images. In the last case of *phantom*, the gain in NSNR with GrAMPA with respect to Scampi is due to the large constant areas, but we notice that even in this case, Scampi gets a better resolution at the edges.

4. Conclusion

We have presented a new probabilistic AMP-based reconstruction algorithm for compressive imaging based on the cosparse analysis model with dual variables. This approach is similar to a finite temperature version of the GrAMPA algorithm, which makes it particularly adapted to natural images. Indeed, for this class of signals, the proposed algorithm *Scampi* reaches state -of-the-art perfomance, overcoming the previous best probabilistic methods. Furthermore, it is highly robust to noise channel uncertainty and does not require near-perfect knowledge to reach optimal performance. However, our experiments show that for cartoon images, consisting of only constant regions separated by instantaneous transitions, the zero temperature model does a better job of enforcing the piecewise-continuous prior on the image.

We have also shown through intensive numerical experiments that when we push the convex optimization TV-AL3 algorithm to its limits via oracle tuning parameter selection, it gives results very close to the proposed, easier to tune, approach. The fundamental reason for this empirical finding is yet to be understood.

Acknowledgments

This work has been supported in part by the ERC under the European Union's 7th Framework Programme Grant Agreement 307087-SPARCS, by the French Ministère de la défense/DGA and by the Swiss National Science Foundation grant number 200021-156672.

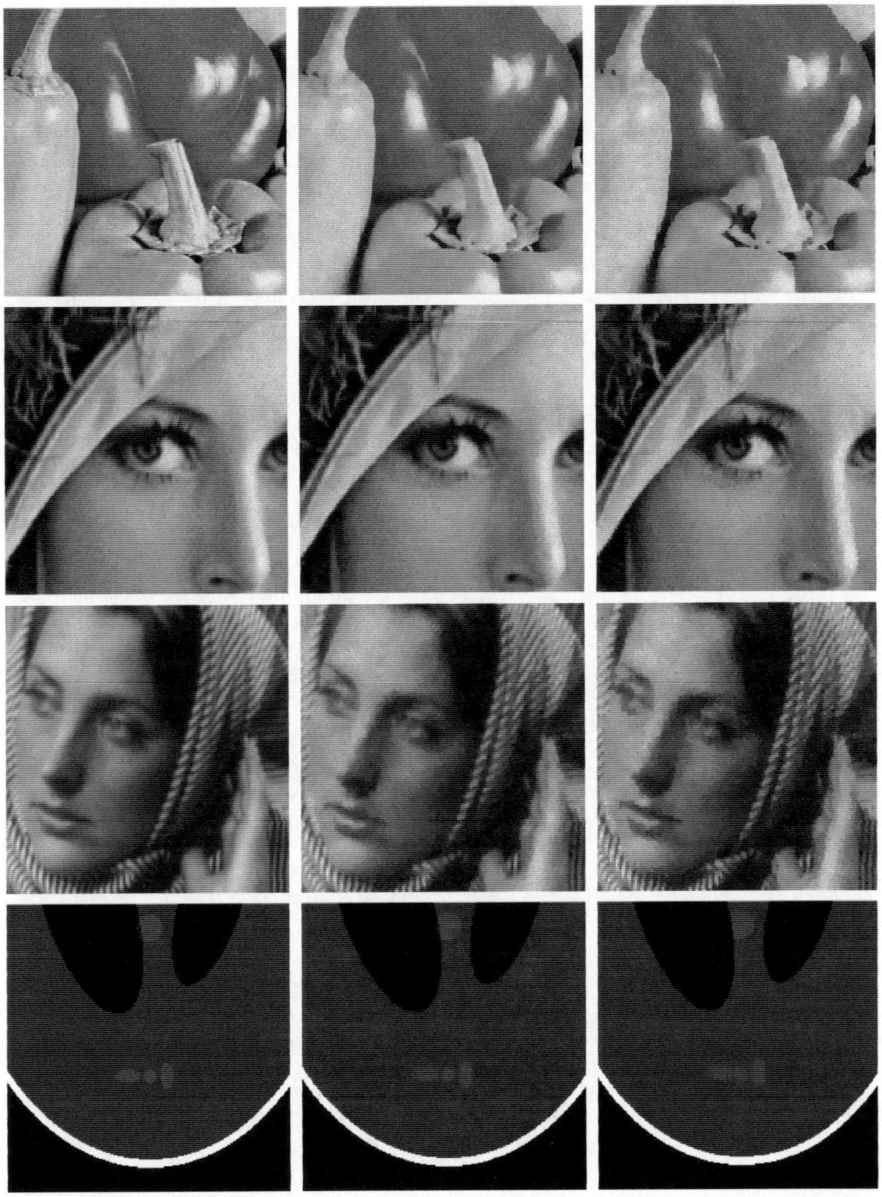

Figure 5. Visual comparison for some 512×512 images: a zoom on the original image (left) is compared with the Scampi (center) and GrAMPA (right) reconstructions for different settings, all with a fixed SNIPE prior parameter $\omega = 0$. From top to bottom are *peppers* at $\alpha = 0.1$ for an ISNR of 20 dB, *Lena* at $\alpha = 0.5$ for an ISNR of 40 dB, *Barbara* at $\alpha = 0.5$ for an ISNR of 30 dB and *phantom* at $\alpha = 0.1$ for an ISNR of 20 dB.

References

[1] Candès E, Romberg J and Tao T 2006 *Communications on Pure and Applied Mathematics* **59** 1207–1223
[2] Candès E J and Tao T 2006 *IEEE Trans. Inform. Theory* **52** 5406
[3] Candès E, Romberg J and Tao T 2006 *IEEE Trans. Inform. Theory* **52** 489–509
[4] Donoho D L 2006 *IEEE Trans. Inform. Theory* **52** 1289
[5] Duarte M F, Davenport M A, Takhar D, Laska J N, Sun T, Kelly K F and Baraniuk R G 2008 *Signal*

Processing Magazine **25** 83–91

[6] Nam S, Davies M, Elad M and Gribonval R 2013 *Applied and Computational Harmonic Analysis* **34** 30 – 56

[7] Tibshirani R 1996 *Journal of the Royal Statistical Society, Series B* **58** 267–288

[8] Chambolle A and Pock T 2011 *Journal of Mathematical Imaging and Vision* **40** 120–145

[9] Li C 2009 *An Efficient Algorithm for Total Variation Regularization with Applications to the Single Pixel Camera and Compressive Sensing* Master's thesis Rice University

[10] Donoho D, Johnstone I and Montanari A 2013 *IEEE Trans. Inform. Theory* **59**

[11] Metzler C A, Maleki A and Baraniuk R G 2014 *arXiv Preprint* **cs.IT/1406.4175**

[12] Tan J, Ma Y and Baron D 2015 *IEEE Trans. Sig. Proc.* **63** 2085–2092

[13] Kang J, Jung H, Lee H N and Kim K 2014 *arXiv Preprint* **cs.IT/1408.3930**

[14] Rangan S 2011 Generalized approximate message passing for estimation with random linear mixing *Proc. IEEE Int. Symp. on Inform. Theory* pp 2168 –2172

[15] Borgerding M A, Schniter P and Rangan S 2013 *arXiv Preprint* **cs.IT/1312.3968**

[16] Donoho D L, Maleki A and Montanari A 2009 *Proc. Natl. Acad. Sci.* **106** 18914–18919

[17] Bayati M and Montanari A 2011 *IEEE Transactions on Information Theory* **57** 764 –785

[18] Donoho D L, Maleki A and Montanari A 2010 Message passing algorithms for compressed sensing: I. motivation and construction *Proc. IEEE Inform. Theory Workshop* pp 1–5

[19] Krzakala F, Mézard M, Sausset F, Sun Y F and Zdeborová L 2012 *Physical Review X* **2** 021005

[20] Krzakala F, Mézard M, Sausset F, Sun Y and Zdeborová L 2012 *J. Stat. Mech.* **P08009**

[21] Javanmard A and Montanari A 2012 Subsampling at information theoretically optimal rates *Proc. IEEE Int. Symp. on Inform. Theory* pp 2431–2435

[22] Barbier J 2015 *Statistical physics and approximate message-passing algorithms for sparse linear estimation problems in signal processing and coding theory* Ph.D. thesis Universite Paris Diderot URL `http://arxiv.org/abs/1511.01650`

[23] Krzakala F, Manoel A, Tramel E W and Zdeborová L 2014 Variational free energies for compressed sensing *Proc. IEEE Int. Symp. on Inform. Theory* pp 1499–1503

[24] Barbier J, Schulke C and Krzakala F 2015 *Journal of Statistical Mechanics: Theory and Experiment* **2015** P05013

Generalized minimum dominating set and application in automatic text summarization

Yi-Zhi Xu and Hai-Jun Zhou

State Key Laboratory of Theoretical Physics, Institute of Theoretical Physics, Chinese Academy of Sciences, Zhong-Guan-Cun East Road 55, Beijing 100190, China

E-mail: `xyz@itp.ac.cn`, `zhouhj@itp.ac.cn`

Abstract. For a graph formed by vertices and weighted edges, a generalized minimum dominating set (MDS) is a vertex set of smallest cardinality such that the summed weight of edges from each outside vertex to vertices in this set is equal to or larger than certain threshold value. This generalized MDS problem reduces to the conventional MDS problem in the limiting case of all the edge weights being equal to the threshold value. We treat the generalized MDS problem in the present paper by a replica-symmetric spin glass theory and derive a set of belief-propagation equations. As a practical application we consider the problem of extracting a set of sentences that best summarize a given input text document. We carry out a preliminary test of the statistical physics-inspired method to this automatic text summarization problem.

1. Introduction

Minimum dominating set (MDS) is a well-known concept in the computer science community (see review [1]). For a given graph, a MDS is just a minimum-sized vertex set such that either a vertex belongs to this set or at least one of its neighbors belongs to this set. In the last few years researchers from the statistical physics community also got quite interested in this concept, as it is closely related to various network problems such as network monitoring, network control, infectious disease suppression, and resource allocation (see, for example, [2, 3, 4, 5, 6, 7, 8, 9, 10] and review [11]). Constructing an exact MDS for a large graph is, generally speaking, an extremely difficult task and it is very likely that no complete algorithm is capable of solving it in an efficient way. On the other hand, by mapping the MDS problem into a spin glass system with local many-body constraints and then treating it by statistical-physics methods, one can estimate with high empirical confidence the sizes of minimum dominating sets for single graph instances [12, 13]. One can also construct close-to-minimum dominating sets quickly through a physics-inspired heuristic algorithm [12, 13], which might be important for many practical applications.

In the present work we extend the statistical-physics approach of [12, 13] to edge-weighted graphs and study a generalized minimum dominating set problem. Our work is motivated by a practical knowledge-mining problem: extracting a set of sentences to best summarize one or more input text documents [14, 15]. We consider a general graph of vertices and edges, each edge connecting two different vertices and bearing one weight or a pair of weights (see Fig. 1). In the context of text summarization, a vertex represents a sentence of some text documents and an edge weight is the similarity between two sentences. Various data-clustering problems

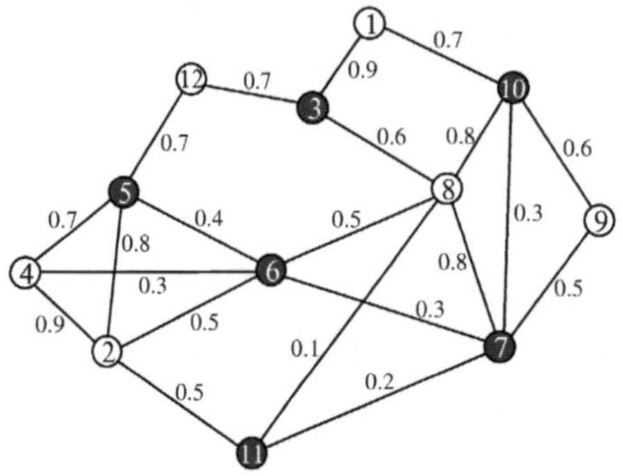

Figure 1. An graph with $N = 12$ vertices and $M = 21$ weighted edges. In this example the two weights $w_{i,j}$ and $w_{j,i}$ of each edge (i, j) are equal $(w_{i,j} = w_{j,i})$, and the threshold value of each vertex is $\theta = 1.0$. The vertex set $\Gamma_0 = \{3, 5, 6, 7, 10, 11\}$ is a generalized minimum dominating set for this graph. The summed weight of edges from every vertex $j \notin \Gamma_0$ to vertices in Γ_0 is equal to or greater than θ.

can also be represented as weighted graphs. Given such a weighted graph, our task is then to construct a minimum-cardinality set Γ_0 of vertices such that if a vertex i is not included in Γ_0, the summed weight of the edges from i to vertices in Γ_0 must reach at least certain threshold value θ. The set Γ_0 is referred to as a (generalized) MDS.

We introduce a spin glass model for this generalized MDS problem in Sec. 2 and then describe a replica-symmetric (RS) mean field theory in Sec. 3. A message-passing algorithm BPD (belief-propagation guided decimation) is outlined in Sec. 4, and is then applied to the automatic text summarization problem in Sec. 5. We conclude this work in Sec. 6 and discuss a way of modifying the spin glass model for better treating the text summarization problem.

2. Constraints and a spin glass model

We consider a generic graph G formed by N vertices with indices $i, j, k, \ldots \in \{1, 2, \ldots, N\}$ and $M = (c/2)N$ edges between pairs of these vertices (Fig. 1). The constant c is the mean vertex degree of the graph (on average a vertex is attached with c edges). Each edge (i, j) is associated with a pair of non-negative weights $w_{i,j}$ and $w_{j,i}$ which may or may not be equal. The meaning of the edge weights depend on the actual context. For example, $w_{i,j}$ may be interpreted as the extent that vertex i represents vertex j; in the symmetric case of $w_{i,j} = w_{j,i}$, we may also interpret $w_{i,j}$ as the similarity between i and j. Two vertices i and j are referred to as mutual neighbors if they are connected by an edge (i, j). The set of neighbors of vertex i is denoted as ∂i, i.e., $\partial i \equiv \{j \,|\, (i, j) \in G\}$.

Given a graph G, we want to construct a vertex set Γ_0 that is as small as possible and at the same time is a good representation of all the other vertices not in this set. Let us assign a state $c_i \in \{0, 1\}$ to each vertex i, $c_i = 1$ if $i \in \Gamma_0$ (referred to as being occupied) and $c_i = 0$ if $i \notin \Gamma_0$ (referred to as being empty). For each vertex $j \notin \Gamma_0$ we require that $\sum_{i \in \partial j} c_i w_{i,j} \geq \theta$, where θ is a fixed threshold value. A vertex j is regarded as being satisfied if it is occupied $(c_j = 1)$ or the condition $\sum_{i \in \partial j} c_i w_{i,j} \geq \theta$ holds, otherwise it is regarded as being unsatisfied. Therefore there are N vertex constraints in the system. A configuration $(c_1, c_2, \ldots, c_N)$ for the whole graph is referred to as a satisfying configuration if and only if it makes all the vertices to be satisfied (Fig. 1). Constructing such a generalized MDS Γ_0, i.e., a satisfying configuration with the smallest number of occupied vertices, is a 0–1 integer programming problem, but as it belongs to the nondeterministic polynomial-hard (NP-hard) computational complexity class, no algorithm is guaranteed to solve it in polynomial time. We now seek to solve it approximately through a statistical physics approach.

Let us introduce a weighted sum $Z(\beta)$ of all the 2^N possible microscopic configurations $(c_1, c_2, \ldots, c_N)$ as

$$Z(\beta) = \sum_{c_1,\ldots,c_N} \prod_{j=1}^{N} \left[\delta_{c_j}^1 e^{-\beta} + \delta_{c_j}^0 \Theta\Big(\sum_{i\in\partial j} c_i w_{i,j} - \theta \Big) \right] , \tag{1}$$

where δ_a^b is the Kronecker symbol ($\delta_a^b = 1$ if $a = b$ and $\delta_a^b = 0$ if $a \neq b$), and $\Theta(x)$ is the Heaviside step function such that $\Theta(x) = 0$ for $x < 0$ and $\Theta(x) = 1$ for $x \geq 0$. In the statistical physics community, $Z(\beta)$ is known as the partition function and the non-negative parameter β is the inverse temperature. Notice a configuration $(c_1, c_2, \ldots, c_N)$ has no contribution to $Z(\beta)$ if it is not a satisfying configuration. If a configuration satisfies all the vertex constraints, it contributes a term $e^{-\beta N_1}$ to $Z(\beta)$, where $N_1 \equiv \sum_{i=1}^N c_i$ is the total number of occupied vertices. As β increases, satisfying configurations with smaller N_1 values become more important for $Z(\beta)$, and at $\beta \to \infty$ the partition function is contributed exclusively by the satisfying configurations with the smallest N_1. For the purpose of constructing a minimum or close-to-minimum dominating set, we are therefore interested in the large-β limit of $Z(\beta)$.

3. Replica-symmetric mean field theory

It is very difficult to compute the partition function $Z(\beta)$ exactly, here we compute it approximately using the replica-symmetric mean field theory of statistical physics. This RS mean field theory can be understood from the angle of Bethe-Peierls approximation [16, 17], it can also be derived through loop expansion of the partition function [18, 19].

3.1. Thermodynamic quantities

We denote by $q_j^{c_j}$ the marginal probability that vertex j is in state $c_j \in \{0, 1\}$. Due to the constraints associated with vertex j and all its neighboring vertices, the state c_j is strongly correlated with those of the neighbors. To write down an approximate expression for $q_j^{c_j}$, let us assume that the states of all the vertices in set ∂j are independent before the constraint of vertex j is enforced. Under this Bethe-Peierls approximation we then obtain that

$$q_j^{c_j} \approx \frac{\delta_{c_j}^1 e^{-\beta} \sum\limits_{\{c_i\,:\,i\in\partial j\}} \prod\limits_{i\in\partial j} q_{i\to j}^{(c_i,1)} + \delta_{c_j}^0 \sum\limits_{\{c_i\,:\,i\in\partial j\}} \Theta\Big(\sum\limits_{i\in\partial j} c_i w_{i,j} - \theta\Big) \prod\limits_{i\in\partial j} q_{i\to j}^{(c_i,0)}}{e^{-\beta} \sum\limits_{\{c_i\,:\,i\in\partial j\}} \prod\limits_{i\in\partial j} q_{i\to j}^{(c_i,1)} + \sum\limits_{\{c_i\,:\,i\in\partial j\}} \Theta\Big(\sum\limits_{i\in\partial j} c_i w_{i,j} - \theta\Big) \prod\limits_{i\in\partial j} q_{i\to j}^{(c_i,0)}} . \tag{2}$$

In the above equation, $q_{i\to j}^{(c_i,c_j)}$ is the joint probability that vertex i has state c_i and its neighboring vertex j has state c_j *when the constraint associated with vertex j is not enforced.* The product $\prod_{i\in\partial j} q_{i\to j}^{(c_i,c_j)}$ is a direct consequence of neglecting the correlations among vertices in ∂j in the absence of vertex j's constraint. The mean fraction $\rho \equiv N_1/N$ of occupied vertices is then obtained through

$$\rho = \frac{1}{N} \sum_{j=1}^{N} q_j^1 , \tag{3}$$

This fraction should be a decreasing function of β.

We can define the free energy of the system as $F(\beta) = -\frac{1}{\beta} \ln Z(\beta)$. Within the RS mean field theory this free energy can be computed through

$$F \equiv N f = \sum_{j=1}^{N} f_j - \sum_{(i,j)\in G} f_{(i,j)} , \tag{4}$$

where f is the free energy density; and f_j and $f_{(i,j)}$ are, respectively, the free energy contribution of a vertex j and an edge (i,j):

$$f_j = -\frac{1}{\beta}\ln\left[e^{-\beta}\sum_{\{c_i\,:\,i\in\partial j\}}\prod_{i\in\partial j} q_{i\to j}^{(c_i,1)} + \sum_{\{c_i\,:\,i\in\partial j\}}\Theta\Big(\sum_{i\in\partial j} c_i w_{i,j} - \theta\Big)\prod_{i\in\partial j} q_{i\to j}^{(c_i,0)}\right], \tag{5a}$$

$$f_{(i,j)} = -\frac{1}{\beta}\ln\left[\sum_{c_i,c_j} q_{i\to j}^{(c_i,c_j)} q_{j\to i}^{(c_j,c_i)}\right]. \tag{5b}$$

The partition function is predominantly contributed by satisfying configurations with number of occupied vertices $N_1 \approx N\rho$, namely $Z(\beta) \approx e^{-\rho\beta N}\Omega(\rho)$ with $\Omega(\rho)$ being the total number of satisfying configurations at occupation density ρ. Then the entropy density $s(\rho) \equiv \frac{1}{N}\ln\Omega(\rho)$ of the system is computed through

$$s = (\rho - f)\beta\,. \tag{6}$$

The entropy density is required to be non-negative by definition. If $s(\rho) < 0$ as ρ decreases below certain value ρ_0, then $\Omega(\rho) = e^{Ns(\rho)} \to 0$ suggests that there is no satisfying configurations with $\rho < \rho_0$. We therefore take the value ρ_0 as the fraction of vertices contained in a minimum dominating set.

3.2. Belief-propagation equation

We need to determine the probabilities $q_{i\to j}^{(c_i,c_j)}$ to compute the thermodynamic densities ρ, f, and s. Following the Bethe-Peierls approximation and similar to Eq. (2), $q_{i\to j}^{(c_i,c_j)}$ is self-consistently determined through

$$q_{i\to j}^{(0,0)} = \frac{1}{z_{i\to j}}\sum_{\{c_k\,:\,k\in\partial i\backslash j\}}\Theta\Big(\sum_{k\in\partial i\backslash j} c_k w_{k,i} - \theta\Big)\prod_{k\in\partial i\backslash j} q_{k\to i}^{(c_k,0)}, \tag{7a}$$

$$q_{i\to j}^{(0,1)} = \frac{1}{z_{i\to j}}\sum_{\{c_k\,:\,k\in\partial i\backslash j\}}\Theta\Big(w_{j,i} + \sum_{k\in\partial i\backslash j} c_k w_{k,i} - \theta\Big)\prod_{k\in\partial i\backslash j} q_{k\to i}^{(c_k,0)}, \tag{7b}$$

$$q_{i\to j}^{(1,0)} = q_{i\to j}^{(1,1)} = \frac{1}{z_{i\to j}}e^{-\beta}\prod_{k\in\partial i\backslash j}\left[q_{k\to i}^{(1,1)} + q_{k\to i}^{(0,1)}\right], \tag{7c}$$

where $\partial i\backslash j$ is the subset of ∂i with vertex j being deleted, and $z_{i\to j}$ is a normalization constant. Equation (7) is called a belief-propagation (BP) equation in the literature. To find a solution to Eq. (7) we iterate this equation on all the edges of the input graph G (see, for example, [12, 13] or [19] for implementing details). However convergence is not guaranteed to achieve. If the reweighting parameter β is small this BP iteration quickly reaches a fixed point; while at large values of β we notice that it usually fails to converge (see next subsection).

3.3. Results on Erdös-Rényi random graphs

We first apply the RS mean field theory to Erdös-Rényi (ER) random graphs. To generate an ER random graph, we select M different pairs of edges uniformly at random from the whole set of $N(N-1)/2$ vertex pairs and then connect each selected pair of vertices by an edge. For N sufficiently large there is no structural correlations in such a random graph, and the typical length of a loop in the graph diverges with N in a logarithmic way.

If the two edge weights of every edge (i,j) are equal to the vertex threshold value θ ($w_{i,j} = w_{j,i} = \theta$), the generalized MDS problem reduces to the conventional MDS problem on an undirected graph, which has been successfully treated in [12]. For example, for ER random

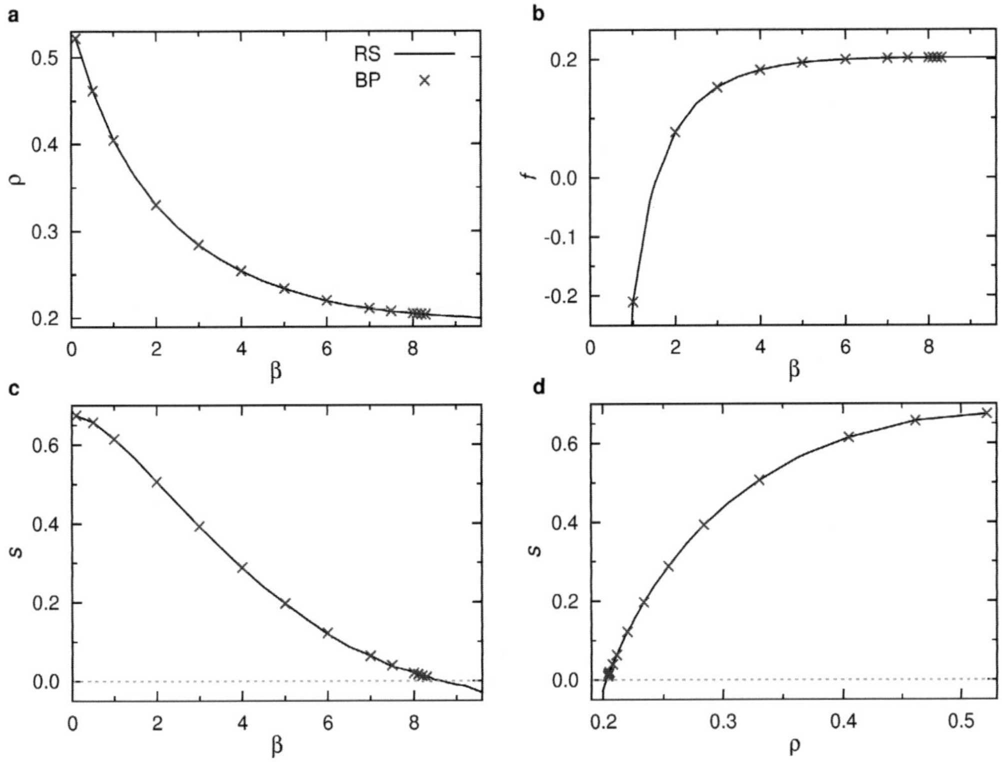

Figure 2. Replica-symmetric mean field results on ER random networks of mean vertex degree $c = 10.0$. The symmetric edge weights are drawn from the set $\{0.4, 0.5, 0.6, 0.7, 0.8, 0.9, 1.0\}$ and the vertex threshold value is $\theta = 1$. The cross symbols (BP) are results obtained by belief-propagation on a single graph instance of size $N = 10^5$, while the solid lines (RS) are ensemble-averaged results obtained by population dynamics simulations. (a) Occupation density ρ versus inverse temperature β; (b) free energy density f versus β; (c) entropy density s versus β; (d) entropy density s as a function of ρ obtained by combining data of (a) and (c).

graphs with mean vertex degree $c = 10.0$ the MDS relative size is $\rho_0 \approx 0.120$ [12]. On the other hand, if the two edge weights of every edge are strongly non-symmetric such that either $w_{i,j} = \theta$ and $w_{j,i} = 0$ (with probability 1/2) or $w_{i,j} = 0$ and $w_{j,i} = \theta$ (also with probability 1/2), the generalized MDS problem reduces to the conventional MDS problem on a directed graph, which again has been successfully treated in [13] (e.g., at $c = 10.0$ the MDS relative size is $\rho_0 \approx 0.195$).

In this paper, as a particular example, we consider a distribution of edge weights with the following properties: (1) the weights of every edge (i, j) are symmetric, so $w_{i,j} = w_{j,i}$; (2) the edge weights of different edges are not correlated but completely independent; (3) for each edge (i, j) its weight $w_{i,j}$ is assigned the value 0.4θ or 1.0θ with probability $1/12$ each and assigned values in the set $\{0.5\theta, 0.6\theta, 0.7\theta, 0.8\theta, 0.9\theta\}$ with equal probability $1/6$ each.

The BP results on the occupation density ρ, the free energy density f, and the entropy density s are shown in Fig. 2 for a single ER random graph of $N = 10^5$ vertices and mean degree $c = 10.0$. The BP iteration for this this graph instance is convergent for $0 \leq \beta \leq 8.3$. The occupation density ρ and the entropy density s both decrease with inverse temperature β. The entropy density as a function of occupation density, $s(\rho)$, approaches zero at $\rho = \rho_0 \approx 0.202$, indicating there is no satisfying configurations at occupation density $\rho < \rho_0$. The BP results

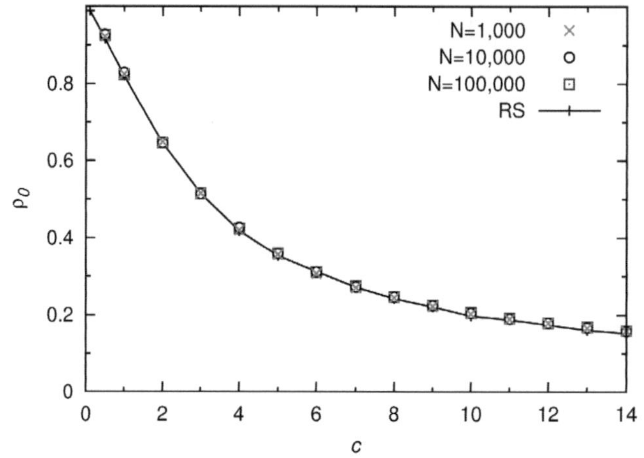

Figure 3. The relative size ρ_0 of minimum dominating sets for ER random graphs of mean vertex degree c. The edge weight distribution for these random graphs are the same as that of Fig. 2, and the vertex threshold value $\theta = 1.0$. The solid-line connected plus symbols are the predictions of the RS mean field theory, while the results obtained by the BPD algorithm at $\beta = 8.0$ are drawn as cross symbols (for graph size $N = 10^3$), circles ($N = 10^4$), and squares ($N = 10^5$). Each BPD data point is the result of a single run on one graph instance.

therefore predict that a MDS for this problem instance must contain at least $0.202N$ vertices.

We can also obtain RS mean field results on the thermodynamic densities by averaging over the whole ensemble of ER random graphs (with $N \to \infty$ and fixed mean vertex degree c). This is achieved by population dynamics simulations [16]. We store a population of probabilities $\{q_{i\to j}^{(c_i,c_j)}\}$ and update this population using Eq. (7), and at the same time compute the densities of thermodynamic quantities. A detailed description on the implementation can be found in section 4.3 of [12]. The ensemble-averaged results for the ER random network ensemble of $c = 10.0$ and $N \to \infty$ are also shown in Fig. 2. These results are in good agreement with the BP results obtained on the single graph instance.

Through the RS population dynamics simulations we can estimate the ensemble-averaged value of ρ_0 (the minimum fraction of occupied vertices) by the equation $s(\rho_0) = 0$. The value of ρ_0 obtained in such a way decreases with mean vertex degree c continuously, see Fig. 3 (solid line).

4. Belief-propagation-guided decimation algorithm

For β sufficiently large, the marginal occupation probability $q_j^{c_j}$ obtained by Eq. (2) tells us the likelihood of each vertex j to belong to a minimum dominating set. This information can serve as a guide for constructing close-to-minimum dominating sets. Based on the BP equation (2) we implement a simple belief-propagation-guided decimation (BPD) algorithm as follows. Starting from an input graph G and an empty vertex set Γ, at each step we (1) iterate the BP equation for a number of repeats and then estimate the occupation probability q_j^1 for all the vertices j not in Γ; and (2) add a tiny fraction (e.g., 1%) of those vertices j with the highest values of q_j^1 into the set Γ and set their state to be $c_j = 1$; (3) then simplify the graph and repeat the operations (1)–(3) on the simplified graph, until Γ becomes a dominating set.

The detailed implementation of this BPD algorithm is the same as described in section 5 of [12]. Here we only need to emphasize one new feature: after a vertex i is newly occupied, the threshold value (say θ_j) of every neighboring vertex j should be updated as $\theta_j \leftarrow (\theta_j - w_{i,j})$, and if this updated θ_j is non-positive then vertex j should be regarded as being satisfied.

For the same graph of Fig. 2, a single trial of this BPD algorithm at $\beta = 8.0$ results in a dominating set of size 21009, which is very close to the predicted MDS size by the RS mean field theory. Equally good performance of the BPD algorithm is also achieved on other ER random graphs with mean vertex degree c ranging from $c = 0.5$ to $c = 14$ (see Fig. 3), suggesting that the BPD algorithm is able to construct a dominating set which is very close to a MDS. We emphasize that in the BPD algorithm we do not require the BP iteration to converge.

5. Application: Automatic text summarization

Automatic text summarization is an important issue in the research field of natural language processing [14]. One is faced with the difficult task of constructing a set of sentences to summarize a text document (or a collection of text documents) in a most informative and efficient way. Here we extend the initial idea of Shen and Li [15] and consider this information retrieval problem as a generalized minimum dominating set problem.

We represent each sentence of an input text document as a vertex and connect two vertices (say i and j) by an weighted edge, with the symmetric edge weight $w_{i,j}$ ($= w_{j,i}$) being equal to the similarity of the two corresponding sentences. Before computing the edge weight a pre-treatment is applied to all the sentences to remove stop-words (such as 'a', 'an', 'at', 'do', 'but', 'of', 'with') and to transform words to their prototypes according to the WordNet dictionary [20] (e.g., 'airier' → 'airy', 'fleshier' → fleshy, 'are' → 'be', 'children' → child, 'looking' → 'look'). There are different ways to measure sentence similarity, here we consider a simple one, the cosine similarity [21]. To compute the cosine similarity, we map each sentence i to a high-dimensional vector $\vec{S}_i$, the k-th element of which is just the number of times the k-th word of the text appears in this sentence. Then the edge weight between vertices i and j is defined as

$$w_{i,j} = \frac{\vec{S}_i \cdot \vec{S}_j}{\sqrt{\vec{S}_i \cdot \vec{S}_i}\sqrt{\vec{S}_j \cdot \vec{S}_j}} . \tag{8}$$

To give a simple example, let us consider a document with only two sentences 'Tom is looking at his children with a smile.' and 'These children are good at singing.'. The word set of this document is {Tom, be, look, child, smile, good, sing}, and the vectors for the two sentences are $\vec{S}_1 = (1,1,1,1,1,0,0)$ and $\vec{S}_2 = (0,1,0,1,0,1,1)$, respectively. The cosine similarity w_{12} between these two sentences is then $w_{12} = \frac{2}{\sqrt{5}\sqrt{4}} \approx 0.447$.

We first test the performance of the BPD algorithm on 32 short English text documents of different lengths (on average a document has 17.7 sentences). We compare the outputs from the BPD algorithm with the key sentences manually selected by the first author. For each text document we denote by B and $\tilde{B}$ the set of key sentences selected by human inspection and by the algorithm, respectively. On average the set B of human inspection contains a fraction $\rho = 0.226$ of the sentences in the input text document. Then we define the coverage ratio R_{cov} and the difference ratio R_{dif} between B and $\tilde{B}$ as

$$R_{cov} = \frac{|B \cap \tilde{B}|}{|B|} , \qquad R_{dif} = \frac{|\tilde{B} - B|}{|\tilde{B}|} , \tag{9}$$

where $(\tilde{B} - B)$ denotes the set of sentences belonging to $\tilde{B}$ but not to B. The ratio R_{cov} quantifies the probability of a manually selected key sentence also being selected by the algorithm, while the ratio R_{dif} quantifies the extent that a sentence selected by the algorithm does not belong to the set of manually selected key sentences.

Table 1. Averaged performances of the BPD algorithm ($\beta = 8.0$), the PR (PageRank) algorithm, and the AP (affinity-propagation) algorithm on 32 English text documents (average number of sentences per document 17.7). For BPD the vertex threshold is set to $\theta = 0.6$ ($\text{BPD}_{0.6}$), $\theta = 0.8$ ($\text{BPD}_{0.8}$) and $\theta = 1.0$ ($\text{BPD}_{1.0}$). For PR the fraction of sentences selected is 25% ($\text{PR}_{25\%}$), 30% ($\text{PR}_{30\%}$), and 40% ($\text{PR}_{40\%}$). For AP the adjustable parameter is set to be $w_{i,i} = 0.0$ ($\text{AP}_{0.0}$) and $w_{i,i} = 0.2$ ($\text{AP}_{0.2}$). ρ is the fraction of representative sentences chosen by the algorithm, and R_{cov} and R_{dif} are two performance measures defined by Eq. (9). The average fraction of representative sentences constructed by human inspection is $\rho = 0.226$.

	$\text{BPD}_{0.6}$	$\text{BPD}_{0.8}$	$\text{BPD}_{1.0}$	$\text{PR}_{25\%}$	$\text{PR}_{30\%}$	$\text{PR}_{40\%}$	$\text{AP}_{0.0}$	$\text{AP}_{0.2}$
ρ	0.44	0.48	0.56	0.27	0.32	0.42	0.17	0.39
R_{cov}	39.9%	47.2%	49.6%	30.0%	41.7%	50.6%	15.6%	39.3%
R_{dif}	79.4%	78.6%	80.2%	74.2%	71.4%	72.2%	76.3%	77.4%

We also apply two other summarization algorithms to the same set of text documents, one is the PageRank (PR) algorithm [22, 23, 24], and the other is the Affinity-Propagation (AP) algorithm [25]. PageRank is based on the idea of random walk on a graph, and it offers an efficient way of measuring vertex significance. The importance P_i of a vertex i is determined by the following self-consistent equation

$$P_i = (1-p)\frac{1}{N} + p * \sum_{j\in\partial i} P_j \frac{w_{j,i}}{\sum_{k\in\partial j} w_{j,k}} , \tag{10}$$

where p is the probability to jump from one vertex to a neighboring vertex (we set $p = 0.85$ following [22]). Those vertices i with high values of P_i are then selected as the representative vertices.

On the other hand, Affinity-Propagation is a clustering algorithm: each vertex either selects a neighboring vertex as its exemplar or serves as an exemplar for some or all of its neighbors [25]. For any pair of vertices i and j, the responsibility $r_{i,j}$ of j to i and the availability $a_{i,j}$ of j to i are determined by the following set of iterative equations:

$$r_{i,j} = w_{i,j} - \max_{k\neq j}\{a_{i,k} + w_{i,k}\} , \tag{11a}$$

$$a_{i,j} = \min\Big[0,\ r_{j,j} + \sum_{k\neq i,j} \max[0,\ r_{k,j}]\Big] , \tag{11b}$$

$$a_{j,j} = \sum_{i\neq j} \max[0,\ r_{i,j}] . \tag{11c}$$

In Eq. (11a) $w_{i,j}$ is the weight of edge (i,j) for $i \neq j$, and $w_{i,i}$ is an adjustable parameter which affects the final number of examplars. We iterate the AP equation (11) on the sentence graph starting from the initial condition of $r_{i,j} = a_{i,j} = 0$ and, after convergence is reached, then consider all the vertices i with positive values of $(r_{i,i} + a_{i,i})$ as the examplar vertices.

For the 32 short text documents used in our preliminary test, the comparative results of Table 1 do not distinguish much the three heuristic algorithms, yet it appears that PageRank performs slightly better than BPD and AP. When the fraction of extracted sentences is $\rho = 0.42$, the coverage ratio reached by PR is $R_{cov} = 0.51$ and the difference ratio is $R_{dif} = 0.72$, while $R_{cov} = 0.40$ and $R_{dif} = 0.79$ for BPD at $\rho = 0.44$ and $R_{cov} = 0.39$ and $R_{dif} = 0.77$ for AP at $\rho = 0.39$.

We then continue to evaluate the performance of the belief-propagation approach on a benchmark set of longer text documents, namely the DUC (Document Understanding

Table 2. Averaged performances of the BPD algorithms $\mathrm{BPD}_\theta^{100}$ ($\theta = 0.6$ or $\theta = 1.0$) and BPD_θ ($\theta = 1.0$) and the PageRank algorithm PR^{100} on the 533 text documents of DUC 2002 [26]. The Precision, Recall, and F-score values are obtained by averaging over the results of individual text documents. The inverse temperature of BPD is fixed to be $\beta = 8.0$.

	PR^{100}	$\mathrm{BPD}_{0.6}^{100}$	$\mathrm{BPD}_{1.0}^{100}$	$\mathrm{BPD}_{1.0}$
Recall	0.455	0.249	0.264	0.727
Precision	0.407	0.396	0.410	0.256
Fscore	0.429	0.303	0.318	0.359

Conference) data set used in [24]. We examine a total number of 533 text documents from the DUC 2002 directory [26]. The average number of sentences per document is about 28 and the average number of words per sentence is about 20.

The DUC data set offers, for each of these text documents, two sets B of representative sentences chosen by two human experts, the total number of words in such a set B being ≈ 100. The PageRank algorithm (PR^{100}) and one version of the BPD algorithm ($\mathrm{BPD}_\theta^{100}$, $\theta = 0.6$ or $\theta = 1.0$) also construct a set $\tilde{B}$ of sentences for each of these documents under the constraint that the total number of words in $\tilde{B}$ should be about 100. In another version of the BPD algorithm (BPD_θ) the restriction on the words number in $\tilde{B}$ is removed. We follow the DUC convention and use the toolkit ROUGE [27] to evaluate the agreement between $\tilde{B}$ and B in terms of Recall, Precision, and F-score:

$$\mathrm{Recall} = \frac{\sum\limits_{\mathrm{word}\in B} \min\big[C(\mathrm{word}), \tilde{C}(\mathrm{word})\big]}{\mathrm{WordsNum}(B)}, \tag{12a}$$

$$\mathrm{Precision} = \frac{\sum\limits_{\mathrm{word}\in B} \min\big[C(\mathrm{word}), \tilde{C}(\mathrm{word})\big]}{\mathrm{WordsNum}(\tilde{B})}, \tag{12b}$$

$$\mathrm{Fscore} = \frac{2 \times \mathrm{Precision} \times \mathrm{Recall}}{\mathrm{Precision} + \mathrm{Recall}}. \tag{12c}$$

where $C(\mathrm{word})$ is the total number of times a given word appears in the summary B, and $\tilde{C}(\mathrm{word})$ is the number of times this word appears in the summary $\tilde{B}$; $\mathrm{WordsNum}(B)$ is the total number of words in the summary B and similarly for $\mathrm{WordsNum}(\tilde{B})$.

The comparative results for the DUC 2002 data set are shown in Table 2. We notice that $\mathrm{BPD}_{1.0}$ ($\theta = 1.0$) has the highest Recall value of 0.727, namely the summary obtained by this algorithm contains most of contents in the summary of human experts, but its Precision value of 0.256 is much lower than that of the PR^{100} algorithm, indicating that the BPD algorithm add more sentences into the summary than the human experts do. In terms of the F-score which balances Recall and Precision (the last row of Table 2) we conclude that PageRank also performs a little bit better than BPD for the DUC 2002 benchmark.

The generalized MDS model for the text summarization problem aims at a complete coverage of an input text document. It is therefore natural that the summary constructed by BPD contains more sentences than the summary constructed by the human experts (which may only choose the sentences that best summarize the key points of a text document). All the tested documents in the present work are rather short, which may make the advantages of the BPD message-passing algorithm difficult to be manifested. More work needs to be done to test the performance of the BPD algorithm on very long text documents.

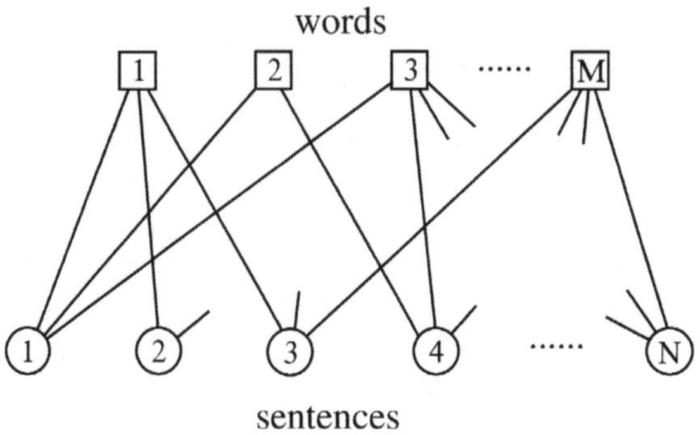

Figure 4. The word–sentence graph representation for a text document. The M words and N sentences of an input text document are denoted by squares and circles, respectively, and a link between a word a and a sentence i is drawn if and only if word a appears in sentence i. To get a set Γ of representative sentences we may require that each word must be connected to at least n ($n \geq 1$) sentences of the set Γ.

6. Outlook

In this paper we presented a replica-symmetric mean field theory for the generalized minimum dominating set problem, and we considered the task of automatic text summarization as such a MDS problem and applied the BPD message-passing algorithm to construct a set of representative sentences for a text document. When tested on a set of short text documents the BPD algorithm has comparable performance as the PageRank and the Affinity-Propagation algorithms. We feel that the BPD approach will be most powerful for extracting sentences out of lengthy text documents (e.g., scientific papers containing thousands of sentences). We hope that our work will stimulate further efforts on this important application.

The belief-propagation based method for the automatic text summarization problem might be improved in various ways. For example, it may not be necessary to perform the decimation step, rather one may run BP on the input sentence graph until convergence (or for a sufficient number of rounds) and then return an adjustable fraction ρ of the sentences i according to their estimated occupation probabilities q_i^1.

One may also convert the text summarization problem to other generalized MDS problems. A particularly simple but potentially useful one can be constructed as follows: we first construct a bi-partite graph formed by words, sentences, and the links between words and sentences (see Fig. 4); we then construct a minimum-sized dominating set of sentences Γ such that every word of the whole bipartite graph must appear in at least n ($n \geq 1$) of the sentences of Γ. Such a generalized MDS problem can be studied by slightly modifying the BP equation Eq. (7). We notice that this alternative construction has the advantage of encouraging diversity in the selected representative sentences.

Acknowledgments

We thank Jin-Hua Zhao and Yusupjan Habibulla for helpful discussions. This research is partially supported by the National Basic Research Program of China (grant number 2013CB932804) and by the National Natural Science Foundation of China (grand numbers 11121403 and 11225526).

References

[1] Haynes T W, Hedetniemi S T and Slater P J 1998 *Fundamentals of Domination in Graphs* (New York: Marcel Dekker)

[2] Echenique P, Gómez-Gardeñes J, Moreno Y and Vázquez A 2005 Distance-*d* covering problems in scale-free networks with degree correlations *Phys. Rev.* E **71** 035102(R)

[3] Dall'Asta L, Pin P and Ramezanpour A 2009 Statistical mechanics of maximal independent sets *Phys. Rev.* E **80** 061136

[4] Dall'Asta L, Pin P and Ramezanpour A 2011 Optimal equilibria of the best shot game *J. Public Economic Theor.* **13** 885–901

[5] Yang Y, Wang J and Motter A E 2012 Network observability transitions *Phys. Rev. Lett.* **109** 258701

[6] Molnár Jr. F, Sreenivasan S, Szymanski B K and Korniss K 2013 Minimum dominating sets in scale-free network ensembles *Sci. Rep.* **3** 1736

[7] Nacher J C and Akutsu T 2013 Analysis on critical nodes in controlling complex networks using dominating sets In *International Conference on Signal-Image Technology & Internet-Based Systems (Kyoto)* 649–654

[8] Takaguchi T, Hasegawa T and Yoshida Y 2014 Suppressing epidemics on networks by exploiting observer nodes *Phys. Rev.* E **90** 012807

[9] Wuchty S 2014 Controllability in protein interaction networks *Proc. Natl. Acad. Sci. USA* **111** 7156–7160

[10] Wang H, Zheng H, Browne F and Wang C 2014 Minimum dominating sets in cell cycle specific protein interaction networks In *Proceedings of International Conference on Bioinformatics and Biomedicine* (IEEE) 25–30

[11] Liu Y Y and Barabási A L 2015 Control principles of complex networks arXiv:1508.05384

[12] Zhao J H, Habibulla Y and Zhou H J 2015 Statistical mechanics of the minimum dominating set problem *J. Stat. Phys.* **159** 1154–1174

[13] Habibulla Y, Zhao J H and Zhou H J 2015 The directed dominating set problem: Generalized leaf removal and belief propagation *Lect. Notes Comput. Sci.* **9130** 78–88

[14] Mani I 1999 *Advances in Automatic Text Summarization* (Cambridge, MA: MIT Press)

[15] Shen C and Li T 2010 Multi-document summarization via the minimum dominating set In *Proceedings of the 23rd International Conference on Computational Linguistics (Beijing)* (Association for Computational Linguistics) 984–992

[16] Mézard M and Parisi G 2001 The bethe lattice spin glass revisited *Eur. Phys. J.* B **20** 217–233

[17] Mézard M and Montanari A 2009 *Information, Physics, and Computation* (New York: Oxford Univ. Press)

[18] Zhou H J and Wang C 2012 Region graph partition function expansion and approximate free energy landscapes: Theory and some numerical results *J. Stat. Phys.* **148** 513–547

[19] Zhou H J 2015 *Spin Glass and Message Passing* (Beijing: Science Press)

[20] Fellbaum C 1998 *WordNet: an electronic lexical database* (Cambridge, MA: MIT Press)

[21] Singhal A 2001 Modern information retrieval: a brief overview *IEEE Data Engineering Bulletin* **24** 35–43

[22] Brin S and Page L 1998 The anatomy of a large-scale hypertextual web search engine *Computer Networks and ISDN Systems* **30** 107–117

[23] Mihalcea R and Tarau P 2004 Textrank: Bringing order into texts In *Preceedings of the Conference on Empirical Methods in Natural Language Processing (Barcelona)* (Association for Computational Linguistics) 404–411

[24] Erkan G and Radev D R 2004 Lexrank: Graph-based lexical centrality as salience in text summarization *J. Artifical Intelligence Res.* **22** 457–479

[25] Frey B J and Dueck D 2007 Clustering by passing messages between data points *Science* **315** 972–976

[26] Document Understanding Conference 2002 http://www-nlpir.nist.gov/projects/duc

[27] Lin C Y 2004 Rouge: a package for automatic evaluation of summaries In *Preceedings of the ACL-04 Workshop: Text Summarization Branches Out (Barcelona)* (Association for Computational Linguistics) 74–81

Journal of Physics: Conference Series **699** (2016) 012015 doi:10.1088/1742-6596/699/1/012015

Performance of a community detection algorithm based on semidefinite programming

Federico Ricci-Tersenghi

Dipartimento di Fisica, INFN–Sezione di Roma1 and CNR–Nanotec, Università La Sapienza, Piazzale Aldo Moro 5, I-00185 Roma, Italy.

E-mail: `federico.ricci@roma1.infn.it`

Adel Javanmard

USC Marshall School of Business, University of Southern California

Andrea Montanari

Department of Electrical Engineering and Department of Statistics, Stanford University

Abstract. The problem of detecting communities in a graph is maybe one the most studied inference problems, given its simplicity and widespread diffusion among several disciplines. A very common benchmark for this problem is the stochastic block model or planted partition problem, where a phase transition takes place in the detection of the planted partition by changing the signal-to-noise ratio. Optimal algorithms for the detection exist which are based on spectral methods, but we show these are extremely sensible to slight modification in the generative model. Recently Javanmard, Montanari and Ricci-Tersenghi [1] have used statistical physics arguments, and numerical simulations to show that finding communities in the stochastic block model via semidefinite programming is quasi optimal. Further, the resulting semidefinite relaxation can be solved efficiently, and is very robust with respect to changes in the generative model. In this paper we study in detail several practical aspects of this new algorithm based on semidefinite programming for the detection of the planted partition. The algorithm turns out to be very fast, allowing the solution of problems with $O(10^5)$ variables in few second on a laptop computer.

1. Introduction and model definition

When dealing with a high-dimensional dataset one often looks for hidden structures, that may be representative of the signal one is trying to extract from the noisy dataset. In the community detection problem, we are asked to find the most significative clustering of the vertices of a graph, such that intra-cluster connections are much more abundant/sparse in the assortative/disassortative case with respect to inter-clusters connections (see Ref. [2] for a review). A common benchmark in this class is provided by the the so-called planted partition problem or stochastic block model (SBM) [3], defined as follows: an undirected graph $G = (V, E)$ is given, where $V = \{1, \ldots, n\}$ is the vertex set and E is the edge set. Vertices are divided in q equal size groups V_k, with $k = 1, \ldots, q$, such that $V = \cup_{k=1}^{q} V_k$, $|V_k| = n/q$ and $V_i \cap V_j = \emptyset$ if

$i \neq j$. The function $\kappa(i)$ returns the cluster vertex i belongs to. Conditioned on the partition $\{V_k\}_{k=1,\ldots,q}$, edges are drawn independent with

$$\mathbb{P}\Big[(i,j) \in E \Big| \{V_k\}\Big] = \begin{cases} c_{\text{in}}/n & \text{if } \kappa(i) = \kappa(j)\,, \\ c_{\text{out}}/n & \text{if } \kappa(i) \neq \kappa(j)\,. \end{cases}$$

The resulting random graph is sparse and has a mean degree equal to $c = [c_{\text{in}} + (q-1)c_{\text{out}}]/q$. We will be mainly concerned with the assortative case, where $c_{\text{in}} > c_{\text{out}}$ holds strictly. The goal is to detect the partition $\{V_k\}$ given the graph G.

The Bayesian approach [4] predicts the existence of a threshold at

$$c_{\text{in}} - c_{\text{out}} = q\sqrt{c}\,,$$

for $q \leq 4$, separating a phase where detecting the partition is impossible from a phase where the detection can be achieved, using the best possible algorithms. We defined the signal-to-noise ratio as

$$\lambda = \frac{c_{\text{in}} - c_{\text{out}}}{q\sqrt{c}}\,,$$

such that the phase transition for $q \leq 4$ takes place at $\lambda_c = 1$. We will mainly be interested in the $q = 2$ case, where the existence of the threshold at $\lambda_c = 1$ has been proved rigorously [5, 6]. In this case the planted partition can be conveniently coded in a vector $\boldsymbol{x}_0 \in \{+1, -1\}^n$, and, calling $\hat{\boldsymbol{x}}(G)$ the estimate of the partition in graph G obtained by any inference procedure, we can quantify the success of the detection algorithm computing the overlap with respect to the planted partition (i.e. the absolute value of the normalized scalar product)

$$Q = \frac{1}{n}|\langle \hat{\boldsymbol{x}}(G), \boldsymbol{x}_0 \rangle|\,.$$

Eventually we can consider also its average value over the ensemble of random graphs, $\mathbb{E}[Q]$. For the SBM Ref. [7] proposed a belief-propagation message-passing algorithm to find the Bayes optimal estimator, which has indeed a non-zero overlap with the planted partition as soon as $\lambda > \lambda_c = 1$.

Spectral methods based on the Laplacian (unnormalized or normalized) are known to be sub-optimal, since they have a threshold which is strictly larger than the optimal one [8], i.e. $\lambda_c^{\text{Lap}} = \sqrt{c/(c-1)} > 1$. However, a new spectral method based on the non-backtracking matrix introduced in Ref. [9] achieves optimality in the detection of the planted partition in the SBM, at the cost of computing the complex spectrum of a non symmetric matrix. Later, such a spectral method has been strongly simplified by showing its similarity to the computation of the spectrum of the so-called Bethe Hessian matrix [10], which is a $n \times n$ symmetric matrix defined as

$$H(r) = (r^2 - 1)\mathbb{1} - rA - D\,,$$

where A is the adjacency matrix, $A_{ij} = A_{ji} = \mathbb{I}[(ij) \in E]$, and D is a diagonal matrix with entries equal to the vertex degrees d_i. In the assortative SBM with $q = 2$, the planted partition is detected by computing the negative eigenvalues of $H(\sqrt{c})$ [10] and the best estimator $\hat{\boldsymbol{x}}^{\text{BH}}(G)$ turns out to be given by the vector of signs of the components of the eigenvector corresponding to the second largest (in absolute value) eigenvalue.

2. Spectral methods versus optimization methods

Although the partition detection based on the Bethe Hessian is optimal for the SBM, it turns out to be not very robust if the generative model departs even slightly from the random graph

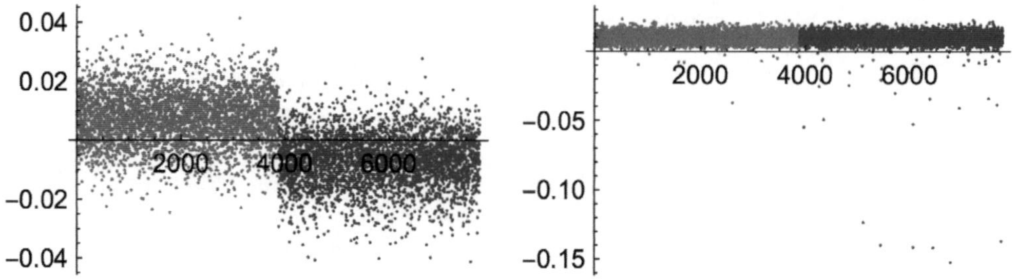

Figure 1. Components of the Bethe Hessian estimator for the planted partition in a graph of 7 755 vertices and 13 348 edges generated according to the SBM and then reduced to its 2-core (left). In the right plot we have used the same graph, where 2 cliques of sizes 3 and 5 has been added. A difference of only 13 edges not generated according to the SBM makes the spectral method completely useless for the detection.

ensemble defined above. Given that the underlying assumption in the SBM is that the graph is locally tree-like, the addition of few cliques may drastically deteriorate the performance of the algorithm based on the Bethe Hessian. Just to give you an idea of how fragile may be the Bethe Hessian based estimator $\hat{\boldsymbol{x}}^{\text{BH}}(G)$ we show in Fig. 1 its components for 2 almost identical graphs. The graph in the left plot is a typical graph from the SBM ensemble with $q = 2$, $c = 3$, $\lambda = 1.2$ and $n = 10^4$ (actually we focus on the 2-core of the graph, for reasons explained below, that has 7 755 vertices and 13 348 edges in this case): the resulting overlap with the planted partition is $Q = 0.59$, in agreement with previous studies [10]. The graph in the right plot is exactly the same graph in the left plot with the addition of *only two cliques* of sizes 3 and 5 (that is just 13 more edges!), but the ability to infer the planted partition from the spectrum of the Bethe Hessian is completely lost, as a consequence of the eigenvector localization. Indeed the resulting overlap is $Q = 0.01$.

Robustness is an important requirement, since no generative model is exact in applications. To this aim, converting the inference problem to an optimization problem is welcome. Obviously the result will depend on the objective function to maximize. One of the most used objective functions for detecting communities in real world graphs is the modularity [11, 12] defined as

$$\sum_{g=1}^{q} \sum_{i,j \in g} \left(A_{ij} - \frac{d_i d_j}{2m} \right) ,$$

where the index g runs over the clusters of vertices in the partition. Modularity measures the excess of intra-cluster connections with respect to a random assignment. Maximizing the modularity is not an easy job, especially because one can find several phase transition in the space of partitions [13, 14], which are likely to affect the the maximum likelihood problem.

In the case of $q = 2$ groups of equal size, the problem of detecting the graph partition is equivalent to the minimum bisection problem, and the objective function to be minimized is just the cut size. Consequently the maximum-likelihood estimator is given by

$$\hat{\boldsymbol{x}}^{\text{ML}}(G) = \underset{\boldsymbol{x} \in \{+1,-1\}^n}{\text{argmax}} \left(\sum_{(ij) \in E} x_i x_j \middle| \sum_i x_i = 0 \right) . \tag{1}$$

However computing the estimator in (1) is a hard problem (NP-complete and non polynomial time approximable in the worst case [15]), because the function to be maximized may have several local maxima, that trap the optimization algorithms.

In order to approximate this problem with a more tractable version it is customary to relax it as semidefinite programming (SDP) over the set of $n \times n$ real and symmetric matrices $\boldsymbol{C}$ [16]

$$\text{maximize} \sum_{i,j} A_{ij} C_{ij} \qquad \text{subject to } \boldsymbol{C} \succeq 0\,,\ C_{ii} = 1\,,\ \sum_j C_{ij} = 0\ \forall i\,. \tag{2}$$

The positive semidefinite condition, $\boldsymbol{C} \succeq 0$, requires all the eigenvalues of $\boldsymbol{C}$ to be non-negative and makes the feasible set convex, thus ensuring the existence of a tractable maximizer of (2).

The maximum-likelihood problem (1) is recovered from the formulation in (2) by enforcing the matrix $\boldsymbol{C}$ to be of rank 1: $C_{ij} = x_i x_j$. So, in general, willing to solve the non-convex problem in (2) with rank 1 matrices, one may search for a solution to the convex problem with generic rank n matrices, and then project back this solution to the space of rank 1 matrices.

A convenient way to represent a $n \times n$ real and symmetric positive semidefinite matrix of rank m is to consider it as a correlation matrix between n real vectors of m components each:

$$C_{ij} = \underline{x}_i \cdot \underline{x}_j\,, \quad \text{with } \underline{x}_i \in \mathbb{R}^m\,,\ \|\underline{x}_i\|^2 = \underline{x}_i \cdot \underline{x}_i = 1\,. \tag{3}$$

3. Our community detection algorithm and it performances

In order to solve problem (2) over the set of rank m matrices we have recently proposed the following procedure [1]. First of all search for a configuration of the n unit-length m-components vectors $\underline{x}_i \in \mathbb{R}^m$, $\|\underline{x}_i\| = 1$, optimizing the following objective function

$$\text{maximize} \sum_{(ij)\in E} \underline{x}_i \cdot \underline{x}_j\,, \qquad \text{subject to } \sum_i \underline{x}_i = \underline{0}\,. \tag{4}$$

Let us call $\underline{\boldsymbol{x}}^* = \{\underline{x}_1^*, \ldots, \underline{x}_n^*\}$ the maximizer. To project back the maximizer to a vector of n reals, we first compute the matrix $\Sigma \in \mathbb{R}^{m\times m}$ measuring the correlations among the m components of the maximizer averaged over the entire graph

$$\Sigma_{jk} = \frac{1}{n} \sum_{i=1}^{n} (\underline{x}_i^*)_j (\underline{x}_i^*)_k\,, \tag{5}$$

whose principal component we call $\underline{v}_1$, and then we project each $\underline{x}_i$ over $\underline{v}_1$. Thus the rank m SDP estimator $\hat{\boldsymbol{x}}^{\text{SDP}}(G)$ has components

$$\hat{x}_i^{\text{SDP}}(G) = \text{sign}(\underline{x}_i \cdot \underline{v}_1)\,. \tag{6}$$

As before we measure the success of our detection algorithm by computing the overlap with respect to the planted partition

$$Q^{\text{SDP}} = \frac{1}{n} |\langle \hat{\boldsymbol{x}}^{\text{SDP}}, \boldsymbol{x}_0 \rangle|\,.$$

In Ref. [1] we have shown that the above algorithm, in the $m \to \infty$ limit, is optimal for synchronization problems defined on dense graphs and almost optimal for solving the SBM. By 'almost optimal' we mean that the overlap Q^{SDP} shows a phase transition at λ_c^{SDP} slightly larger than the optimal $\lambda_c = 1$: for example for $c = 3$ we have $\lambda_c^{\text{SDP}} \simeq 1.017$, with $Q^{\text{SDP}} > 0$ for $\lambda > \lambda_c^{\text{SDP}}$. Just for comparison, we remind that for $c = 3$ the spectral methods based on the Laplacian are unable to detect the planted partition as long as $\lambda < \sqrt{3/2} \simeq 1.22$.

The algorithm we use to find the maximizer in (4) is nothing but a zero temperature dynamics for a model with m-component spins $\underline{x}_i$ placed on the vertices of the graph G, with the addition of an external field that self adapt in order to keep the global magnetization null. In practice at each step of the algorithm we update spins in a random order aligning each spin to its local field

$$\underline{x}_i^{(t+1)} = \frac{\sum_{j:(ij)\in E} \underline{x}_j^{(t)} - \underline{M}(t)}{\| \sum_{j:(ij)\in E} \underline{x}_j^{(t)} - \underline{M}(t)\|},$$

with the global magnetization being $\underline{M}(t) = \sum_i \underline{x}_i$. We always check that the global magnetization becomes very small at large times, $\lim_{t\to\infty} \underline{M}(t) = \underline{M}(\infty) \ll 1$. As a stopping criterion we check the largest variation in a spin during the last step, $\Delta_{\max} = \max_i \|\underline{x}_i^{(t+1)} - \underline{x}_i^{(t)}\|$, and we stop when $\Delta_{\max} < \varepsilon$, with $\varepsilon = 10^{-3}$ or 10^{-4} (the specific value is rather irrelevant, being the results independent of ε as long as $\varepsilon \lesssim 10^{-3}$). We call t_{conv} the number of steps required to meet the stopping criterion.

The main parameter in this algorithm is the number of components m, and we expect the behavior of the algorithm to depend strongly on m at least close to the critical point $\lambda_c = 1$. Indeed for $m = 1$ the maximization of the likelihood is a NP-hard problem. We expect a greedy dynamics, as the one we are using, to get stuck in some local maxima, while for $m \to \infty$ the problem is convex and thus the greedy dynamics should be able to reach the maximum.

Given that each step of the dynamics requires $O(m\,n)$ operations, an interesting aspect to study is the minimal value of m that allows the algorithm to get close to the $m = \infty$ maximizer, such that the quasi optimal solution of Ref. [1] can then be obtained through the projection in (6). On general grounds, since the SDP in (2) has n constraints, one can show [17] that for $m \geq 2\sqrt{n}$ the objective function in (4) has no local maxima, which are not global maxima. However, we expect $\sqrt{n}$ to be a loose upper bound for the optimal value of m on random instances. Indeed, for $\lambda = 0$ the model we are studying is close to an m-components spin glass on a random graph, whose ground state (i.e. the maximizer) has a number of non-zero components (i.e. the rank) growing roughly as n^{μ}, with $1/3 \lesssim \mu < 2/5$ depending on the mean degree c [18]. From the recent work [19] we known that the maximum of the objective function on rank m solutions deviates for the corresponding $m = \infty$ value by no more than $O(1/m)$ terms. We will see that actually the convergence on the SDP estimator $\hat{\boldsymbol{x}}^{\text{SDP}}$ is much faster.

In Ref. [1] we have studied the effect of changing m in solving the SBM close to the critical value $\lambda_c = 1$. The kind of results we got are illustrated in Fig. 2 for graphs generated according to the SBM with $c = 5$ and $\lambda = 1$. We plot the histogram of the convergence times, t_{conv}, and we see that running the algorithm with $m = 20$ and $m = 40$ leads essentially to the same dynamics and the same convergence times for $n = 1000$ and $n = 2000$. On the contrary for $n = 8000$ convergence times with $m = 20$ are sensibly larger and much more disperse than with $m = 40$ (notice we are plotting $\log(t_{\text{conv}})$). Our conclusion is that the objective function we are climbing with the greedy dynamics is less smooth for $m = 20$ than for $m = 40$.

Although the dynamics get slower (i.e. t_{conv} increases) reducing the value of m, because the objective function to be maximized gets rougher, each single step of the algorithm becomes much more economic, and so it is worth asking how much the maximizer reached by running the algorithm with a “too small” value of m is informative for the goal of detecting communities.

The graphs that we use are generated according to the SBM with $c = 3$, and subsequently they are pruned by eliminating recursively dangling ends. Equivalently we reduce the graph to its 2-core, and this has the advantage of reducing the noise level, without changing the complexity of the problem. Indeed any removed tree joins the core on a single vertex and the optimal solution is to assign all tree vertices to the same cluster the core vertex belongs to.

In order to study the robustness of our approach, we add cliques according to the following rule: for each vertex, with probability p, we add a clique among all its neighbour vertices. This superimposed structure is well justified since in many real world networks, especially social networks, structures of this kind are very common (two friends of a person tend to be also friends). But from the point of view of the SBM this is an additional perturbation which may

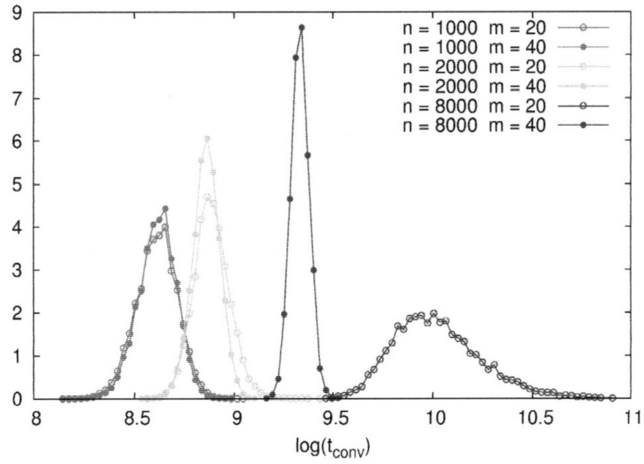

Figure 2. Distribution of convergence times for our greedy algorithm measured on many graphs generated according to the SBM with $c = 5$ and $\lambda = 1$. The typical running time grows mildly with the problem size n, but if m is not large enough the convergence time becomes much larger and sample to sample fluctuations much more severe.

degrade the performance of detecting algorithms. Indeed this is what we have shown preliminary in Fig. 1 where the graph of the right plot has been generated with a very small $p = 10^{-4}$ and still the effect on the Bethe Hessian algorithm is dramatic.

In order to get statistics on a single graph, we run the algorithm with 100 different *clones*. Each clone starts from a different random initial condition and evolve according to the greedy algorithm illustrated above. Remind that the algorithm, although greedy, is not deterministic, because variables are updated in a random order. At all the times when we take measurements we compute the SDP estimator $\hat{\boldsymbol{x}}^{\text{SDP}}$ and the corresponding overlap Q^{SDP} for each of the 100 clones.

In Fig. 3 we show the robustness of our algorithm with respect to the addition of cliques. We consider a graph generated initially with $n = 40\,000$ nodes at $\lambda = 1.1$: the resulting core has 30 597 vertices, 53 161 edges for $p = 0$ and 54 734 edges for $p = 0.01$. In Fig. 3 we plot Q^{SDP} as a function of the number of iteration of our algorithm, for $m = 16$ and $m = 64$. The horizontal lines report the value of the overlap q achieved by the Bethe Hessian algorithm, which is very close to the Bayes optimal value, according to Ref. [10]. We observe that for $p = 0$ our algorithm reaches a higher value of the overlap with respect to the Bethe Hessian result, both for $m = 16$ and $m = 64$. More importantly the addition of cliques ($p = 0.01$ case) deteriorates very little the performances of our algorithm, while the spectral method return a random guess with $Q \simeq 0$. By virtue of robustness of our algorithm, hereafter we are going to study only the $p = 0$ case.

In Fig. 4 we report the results for Q^{SDP} obtained on the same graph as in Fig. 3 with many more values of m. Error bars represent the standard deviation over the 100 clones. We observe that, while for $m = 1, 2$ the algorithm is unable to detect any significant partition, already for $m = 3$ the signal is substantial and eventually for $m \geq 8$ the algorithm achieves the optimal value for Q^{SDP}. Being the graph size quite large (30 597 vertices and 53 161 edges) is surprising that our algorithm works so nicely also for values of m as small as $m = 8$.

Let us focus on values $m \geq 8$ and let us compute the real running time, that is the wall clock time our algorithm takes to maximize the objective function. This time is not exactly proportional to mn because all the operations among the m-components vectors (e.g. sums and scalar products) can be highly optimized by modern compilers, and so we expect the real

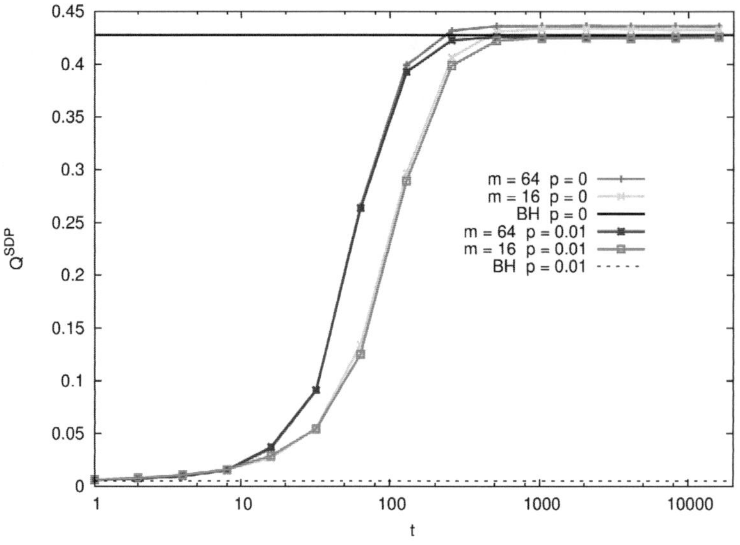

Figure 3. Overlap with the planted partition obtained from the configurations at time t in our algorithm. We report the average over the 100 clones. The comparison with the overlap achieved by the Bethe Hessian algorithm is favorable, especially when a small fraction of cliques is added, and the prediction from the Bethe Hessian is equivalent to a random guess. The algorithm run with $m = 64$ and $m = 16$ requires different running times, but achieves essentially the same inference accuracy.

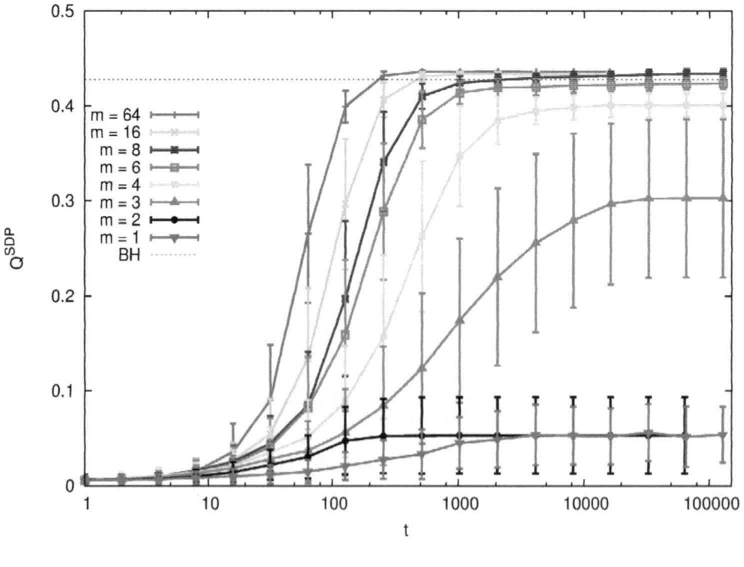

Figure 4. Dependence on m of the inferred overlap Q^{SDP} for $\lambda = 1.1$. Error bars represent standard deviation in the population of 100 clones.

running time to grow sublinearly with m. In Fig. 5 we report the growth of Q^{SDP} for 2 graphs of initial size $n = 10^5$ and $\lambda = 1.1$ (left) and $\lambda = 1.2$ (right); the sizes of their two cores are 77 250 nodes, 132 578 edges, and 77 370 nodes, 132 890 edges respectively. Error bars represent again the standard deviation over the 100 clones, while the CPU time has been measured for the

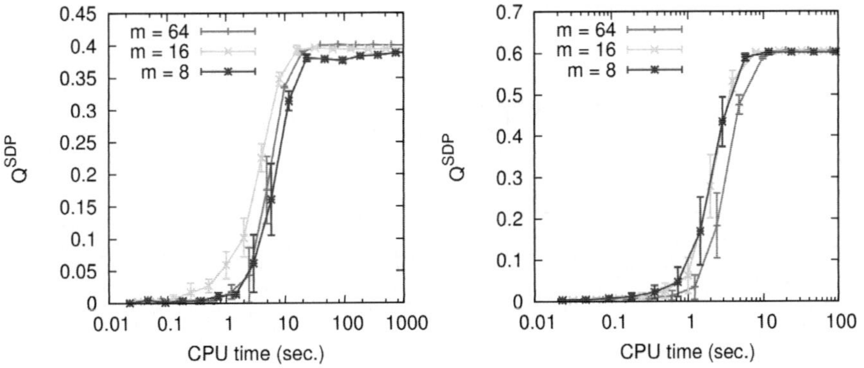

Figure 5. Inferred overlap by our SDP-based greedy algorithm as a function of the real wall clock time. The two problems have been generated according to the SBM with $n = 10^5$, $c = 3$ and $\lambda = 1.1$ (left) and $\lambda = 1.2$ (right). Times have been measured in seconds on a personal laptop with a 2 GHz Intel Core i7 processor. Dependence on m in running times is very weak.

simulation of a single clone on a personal laptop with a 2 GHz Intel Core i7 processor. Looking at Fig. 5 we notice that the differences in the real running time by varying the value of m are not very significant, and so maybe working with a moderate value of m is preferred, especially because the actual running time is very small: few seconds on a laptop to solve a problem with about 10^5 variables!

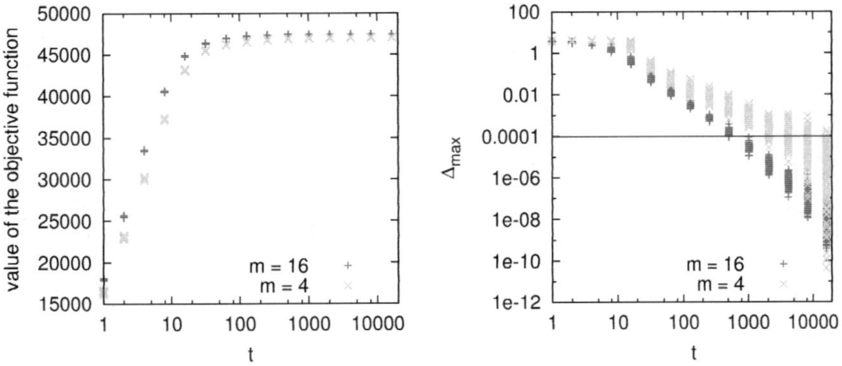

Figure 6. The objective function and the maximum variation in a variable during the last step, $\Delta_{\max}$, as a function of the number of algorithm steps. At each time we plot 100 points corresponding to different clones (although in the left panel they are hardly visible, because almost perfectly superimposed). The problem has size $n = 40\,000$ and $\lambda = 1.1$ (and is the same studied in Fig. 7).

In Figs. 3, 4 and 5 we have always considered the overlap with the planted partition; however in practice such an overlap is unknown. The only information available to us are the value of the objective function, $\Delta_{\max}$ and the distances between clones (if we run more than one clone). The first two quantities are shown in Fig. 6 for a graph generated with $n = 40\,000$ and $\lambda = 1.1$ (at each time we report the 100 measurements taken in different clones). From the previous analysis we known that $m = 16$ may be a good value, while $m = 4$ is definitely too small (see green and cyan lines in Fig. 4). However looking at the data in Fig. 6 we do not see large differences in the objective function and only the fluctuations in t_{conv}, given by the time where $\Delta_{\max} = 10^{-4}$

(blue horizontal line in the right panel), are suggesting $m = 4$ may be too small.

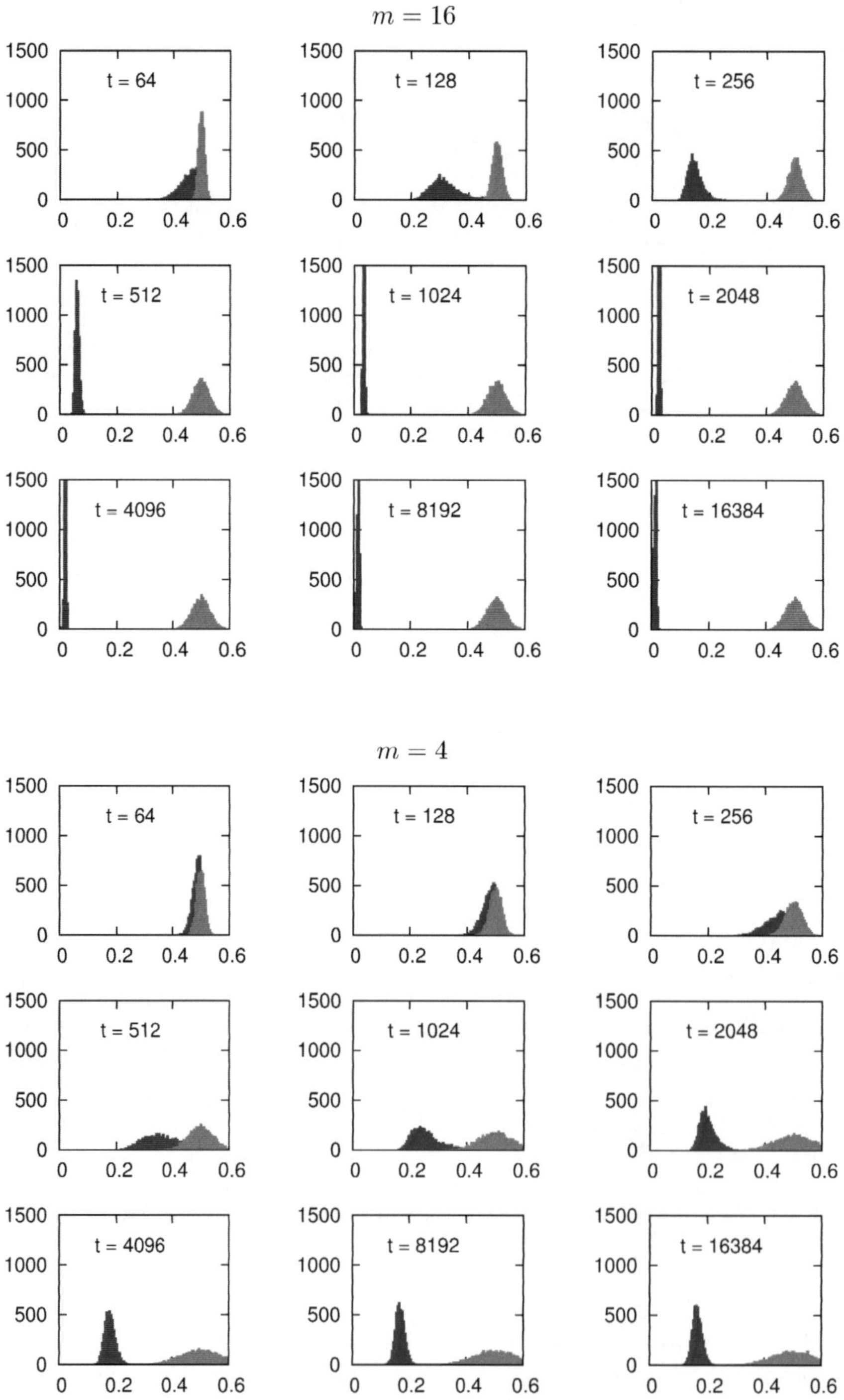

Figure 7. Histograms of the 4950 distances among the 100 clones before (red) and after (blue) the eliminating the rotational symmetry. Each panel is for a different algorithm time; upper panels are for $m = 16$ and lower ones are for $m = 4$.

Given that any clone eventually gets stuck in a maxima (either local or global), we would like to understand how likely the clones have reached the global maximum, without the need to run the algorithm with a larger value of m (that would make convergence to the global maximum easier). To this end we measure the distances between any pair of clones, by measuring

$$d = \frac{1}{2}\left(1 - \frac{1}{n}\sum_{i=1}^{n} \underline{x}_i^{(1)} \cdot \underline{x}_i^{(2)}\right),$$

where $\underline{\boldsymbol{x}}^{(1)}$ and $\underline{\boldsymbol{x}}^{(2)}$ are their configurations. The idea is that if all the clones have reached the global maximum they are all very close by, while if the objective function is too rough and they get stuck in local maxima, then their distances remain strictly positive. There is one caveat in computing straightforwardly the distance between the clones: since the objective function is invariant under a rotation $R : \mathbb{R}^m \to \mathbb{R}^m$ applied to all the n vectors and since the clones start from random configurations, we expect the clones to remain as orthogonal as possible, with a value of the distance $d \simeq 1/2$. In Fig. 7 we plot in red the histograms of the 4950 distances between the 100 clones, measured during the running of the algorithm with $m = 16$ (above) and $m = 4$ (below). Indeed, we see that the peak of the distributions remain close to $d = 1/2$ and provide no information at all about the real closeness of the clones. In order to get a meaningful information we should break the rotational invariance, by rotating the configuration of each clone and minimize their distances. We have done this in the following way: for each clone, we have computed the empirical correlation function Σ defined in (5), the corresponding principal component $\underline{v}_1$ and we have rotated the configuration such that $\underline{v}_1 \parallel (1, 0, \ldots, 0)$ and all distances are smaller than $1/2$. The resulting distances are shown in Fig. 7 with the blue histograms: it is now clear that running the algorithm with $m = 16$ makes all the clones converge to the same optimal configuration, while in the $m = 4$ case the clones get stuck in local maxima and their distances converge to strictly positive values.

4. Conclusions and perspectives

We studied how the algorithm based on SDP for community detection introduced in Ref. [1] works on very large graphs generated according to the SBM. We have shown the algorithm is very robust with respect to variations in the generative model: adding a good number of cliques only slightly degrades the algorithm performances (in contrast with spectral methods whose accuracy drops dramatically). The present work demonstrates that surprisingly small values of m may be sufficient for detecting the partition even in very large problems. Running few clones (even just 2 clones) allows one to understand whether the global maximum of the objective function is achieved. Consequently one can choose whether to rerun the algorithm with a larger value for m.

There are many possible extensions of our work. We just mention the most straightforward ones. We have made some preliminary runs of the algorithm with $q = 3$ and observed that 3 clusters of equal (or almost equal) size can be detected with a very good accuracy. However a systematic study of the performances of our algorithm for $q > 2$ is still missing. Given that for $q > 4$ the SBM undergoes a first order phase transition, it would be very interesting to study how the SDP relaxation in general and our algorithm in particular perform in detecting communities in that case. We think our algorithm can be also adapted to detect communities of different sizes by modifying the constraint of zero global magnetization; this modification is relevant to study real world problems.

Acknowledgments

A.J. and A.M. were partially supported by NSF grants CCF-1319979 and DMS-1106627 and the AFOSR grant FA9550-13-1-0036.

References

[1] Javanmard A, Montanari A and Ricci-Tersenghi F 2015 *Submitted*
[2] Fortunato S 2010 *Physics Reports* **486** 75–174
[3] Holland P W, Laskey K B and Leinhardt S 1983 *Social networks* **5** 109–137
[4] Decelle A, Krzakala F, Moore C and Zdeborová L 2011 *Physical Review E* **84** 066106
[5] Mossel E, Neeman J and Sly A 2013 *arXiv preprint arXiv:1311.4115*
[6] Massoulié L 2014 *Proceedings of the 46th Annual ACM Symposium on Theory of Computing* (ACM) 694–703
[7] Decelle A, Krzakala F, Moore C and Zdeborová L 2011 *Physical Review Letters* **107** 065701
[8] Kawamoto T and Kabashima Y 2015 *Phys. Rev. E* **91**(6) 062803
[9] Krzakala F, Moore C, Mossel E, Neeman J, Sly A, Zdeborová L and Zhang P 2013 *Proceedings of the National Academy of Sciences* **110** 20935–20940
[10] Saade A, Krzakala F and Zdeborová L 2014 *Advances in Neural Information Processing Systems* 406–414
[11] Clauset A, Newman M E and Moore C 2004 *Physical review E* **70** 066111
[12] Newman M E 2006 *Proceedings of the National Academy of Sciences* **103** 8577–8582
[13] Zhang P and Moore C 2014 *Proceedings of the National Academy of Sciences* **111** 18144–18149
[14] Schülke C and Ricci-Tersenghi F 2015 *Physical Review E* **92** 042804
[15] Arora S, Berger E, Hazan E, Kindler G and Safra M 2005 *Foundations of Computer Science, 2005. FOCS 2005. 46th Annual IEEE Symposium on* (IEEE) 206–215
[16] Goemans M X and Williamson D P 1995 *Journal of the ACM (JACM)* **42** 1115–1145
[17] Burer S and Monteiro R D 2003 *Mathematical Programming* **95** 329–357
[18] Braun A and Aspelmeier T 2006 *Physical Review B* **74** 144205
[19] Montanari A and Sen S 2015 *arXiv preprint arXiv:1504.05910*

Journal of Physics: Conference Series **699** (2016) 012016 doi:10.1088/1742-6596/699/1/012016

Exact Algorithms for Isotonic Regression and Related

Yao-Liang Yu and Eric P. Xing
Machine Learning Department, Carnegie Mellon University, Pittsburgh, PA, 15213, USA
E-mail: {yaoliang, epxing}@cs.cmu.edu

Abstract. Statistical estimation under order restrictions, also known as isotonic regression, has been extensively studied, with many important practical applications. The same order restrictions also appear *implicitly* in sparse estimation, where intuitively we should shrink variables starting from smaller ones. Inspired by the achievements in both fields, we first propose the GPAV algorithm for solving problems with order restrictions. We study its theoretical properties, present an online linear time implementation, and prove a converse theorem to pinpoint the exact correctness condition. When specialized to the proximity operator of an order restricted regularization function, GPAV recovers, as special cases, many existing algorithms, and also leads to many new extensions that even involve nonconvex functions.

1. Introduction

As (data) scientists, how do we argue global warming is not bogus? One established way is to perform statistical hypothesis testing using say the likelihood ratio: Given a collection of everyday temperatures over a certain time period, we can estimate the yearly mean temperatures under the null hypothesis that global warming is not happening (*i.e.*, constant mean), and also the mean temperatures under the alternative hypothesis that global warming is indeed happening (*i.e.*, monotonically increasing mean). Adopting an appropriate statistical temperature model, the latter would require maximizing the likelihood function subject to the *isotonic* constraint, *i.e.*, the estimated yearly mean temperatures should be monotonically increasing. This problem is widely known as isotonic regression and has been extensively studied in the 1970s, culminating in the excellent books [1, 2], with many illuminating results and applications. See §2 for some examples.

More generally, statistical estimation under shape constraints (such as monotonicity, convexity) has been an important topic in statistics since the seminal works [3, 4]. A large part of the theory focuses on studying the asymptotic distribution of estimators (parametric or nonparametric) under shape constraint, which are generally different from those obtained without shape constraint and are rather difficult to derive. An excellent summary in this direction is the recent book [5]. Computationally, how to obtain the estimator under shape constraint has also attracted lots of attention. The pool-adjacent-violators (PAV) algorithm [6], essentially the Euclidean projection onto the isotonic cone, is one of the early achievements in this field and is still widely used today.

Order constraints also implicitly appear in sparse estimation problems. If the true estimator is sparse or can be well approximated by a sparse vector, it would be beneficial to "sparsify" our estimator, effectively reducing model complexity hence potentially improving generalization. "Sparsification" is usually executed through the *proximity operator* of carefully designed sparse

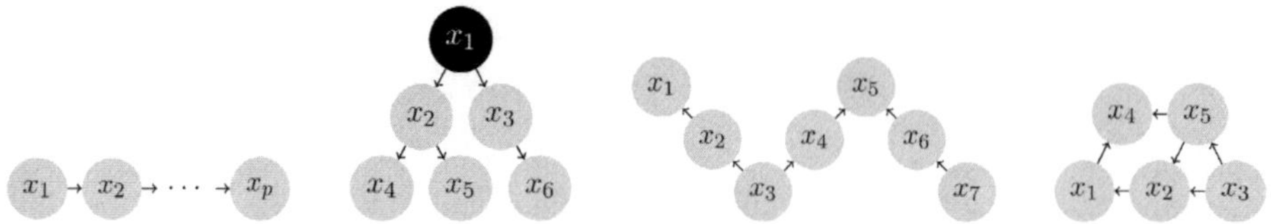

(a) Chain. (b) Rooted tree. (c) "Soft" chain. (d) DAG.

Figure 1: Partial orders induced by a chain, a rooted tree, a "soft" chain, and a DAG.

regularizers, among which the ℓ_1 norm, along with its soft-shrinkage operator, is perhaps the most famous. Intuitively, if we were to shrink a subset of variables to zero, we should start with the smaller ones, since larger estimates are more likely to be genuinely nonzero. Indeed, almost all sparse regularizers that we are aware of obey this principle and *implicitly* enforce an order restriction on the output. More recent sparse regularizers, such as the k-support norm [7], OSCAR [8], and the sorted ℓ_1 norm [9] even made the order restriction explicit. Computationally, the order constraint complicates the computation of the proximity operator and is addressed case by case in many recent works [7, 9–13]. The connection to the field of isotonic regression, in particular, the PAV algorithm, is not always explored.

The main goal of this work is to provide a unified treatment of the various computational algorithms involving order constraints. Based on a surprisingly simple yet general result (Theorem 1 below) we develop in §3 the generalized pool-adjacent-violators (GPAV) algorithm, which is a strict generalization of PAV, hence making the connection to the field of isotonic regression explicit. We also prove a converse theorem that pinpoints exactly when GPAV is provably correct, which, to our best knowledge, is the first result of its kind. An online implementation is provided in §4 to bring down the time complexity from quadratic to linear. Then, in §5 we specialize the online GPAV to the problem of computing the proximity operator of functions involving order restrictions. One nice feature of online GPAV is that it can handle k-piecewise convex functions (that are generally nonconvex). In particular, it also computes the proximity operator for all 2-convex functions *exactly, globally* in *linear time.*

In this work we focus exclusively on the linear model (namely the functions g_i in Equation (1) below are univariate). However, extensions to additive models (where g_i can be multivariate functions of a specific form) can be pursued as in the work of Bacchetti [14] and Fang and Meinshausen [15].

2. Problem definition

We are interested in solving the following problem:

$$\min_{\mathbf{x}\in\mathcal{K}} \sum_{i=1}^{p} g_i(x_i), \tag{1}$$

where $g_i : \mathbb{R} \to \mathbb{R} \cup \{\infty\}$ are *univariate* functions and $\mathcal{K}$ is an isotonic *cone* induced by some partial order $\preceq$ on the set $\{1, \ldots, p\}$:

$$\mathcal{K} := \{\mathbf{x} \in \mathbb{R}^p : x_i \leq x_j \text{ for all comparable pairs } i \preceq j\}. \tag{2}$$

Equivalently, since any partial order can be represented as a directed acyclic graph (DAG), we can define $\mathcal{K}$ through a DAG too. In particular, the following special case where $\mathcal{K}$ is induced by a total order (equivalently a chain graph) motivated the development historically:

$$\mathcal{K}_c := \{\mathbf{x} \in \mathbb{R}^p : x_1 \geq x_2 \geq \cdots \geq x_p\}. \tag{3}$$

There are of course many other possibilities, see Figure 1. We will address the first two cases in Figure 1 and leave the other two cases for future work.

The order constraint modeled by the isotonic cone in (2) has many practical applications. Let us mention a few for the sake of motivation.

Example 1 (Exponential Family Parameter Estimation [1, 6, 16, 17]). *Consider the exponential family (w.r.t. some dominating measure)* $p(x|\theta) \propto \exp(T(x)\theta - f(\theta))$*, where* $T(x)$ *is the sufficient statistic and* $f(\cdot)$ *is the log-partition function. Fix the unknown parameters* θ*. For estimation we take* n_i *observations for each parameter:* $x_{i,j}, i = 1, \ldots, p, j = 1, \ldots, n_i$*. Then the (restricted) maximum likelihood estimator is:*

$$\max_{\theta \in \mathcal{K}} \sum_{i=1}^{p} n_i [\theta_i \bar{T}(\mathbf{x}_i) - f(\theta_i)], \tag{4}$$

where $\bar{T}(\mathbf{x}_i) = \frac{1}{n_i} \sum_{j=1}^{n_i} T(x_{ij})$*. Surprisingly, as shown in [1], the solution depends only mildly on the log-partition function* f*. We refer to the excellent book of Barlow et al. [1] for many eye-opening applications under this setting.*

Example 2 (Inventory Management [18–21]). *Consider an inventory system where a single product is produced and assembled in multiple sites and stages. Once the product is completely assembled it can be sold to the customer which has certain demands. There is a production cost and a storage cost in each site, and each site depends on its "ancestor" sites, that is, in order for site* i *to produce its part it will need a certain amount of parts from say site* j*. Obviously, this quantity cannot exceed the storage in site* j*, hence creating a natural order restriction. An optimal inventory policy can then be formulated and found by solving problem* (1)*, where the functions* g_i *depend on the production cost, storage cost, and demand.*

Example 3 (Sparsity [7–13]). *Sparsity has been recognized as one of the key structures that allow statistical inference in high dimensions, and is usually forced through a penalty function such as the* ℓ_1 *norm. Consider the ideal orthogonal design case:*

$$\min_{\mathbf{x} \in \mathbb{R}^p} \tfrac{1}{2}\|\mathbf{x} - \mathbf{y}\|^2 + f(\mathbf{x}), \tag{5}$$

i.e. *we are trying to sparsify the input vector* $\mathbf{y}$*. Intuitively, a bigger entry should be thresholded to zero only after the smaller entries, because the former is more likely to be genuinely nonzero than the latter. This intuition in turn puts an* implicit *order restriction on the penalty function* f*. Early penalty functions like the* ℓ_1 *norm are simply permutation-invariant, but more recent works [7, 8, 10, 13] have made the order restriction explicit. A similar setting was also considered in [9] for controlling the false discovery rate in multiple hypothesis testing problems. However, the connection to the isotonic cone is not always recognized or exploited.*

Problem (1) is very special in the sense that its objective function is separable in each coordinate, however, the isotonic cone $\mathcal{K}$ couples all coordinates. If the functions g_i are convex, (1) can be solved using for instance the iterative projected gradient algorithm, provided that we know the Euclidean projection onto $\mathcal{K}$. Instead, the GPAV algorithm we present below is based on a different idea (and will prove more useful later). First note that (1) would be easy to solve if we ignore the isotonic constraint $\mathbf{x} \in \mathcal{K}$: we simply minimize the univariate functions g_i separately (or even in parallel). Of course, the resulting *unconstrained* minimizer, denoted generically as $\mathbf{z}$ throughout, may not satisfy the isotonic constraint. The GPAV algorithm then "pools" the functions g_i in (1) so that next time the unconstrained minimizer $\mathbf{z}$ violates *strictly* less order constraints (if any). The "pooling" is intelligently performed so that the (constrained) minimum value of (1) is still maintained. Therefore after at most finitely many steps the unconstrained minimizer $\mathbf{z}$ will automatically satisfy the isotonic constraint hence be a *bona fide* minimizer.

Let us point out the convenience of working with general abstract functions g_i. Suppose we have an additional nonnegative constraint in (1), *i.e.*, $x_1 \geq x_2 \geq \cdots \geq x_p \geq 0$, which is clearly equivalent as $x_1 \geq x_2 \geq \cdots \geq x_p, x_p \geq 0$. Thus, upon redefining $g_p(x_p) = \infty$ if $x_p < 0$ (*i.e.*, restricting the effective domain of g_p to $\mathbb{R}_+$), we reduce back to (1), but this time without any *explicit* nonnegative constraint. Clearly, other interval constraints under any isotonic cone can be dealt with similarly.

3. The generalized pool-adjacent-violator algorithm

We present in this section the generalized pool-adjacent-violators (GPAV) algorithm. The presentation is kept general and abstract so that we can see the idea more clearly, without being distracted by the non-essential details. The generality will be felt when we apply the results in §5 to recover various existing results and also to uncover many new ones.

As we mentioned in the previous subsection, GPAV "pools" the functions g_i so that the unconstrained minimizer $\mathbf{z}$ will satisfy the isotonic constraint eventually. Take any unconstrained minimizer[1] $\mathbf{z}$ of the separable function $\sum_{i=1}^{p} g_i(x_i)$, and consider the violator set

$$\{j : z_j \leq z_i \text{ for some } i \text{ that is an immediate predecessor of } j \text{ w.r.t. the partial order } \mathcal{K}\}. \quad (6)$$

Order the violators by their z-values. We have the following result for pooling:

Theorem 1. *Let $\mathcal{K}$ be induced by a rooted tree (Figure 1b). Suppose for all i the closed function $g_i : \mathbb{R} \to \mathbb{R} \cup \{\infty\}$ is convex[2]. Consider any minimum violator z_j and its predecessor z_i, then there exists a constrained minimizer*

$$\mathbf{y}^\star \in \arg\min_{\mathbf{x} \in \mathcal{K}} \sum_{i=1}^{p} g_i(x_i) \quad (7)$$

such that $y_i^\star = y_j^\star$.

Note that in a rooted tree, every node has exactly one predecessor (except the root which has none). This condition is needed as otherwise it can be tricky to pool which predecessor with the violator. Therefore, whenever the *unconstrained* minimizer $\mathbf{z}$ violates any order constraint, we obtain crucial information about the *constrained* minimizer $\mathbf{y}^\star$. Thanks to separability, the unconstrained minimizer $\mathbf{z}$ may be found cheaply (perhaps even in closed-form).

Theorem 1 is known in special cases: It was first discovered in Ayer et al. [6] where $\mathcal{K} = \mathcal{K}_c$ is induced by a chain (a special rooted tree) and each g_i is quadratic. Strömberg [22] considered the same chain order but allowed g_i to be arbitrary convex functions, see also [23, 24]. W. A. Thompson [17] considered the general rooted tree order but only with specific g_i functions. Our proof (omitted here) also differs substantially from [17]. Orders induced by an arbitrary DAG has also been considered [*e.g.* 1, 2], but the pooling strategies are very complicated, sometimes with unknown complexity.

Theorem 1 immediately suggests an efficient algorithm for solving (7), when $\mathcal{K}$ is induced by a rooted tree. In each iteration we find a minimum violator and perform the pooling, namely, set $g_i(\cdot) \leftarrow g_j(\cdot) \leftarrow \frac{1}{2}[g_i(\cdot) + g_j(\cdot)]$. Since there are at most $O(p)$ violators, and each pooling will reduce the problem size by 1, the total complexity is $O(p^2)$, assuming that finding the unconstrained minimizer $\mathbf{z}$ takes $O(p)$ time. Note that if all functions g_i are quadratic, then it is possible to use advanced data structures to improve the complexity to $O(p \log p)$ [25].

We complement Theorem 1 with a strong converse, which, to the best of our knowledge, is the first of its kind. The choices of the functions g_i below are motivated by the applications in §5.

[1] The existence of minimizers is always assumed in this work, for simplicity.
[2] In fact, we only need the weaker quasiconvexity (unimodality) here.

Theorem 2. *Fix $\lambda > \mathbf{0}$ and consider applying Theorem 1 to solve (1) with $g_i(x) := w_i(x - y_i)^2 + \lambda_i f(x)$. If it always leads to a correct minimizer for any $\mathbf{w} \in \mathbb{R}^p_+, \mathbf{y} \in \mathbb{R}^p$, $p \geq 2$, then the univariate function f must be convex.*

Thus, for nonconvex functions g_i, we cannot directly apply GPAV for solving (1), but see §5 for some exceptions.

4. Linear time algorithm for the chain graph

When the isotonic cone $\mathcal{K}$ is induced by a chain (a special rooted tree), the complexity automatically reduces from $O(p^2)$ for the general rooted tree order to $O(p)$, which is clearly the best possible. This is because in the chain case finding the minimum violator among all violators can be incrementally done in $O(1)$ time. Here we give an online view of this known fact, based on which we will extend the algorithm to piecewise convex functions in §5.

The idea is very simple: we pretend that the input functions $g_1, \ldots, g_p$ are revealed to us one by one, but each time we must solve the current *constrained* problem before the next function is revealed.

To start, we get the first function g_1 and we solve $u_1 \in \arg\min_x g_1(x)$ in $O(1)$ time, completing the first step and enabling the next function g_2 to be revealed. Now suppose we have found[3]

$$(u_1, \ldots, u_j) \in \underset{x_1 \leq \cdots \leq x_j}{\arg\min} \sum_{i=1}^{j} g_i(x_i), \tag{8}$$

and we have kept a partition $0 = t_0 < t_1 < \ldots < t_{\ell_j} = j$ of the set $\{0, 1, \ldots, j\}$ so that for each $s = 1, \ldots, \ell_j$, $u_{t_{s-1}+1} = \cdots = u_{t_s}$. Then the next function g_{j+1} is revealed and we need to solve

$$\underset{x_1 \leq \cdots \leq x_{j+1}}{\arg\min} \sum_{i=1}^{j+1} g_i(x_i). \tag{9}$$

The key is that we need *not* solve (9) from scratch. Indeed, after finding the unconstrained candidate

$$u_{j+1} \in \underset{x}{\arg\min} \quad g_{j+1}(x) \tag{10}$$

in $O(1)$ time, we realize that, augmented with our previous solution $(u_1, \ldots, u_j)$, $\mathbf{u} := (u_1, \ldots, u_{j+1})$ is in fact obtainable from applying the pooling procedures suggested in Theorem 1 to the problem (9). Of course, $\mathbf{u}$ need not be the minimizer of (9) since the (only new) order constraint $u_j \leq u_{j+1}$ may not be satisfied yet. So we continue applying Theorem 1 on top of $\mathbf{u}$ to finish the business (picking violators in any order we like). Inductively, after the p-th function is revealed, we will have $\mathbf{u}$ that is a minimizer of (7).

The partition list $0 = t_0 < t_1 < \ldots < t_{\ell_j} = j$ is maintained so that we can resume and complete the j-th round in say $O(\gamma_{j+1})$ time, instead of the brute-force $O(j)$ time, where γ_{j+1} is the number of poolings we performed in step j. Indeed, all terms fall into the interval $[t_{s-1}+1, t_s]$ have been pulled together hence require only one representative. More crucially, we always have the inequality

$$\ell_{j+1} \leq \ell_j - \gamma_{j+1} + 1, \text{ and } \gamma_{j+1} \leq \ell_j. \tag{11}$$

[3] If not for notational clarity, we should put a superscript on u_j, reflecting its possible change in later iterations than j.

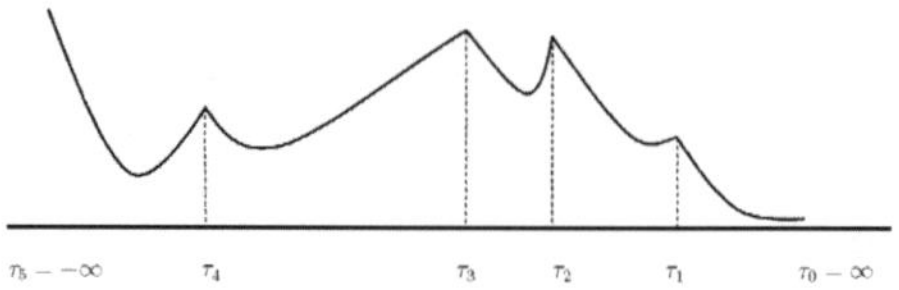

Figure 2: An example k-convex function.

Basically, this means that if we spend a lot of time (large γ_{j+1}) in pulling components in the j-th iteration, then we would have few components (small ℓ_{j+1}) to work with in the $(j+1)$-th iteration, which in turn bounds the time spent in the $(j+1)$-th iteration. Simply put, previous hard work eventually pays out in the future, which is the essence of amortized complexity analysis [26]. After the last function g_p is handled, we can count the total time consumed, using the bound in (11):

$$O\Big(\sum_{i=1}^{p} \gamma_{j+1}\Big) \le O\Big(\sum_{i=1}^{p} (\ell_j - \ell_{j+1} + 1)\Big) = O(p). \tag{12}$$

Perhaps more importantly, we observe that the above online update not only solves (7) in linear time, it actually does much more: it simultaneously solves all the subproblems (8) for all $j = 1, \dots, p$ in linear time. Let us create a vector $\mathsf{r} \in \mathbb{R}^p$ to record the minimum value in (8). Similarly, we create a vector $\mathsf{l} \in \mathbb{R}^p$ that records in linear time the minimum values in the similar subproblems:

$$(v_j, \dots, v_p) \in \operatorname*{arg\,min}_{x_j \le \dots \le x_p} \sum_{i=j}^{p} g_i(x_i). \tag{13}$$

Then these two records will allow us to apply Theorem 1 on even piecewise convex functions g_i (that need not be convex globally). This will be detailed below.

5. Extension to piece-wise convex functions

In this section we apply the online implementation in §4 to the following problem:

$$\min_{\mathbf{x} \in \mathcal{K}_c} \sum_{i=1}^{p} w_i x_i^2 - m_i x_i + \lambda_i f(x_i), \tag{14}$$

where $f : \mathbb{R} \to \mathbb{R} \cup \{\infty\}$ is some univariate closed function, $\mathbf{w}, \boldsymbol{\lambda} \in \mathbb{R}_+^p$, and $\mathcal{K}_c$ is induced by a total order (see Figure 1a). Note that by setting $m_i = 2w_i y_i, \lambda_i \equiv \lambda$ we recover the familiar problem $\min_{\mathbf{x} \in \mathcal{K}_c} \sum_{i=1}^{p} w_i (x_i - y_i)^2 + \lambda f(x_i)$ —a constrained proximity operator of f. Setting $f \equiv 0$ recovers the isotonic regression problem in [1].

We can actually allow the function f to be nonconvex. Let us recall:

Definition 1 (k-convexity). *$f : \mathbb{R} \to \mathbb{R} \cup \{\infty\}$ is called k-convex if there exists $\infty = \tau_0 \ge \tau_1 \ge \cdots \ge \tau_k = -\infty$ such that f restricted to $[\tau_i, \tau_{i+1}]$ is convex for all i.*

Clearly, 1-convexity corresponds to the usual convexity. With $k > 1$, we can start to generate many interesting nonconvex functions, see Figure 2 for an illustration.

Example 4. *The capped absolute function [e.g. 27, 28] $|x|_\rho = \min\{|x|/\rho, 1\}$ is 2-convex (when restricted to $\mathbb{R}_+$) with $\tau_1 = \rho$. Note that $\rho|x|_\rho \to |x|$ when $\rho \to \infty$ while $|x|_\rho \to \mathbb{1}_{x \ne 0}$ when $\rho \to 0$. More generally, for any convex function f, we can define its capped cousin $f_\rho = \min\{f/\rho, 1\}$, which is 2-convex with $\tau_1 = \rho$. The capped functions are preferred in sparse estimation problems for their ability to reduce estimation bias, as compared to the convex alternatives [27, 28].*

One attractive property of k-convex functions is its computational tractability: we can chop it into convex pieces and consider each piece in turns. Indeed, if in (14) the function f is k-convex, with $\tau_1, \ldots, \tau_k$ being given, we can then consider all possible partitions $0 = j_0 < j_1 < \cdots < j_k = p$. For all $s = 1, \ldots, k$ and for all $j_{s-1} + 1 \leq i \leq j_s$, by defining

$$f_i(x_i) = \begin{cases} \lambda_i f(x_i), & \tau_{s-1} \geq x_i \geq \tau_s \\ \infty, & \text{otherwise} \end{cases}, \tag{15}$$

we can "reduce" the nonconvex problem (14) with a k-convex f into the more general problem (1) where all g_i are convex. Since the online implementation in Section 4 solves (1) in linear time, we can simply enumerate all possible partitions $0 = j_0 < j_1 < \cdots < j_k = p$, each yielding a convex problem in the form of (1). Finally, taking the minimum among these $\binom{p}{k-1}$ similar problems yields the (global) minimizer of (14). If k is a constant then we will have a polynomial time algorithm for solving the nonconvex problem (14) (where nonconvexity comes form f). Note that any function can be (uniformly) approximated by a k-convex function. Intuitively, the more nonconvex f is, i.e. a bigger k is needed for a decent approximation, the more time-consuming it is computationally, i.e., more subproblems to be considered (on the order of p^{k-1}). While it may be possible to prune out many of these subproblems, we will simply restrict to $k \leq 2$ below, a choice motivated by Example 4.

Let f be a 2-convex function from now on, with $\tau_1 = \tau$, i.e., f is separately convex on $(-\infty, \tau]$ and $[\tau, \infty)$ but need not be convex on $\mathbb{R}$, see e.g. Example 4. Then, to solve our main problem in (14), we need only find the optimal split

$$x_1 \geq \cdots \geq x_j \geq \tau \geq x_{j+1} \geq \cdots \geq x_p, \tag{16}$$

among all possible positions of τ (namely j). Moreover, given the position of τ, we can convert (14) into two *unrelated* problems in the form of (1), which we already have a linear time algorithm in §4:

$$\operatorname*{arg\,min}_{x_1 \geq \cdots \geq x_j} \sum_{i=1}^{j} w_i x_i^2 - m_i x_i + f_i(x_i), \qquad \operatorname*{arg\,min}_{x_{j+1} \geq \cdots \geq x_p} \sum_{i=j+1}^{p} w_i x_i^2 - m_i x_i + h_i(x_i), \tag{17}$$

where the functions f_i and h_i are defined as:

$$f_i(x_i) = \begin{cases} \lambda_i f(x_i), & x_i \geq \tau \\ \infty, & \text{otherwise} \end{cases}, \qquad h_i(x_i) = \begin{cases} \lambda_i f(x_i), & x_i \leq \tau \\ \infty, & \text{otherwise} \end{cases}. \tag{18}$$

Note that both f_i and h_i are convex thanks to the 2-convexity of f, and more importantly they do not depend on the position j. In other words, applying the online implementation in §4 we would be able to find the minimal values in (17) in linear time for all j simultaneously. These are exactly the records l and r we mentioned at the end of §4. Then the optimal position for τ in (16) is given as

$$j^\star \in \operatorname*{arg\,min}_{0 \leq t \leq p} \ \mathsf{l}_t + \mathsf{r}_{p-t}, \tag{19}$$

where $\mathsf{l}_0 = \mathsf{r}_0 = 0$. After having $j^\star$ we can just re-solve (17) with j replaced by $j^\star$. It is clear that the overall time is still linear $O(p)$, a big improvement compared to the naive $\binom{p}{1} \cdot O(p) = O(p^2)$ time (obtained by calling the online implementation p times).

The above acceleration trick can be generalized to handle a k-convex function f in (14), with the overall complexity $\binom{p}{k-1} + O(kp^{\min\{2,k-1\}} + p) = O(p^{k-1} + p)$ as opposed to the naive $\binom{p}{k-1} \cdot O(p) = O(p^k)$ time. We omit the rather technical (but less insightful) details.

Example 5. *Bogdan et al. [9] considered the following sorted ℓ_1 norm for false discovery rate (FDR) control:*

$$\|\mathbf{x}\|_{\mathsf{SLOPE}} = \sum_{i=1}^{p} \lambda_i |x|_{(i)}. \tag{20}$$

The main computation involved is again (14) with $f = |\cdot|$. A linear time algorithm in this case is provided in [9]. However, our treatment allows dealing simultaneously with any k-convex f. In particular, our algorithm covers the capped and sorted ℓ_1 norm that is known to yield less statistical bias.

6. Conclusion

In this work we explored the connection between statistical estimation under order restriction and sparse regularization in high dimensional statistical inference. We presented a general pooling principle to solve the order constrained optimization problem and proved a converse theorem to reveal the exact correctness condition. An online linear time implementation for the special chain graph was developed, and further extended to deal with piecewise convex functions. We will apply our algorithm to shrinkage parameter estimation and false discovery control problems.

Acknowledgments

The present study was financially supported by NIH Grant R01GM114311. We thank Adams Wei Yu for some helpful discussions.

References

[1] R. E. Barlow, D. J. Bartholomew, J. M. Bremner, and H. D. Brunk. *Statistical inference under order restrictions: The theory and application of isotonic regression.* John Wiley & Sons, 1972.

[2] Tim Robertson, F. T. Wright, and R. L. Dykstra. *Order Restricted Statistical Inference.* John Wiley & Sons, 1988.

[3] Ulf Grenander. "On the theory of mortality measurement: II." *Scandinavian Actuarial Journal*, vol. 39, no. 2 (1956), pp. 125–153.

[4] Herman Chernoff. "Estimation of the mode." *Annals of the Institute of Statistical Mathematics*, vol. 16, no. 1 (1964), pp. 31–41.

[5] Piet Groeneboom and Geurt Jongbloed. *Nonparametric Estimation under Shape Constraints: Estimators, Algorithms and Asymptotics.* Cambridge University Press, 2014.

[6] Miriam Ayer, H. D. Brunk, G. M. Ewing, W. T. Reid, and Edward Silverman. "An Empirical Distribution Function for Sampling with Incomplete Information." *The Annals of Mathematical Statistics*, vol. 26, no. 4 (1955), pp. 641–647.

[7] Andreas Argyriou, Rina Foygel, and Nathan Srebro. "Sparse Prediction with the k-Support Norm." In: *Proceedings of the 25th Advances in Neural Information Processing Systems (NIPS-12).* Ed. by F. Pereira, C.J.C. Burges, L. Bottou, and K.Q. Weinberger. Curran Associates, Inc., 2012, pp. 1457–1465.

[8] Howard Bondell and Brian Reich. "Simultaneous Regression shrinkage, variable selection, and supervised clustering of predictors with oscar." *Biometrics*, vol. 64, no. 1 (2008), pp. 115–123.

[9] Małgorzata Bogdan, Ewout van den Berg, Chiara Sabatti, Weijie Su, and Emmanuel J. Candès. "SLOPE – Adaptive Variable Selection via Convex Optimization." *The Annals of Applied Statistics*, vol. 9, no. 3 (2015), pp. 1103–1140.

[10] Andrew M. McDonald, Massimiliano Pontil, and Dimitris Stamos. "Spectral k-Support Norm Regularization." In: *Proceedings of the 27th Advances in Neural Information Processing Systems (NIPS-14).* Ed. by Z. Ghahramani, M. Welling, C. Cortes, N.D. Lawrence, and K.Q. Weinberger. Curran Associates, Inc., 2014, pp. 3644–3652.

[11] Leon Wenliang Zhong and James T. Kwok. "Efficient Sparse Modeling with Automatic Feature Grouping." In: *Proceedings of the 28th International Conference on Machine Learning (ICML-11)*. Ed. by Lise Getoor and Tobias Scheffer. Bellevue, Washington, USA: ACM, 2011, pp. 9–16.

[12] Jun Liu, Liang Sun, and Jieping Ye. "Projection on A Nonnegative Max-Heap." In: *Proceedings of the 24th Advances in Neural Information Processing Systems (NIPS-11)*. Ed. by J. Shawe-Taylor, R.S. Zemel, P.L. Bartlett, F. Pereira, and K.Q. Weinberger. Curran Associates, Inc., 2011, pp. 487–495.

[13] Soumyadeep Chatterjee, Sheng Chen, and Arindam Banerjee. "Generalized Dantzig Selector: Application to the k-support norm." In: *Proceedings of the 27th Advances in Neural Information Processing Systems (NIPS-14)*. Ed. by Z. Ghahramani, M. Welling, C. Cortes, N.D. Lawrence, and K.Q. Weinberger. Curran Associates, Inc., 2014, pp. 1934–1942.

[14] Peter Bacchetti. "Additive Isotonic Models." *Journal of the American Statistical Association*, vol. 84, no. 405 (1989), pp. 289–294.

[15] Zhou Fang and Nicolai Meinshausen. "LASSO isotone for high-dimensional additive isotonic regression." *Journal of Computational and Graphical Statistics*, vol. 21, no. 1 (2012), pp. 72–91.

[16] R. E. Barlow and H. D. Brunk. "The isotonic regression problem and its dual." *Journal of the American Statistical Association*, vol. 67, no. 337 (1972), pp. 140–147.

[17] Jr. W. A. Thompson. "The Problem of Negative Estimates of Variance Components." *The Annals of Mathematical Statistics*, vol. 33, no. 1 (1962), pp. 273–289.

[18] Peter L. Jackson and Robin O. Roundy. "Minimizing Separable Convex Objectives on Arbitrarily Directed Trees of Variable Upper Bound Constraints." *Mathematics of Operations Research*, vol. 16, no. 3 (1991), pp. 504–533.

[19] William L. Maxwell and John A. Muckstadt. "Establishing Consistent and Realistic Reorder Intervals in Production-Distribution Systems." *Operations Research*, vol. 33, no. 6 (1985), pp. 1316–1341.

[20] Robin O. Roundy. "A 98%-Effective Lot-Sizing Rule for a Multi-Product, Multi-Stage Production/Inventory System." *Mathematics of Operations Research*, vol. 11, no. 4 (1986), pp. 699–727.

[21] Jr Arthur F. Veinott. "Least d-Majorized Network Flows with Inventory and Statistical Applications." *Management Science*, vol. 17, no. 9 (1971), pp. 547–567.

[22] Ulf Strömberg. "An algorithm for isotonic regression with arbitrary convex distance function." *Computational Statistics & Data Analysis*, vol. 11, no. 2 (1991), pp. 205–219.

[23] Michael J. Best, Nilotpal Chakravarti, and Vasant A. Ubhaya. "Minimizing Separable Convex Functions Subject to Simple Chain Constraints." *SIAM Journal on Optimization*, vol. 10, no. 3 (2000), pp. 658–672.

[24] Ravindra K. Ahuja, Dorit S. Hochbaum, and James B. Orlin. "Solving the convex cost integer dual network flow problem." *Management Science*, vol. 49, no. 7 (2003), pp. 950–964.

[25] P. M. Pardalos and G. Xue. "Algorithms for a class of isotonic regression problems." *Algorithmica*, vol. 23, no. 3 (1999), pp. 211–222.

[26] Robert Endre Tarjan. "Amortized Computational Complexity." *SIAM Journal on Algebraic Discrete Methods*, vol. 6, no. 2 (1985), pp. 306–318.

[27] Jianqing Fan. "Comments on "Wavelets in Statistics: A Review"." *Journal of the Italian Statistical Association*, vol. 6, no. 2 (1997), pp. 131–138.

[28] Tong Zhang. "Multi-stage convex relaxation for feature selection." *Bernoulli*, vol. 19, no. 5B (2013), pp. 2277–2293.

Journal of Physics: Conference Series **699** (2016) 012017 doi:10.1088/1742-6596/699/1/012017

Sparse approximation problem: how rapid simulated annealing succeeds and fails

Tomoyuki Obuchi[1] **and Yoshiyuki Kabashima**

Interdisciplinary Graduate School of Science and Engineering, Tokyo Institute of Technology, Yokohama, Kanagawa 226-8502, Japan

E-mail: [1]obuchi@sp.dis.titech.ac.jp

Abstract. Information processing techniques based on sparseness have been actively studied in several disciplines. Among them, a mathematical framework to approximately express a given dataset by a combination of a small number of basis vectors of an overcomplete basis is termed the *sparse approximation*. In this paper, we apply simulated annealing, a metaheuristic algorithm for general optimization problems, to sparse approximation in the situation where the given data have a planted sparse representation and noise is present. The result in the noiseless case shows that our simulated annealing works well in a reasonable parameter region: the planted solution is found fairly rapidly. This is true even in the case where a common relaxation of the sparse approximation problem, the ℓ_1-relaxation, is ineffective. On the other hand, when the dimensionality of the data is close to the number of non-zero components, another metastable state emerges, and our algorithm fails to find the planted solution. This phenomenon is associated with a first-order phase transition. In the case of very strong noise, it is no longer meaningful to search for the planted solution. In this situation, our algorithm determines a solution with close-to-minimum distortion fairly quickly.

1. Introduction

The success of compressed sensing [1, 2, 3, 4] has triggered interest in the utilization of sparseness in signal processing techniques [5, 6, 7, 8, 9, 10, 11, 12, 13]. Sparseness is the property whereby data can be represented, on a proper basis, by some combination of a small number of non-zero components. This property is useful for practical applications such as data compression and data reconstruction from a small number of observations, the latter of which is simply compressed sensing.

Usually, obtaining a sparse representation from a given dataset is formulated as an optimization problem. We refer to this as the sparse approximation problem [14, 15, 16, 17, 18]. This is sometimes recast in a probabilistic formulation by statistical physicists [10, 12, 13] using Bayesian techniques or statistical mechanics. In this paper, we employ such a probabilistic formulation to search for an "optimal" solution with the minimum distortion between the given and reconstructed data.

Unlike previous approaches, we do not use a message passing algorithm [10]. Instead, we use the well-known "simulated annealing" (SA) heuristic. The motivation for using SA comes from our recent theoretical analysis of "entropy" [12, 13], which is the exponential rate of the number of combinations of non-zero components yielding a given level of distortion. Our analysis indicates that entropy exhibits some nice analytical properties, in contrast to other optimizations such as the k-satisfiability problem [19, 20]. This implies a simple structure of the "phase space," the space of possible combinations of non-zero components, and hence SA is expected to work well. Based on this expectation, the actual performance of SA is reported through numerical experiments.

2. Formulation and algorithm

2.1. Combinatorial optimization formulation

Let us suppose a signal vector $\boldsymbol{y} \in \mathbb{R}^M$ is generated from an appropriate sparse representation or a planted solution, $\hat{\boldsymbol{x}}$, through

$$\boldsymbol{y} = A\hat{\boldsymbol{x}} + \boldsymbol{\xi}, \tag{1}$$

where $A = \{\boldsymbol{a}_i\}_{i=1}^N \in \mathbb{R}^{M \times N}$ is an overcomplete matrix with $M < N$ called a dictionary; $\boldsymbol{\xi} \in \mathbb{R}^M$ is a noise vector, each component of which is drawn from the zero-mean normal distribution with variance σ_ξ^2, $\mathcal{N}(0, \sigma_\xi^2)$. The appropriate representation $\hat{\boldsymbol{x}}$ is "sparse," and is assumed to be generated from the following Bernoulli–Gaussian distribution:

$$P(\hat{x}_i) = \hat{\rho}\frac{e^{-\frac{1}{2\sigma_x^2}\hat{x}_i^2}}{\sqrt{2\pi\sigma_x^2}} + (1-\hat{\rho})\delta(\hat{x}_i). \tag{2}$$

We also assume that each component of A is independent and identically distributed from $\mathcal{N}(0, 1/N)$. The aspect ratio of the matrix, $\alpha = M/N$, is assumed to lie within $(0, 1)$.

The sparse approximation problem is to approximate $\boldsymbol{y}$ by a linear combination of a restricted number of column vectors of A. There are various formulations, one of which is based on the following optimization:

$$\boldsymbol{x}^* = \arg\min_{\boldsymbol{x}} \{\mathcal{E}(\boldsymbol{x}|\boldsymbol{y}, A)\} \text{ subject to } ||\boldsymbol{x}||_0 \le N\rho, \tag{3}$$

where $||\boldsymbol{x}||_k = (\sum_i |x_i|^k)^{1/k}$ denotes the ℓ_k norm and the ℓ_0 norm is equal to the number of non-zero components of $\boldsymbol{x}$; the parameter $\rho(< \alpha)$ controls the sparseness of $\boldsymbol{x}$ (and is called the sparsity in this paper); and $\mathcal{E}$ denotes the distortion between $\boldsymbol{y}$ and a reconstructed signal through a representation $\boldsymbol{x}$:

$$\mathcal{E}(\boldsymbol{x}|\boldsymbol{y}, A) = \frac{1}{2}||\boldsymbol{y} - A\boldsymbol{x}||_2^2. \tag{4}$$

2.2. Probabilistic formulation

Eq. (3) is a commonly used formulation, but it has the limitation that it only provides the information of the minimum-distortion solution. To get a wider perspective, it is better to treat *all* possible combinations of the column vectors. For this, we use a probabilistic formulation. Suppose that a binary vector $\boldsymbol{c} = \{c_i = 0, 1\}_{i=1}^N$, which we call the sparse weight, represents the column vectors used to represent $\boldsymbol{y}$: if $c_i = 1$, the ith column of A, $\boldsymbol{a}_i$, is used; if $c_i = 0$, it is not. Once these columns have been determined by $\boldsymbol{c}$, the optimal coefficients of the chosen columns are evaluated by solving

$$\boldsymbol{x}(\boldsymbol{c}) = \arg\min_{\boldsymbol{x}} ||\boldsymbol{y} - A(\boldsymbol{c} \circ \boldsymbol{x})||_2^2, \tag{5}$$

where $(\boldsymbol{c} \circ \boldsymbol{x})_i = c_i x_i$ represents the Hadamard product. The corresponding distortion is

$$\mathcal{E}(\boldsymbol{c}|\boldsymbol{y}, A) = M\epsilon(\boldsymbol{c}|\boldsymbol{y}, A) = \frac{1}{2}||\boldsymbol{y} - A(\boldsymbol{c} \circ \boldsymbol{x}(\boldsymbol{c}))||_2^2. \tag{6}$$

The components of $\boldsymbol{x}(\boldsymbol{c})$ for the zero components of $\boldsymbol{c}$ are actually indefinite, and we set them to be zeros. The definite part of $\boldsymbol{x}(\boldsymbol{c})$, which we denote $\tilde{\boldsymbol{x}}(\boldsymbol{c})$, has the compact analytic form

$$\tilde{\boldsymbol{x}}(\boldsymbol{c}) = \left(\tilde{A}^{\mathrm{T}}(\boldsymbol{c})\tilde{A}(\boldsymbol{c})\right)^{-1}\tilde{A}^{\mathrm{T}}(\boldsymbol{c})\boldsymbol{y}, \tag{7}$$

where $\tilde{A}(\boldsymbol{c})$ denotes the submatrix of columns chosen by $\boldsymbol{c}$.

Let us regard $\mathcal{E}$ as an "energy," and introduce an "inverse temperature" μ. A Gibbs–Boltzmann distribution is thus defined as

$$P(\boldsymbol{c}|\mu;\boldsymbol{y},A) = \frac{1}{G(\mu;\boldsymbol{y},A)}\delta\left(\sum_i c_i - N\rho\right)e^{-\mu\mathcal{E}(\boldsymbol{c}|\boldsymbol{y},A)}, \tag{8}$$

where G is the "partition function"

$$G(\mu;\boldsymbol{y},A) = \sum_{\boldsymbol{c}}\delta\left(\sum_i c_i - N\rho\right)e^{-\mu\mathcal{E}(\boldsymbol{c}|\boldsymbol{y},A)}. \tag{9}$$

Our strategy is to generate $\boldsymbol{c}$ according to eq. (8). Changing μ allows us to sample different sparse solutions with different distortion values.

This formulation provides several options to treat the sparse approximation problem. For example, sampling in $\mu < \infty$ produces solutions with distortion greater than the minimum. These "finite temperature" solutions may be more suitable for capturing the planted solution $\hat{\boldsymbol{x}}$ than $\boldsymbol{x}^*$ in eq. (3) in the presence of noise $\sigma_\xi > 0$, as suggested in [13].

In this probabilistic formulation, the optimization (3) is recovered as a sampling problem in the limit $\mu \to \infty$. We pursue this direction in this paper, and propose an algorithm to solve eq. (3). The performance of our technique is examined in numerical experiments, and is compared with some known analytical results [7, 12, 13].

2.3. Simulated annealing

Our algorithm is a variant of SA, which is a metaheuristic to find the global minimum of a cost function. The outline of the algorithm is as follows: starting from a random initial configuration of $\boldsymbol{c}$ at very high temperature $T = 1/\mu \gg 1$, the algorithm randomly updates the configuration $\boldsymbol{c} \to \boldsymbol{c}'$ in a Monte-Carlo (MC) manner, while gradually decreasing the temperature. Eventually, the temperature becomes very low, $T \approx 0$, and the configuration is no longer updated. This final configuration is expected to be very close (or identical) to the true solution, *i.e.*, $\boldsymbol{x}^* \approx \boldsymbol{c} \circ \boldsymbol{x}(\boldsymbol{c})$.

The Metropolis criterion is adopted in our simulation: an MC move $\boldsymbol{c} \to \boldsymbol{c}'$ is judged to be accepted or not according to the probability

$$p_{\mathrm{accept}}(\boldsymbol{c} \to \boldsymbol{c}') = \max(1, e^{-\mu(\mathcal{E}(\boldsymbol{c}')-\mathcal{E}(\boldsymbol{c}))}). \tag{10}$$

For a fixed value of ρ, the configurations generated by the algorithm should always satisfy $\sum_i c_i = N\rho$. Given an initial configuration satisfying this condition, we generate trial moves $\boldsymbol{c} \to \boldsymbol{c}'$ by "pair flipping" two sparse weights, one equal to 0 and the other equal to 1. Namely, choosing an index i of the sparse weight from ONES $\equiv \{k|c_k = 1\}$ and another index j from ZEROS $\equiv \{k|c_k = 0\}$, we set $\boldsymbol{c}' = \boldsymbol{c}$, except for the counterpart of $(c_i, c_j) = (1, 0)$, which is given as $(c_i', c_j') = (0, 1)$.

The pseudo-code of our MC algorithm is given in Alg. 1, and that of our SA procedure is presented in Alg. 2. The lines marked with # are not necessarily needed for SA, but have been inserted for later convenience. For Alg. 2, we have a set of inverse temperature points $\{\mu_a\}_a^{N_\mu}$ arranged in ascending order $(0 =)\mu_1 < \mu_2 < \cdots < \mu_{N_\mu}(\gg 1)$ and the waiting times

Algorithm 1 MC update with pair flipping

procedure MCPF($\boldsymbol{c}, \mu, \boldsymbol{y}, A$) ▷ MC routine with pair flipping
  ONES $\leftarrow \{k|c_k = 1\}$, ZEROS $\leftarrow \{k|c_k = 0\}$
  randomly choose i from ONES and j from ZEROS
  $\boldsymbol{c}' \leftarrow \boldsymbol{c}$
  $(c'_i, c'_j) \leftarrow (0, 1)$
  $(\mathcal{E}, \mathcal{E}') \leftarrow (\mathcal{E}(\boldsymbol{c}|\boldsymbol{y}, A), \mathcal{E}(\boldsymbol{c}'|\boldsymbol{y}, A))$ ▷ Calculate energy
  $p_{\text{accept}} \leftarrow \max(1, e^{-\mu(\mathcal{E}'-\mathcal{E})})$
  generate a random number $r \in [0, 1]$
  if $r < p_{\text{accept}}$ **then**
    $\boldsymbol{c} \leftarrow \boldsymbol{c}'$
  end if
  return $\boldsymbol{c}$
end procedure

Algorithm 2 SA for sparse approximation problem

procedure SA($\{\mu_a, \tau_a\}_{a=1}^{N_\mu}, \rho, \boldsymbol{y}, A$)
  Generate a random initial configuration $\boldsymbol{c}$ with $\sum_i c_i = N\rho$
  for $a = 1 : N_\mu$ **do** ▷ Changing temperature
    for $t = 1 : \tau_a$ **do** ▷ Sampling at $\mu = \mu_a$
      for $i = 1 : N$ **do** ▷ Extensive number of MC updates
        $\boldsymbol{c} \leftarrow$ MCPF($\boldsymbol{c}, \mu_a, \boldsymbol{y}, A$)
      end for
      # Calculate energy $\epsilon_t = \epsilon(\boldsymbol{c}_t|\boldsymbol{y}, A)$ of the current configuration $\boldsymbol{c}_t = \boldsymbol{c}$
    end for
    # Calculate the average energy $\epsilon_a = (1/\tau_a)\sum_{t=1}^{\tau_a} \epsilon_t$
  end for
  return $\boldsymbol{c}$
end procedure

$\{\tau_a\}_a$ at those points. Hence, as the algorithm proceeds, the temperature of the system $T = 1/\mu$ decreases step by step. It is known that, if the rate of decrease of the temperature obeys

$$T(t) > \frac{A(N)}{\log(t+2)} \tag{11}$$

for some time-independent constant $A(N)$, then the output of SA is guaranteed to be optimal [21]. This is a very slow schedule of decrease for T, and is generally overcautious so as to include the worst-case scenario. Faster schedules are known to work in practical situations. We report the results for such a rapid annealing below.

3. Results

We now present the results of SA according to eq. (8). System sizes of $N = 100, 200$, and 400 will be examined. The annealing schedule is fixed as

$$\mu_a = \mu_0 + r^{a-1} - 1, \ \tau_a = \tau, \ (a = 1, \cdots, 100). \tag{12}$$

We set $\tau = 5$, $\mu_0 = 10^{-8}$, and $r = 1.1$ as default parameter values. Thus, the maximum value of μ is $\mu_{100} \approx 1.3 \times 10^4$.

Journal of Physics: Conference Series **699** (2016) 012017 doi:10.1088/1742-6596/699/1/012017

To determine whether the SA process is proceeding well, we consider the lines marked with # in Alg. 2. If our schedule is sufficiently slow, the $\boldsymbol{c}_t$ obtained during annealing are typical samples from eq. (8), and the values of physical quantities of typical samples should be very close (in the limit $N \to \infty$ almost surely identical) to the thermal averages. This implies the following relation

$$\epsilon_a = \frac{1}{\tau_a}\sum_{t}^{\tau_a} \epsilon(\boldsymbol{c}_t) \approx \langle \epsilon(\boldsymbol{c}|\boldsymbol{y}, A)\rangle_{\mu_a}, \tag{13}$$

where $\langle \cdots \rangle_{\mu}$ denotes the average over eq. (8) with the inverse temperature μ. Fortunately, the right-hand side is analytically assessed in the present case with a random matrix dictionary A in the $N \to \infty$ limit [12, 13]. Hence, we compare $\mathcal{E}_a$ with the analytically evaluated $\langle \mathcal{E}\rangle_{\mu_a}$ to determine how well our annealing process follows the correctly distributed samples from (8). For clarity of comparison, we take an average over a different $N_{\text{samp}} = 100$ samples of $\hat{\boldsymbol{x}}, \boldsymbol{\xi}, A$ in the numerical experiments. The error bar is given by the standard deviation among those samples divided by $\sqrt{N_{\text{samp}} - 1}$.

As well as the distortion, we calculate the mean squared error (MSE) between the planted and inferred representations

$$\mathcal{M}(\boldsymbol{c}) = \frac{1}{N}||\hat{\boldsymbol{x}} - \boldsymbol{c} \circ \boldsymbol{x}(\boldsymbol{c})||_2^2. \tag{14}$$

This metric provides direct information about the reconstruction of the planted solution.

3.1. Noiseless case

Let us start with the noiseless case $\sigma_\xi^2 = 0$. In this subsection, we fix $\sigma_x^2 = 1$.

The easy reconstruction region, where α is sufficiently larger than $\hat{\rho}$, is a good starting point. Fig. 1 plots ϵ (left, middle) and the MSE $\mathcal{M}$ (right) against the temperature for $\alpha = 0.8, \rho = 0.4$, and $\hat{\rho} = 0.2$. The numerical results show a fairly good agreement with the analytical curve.

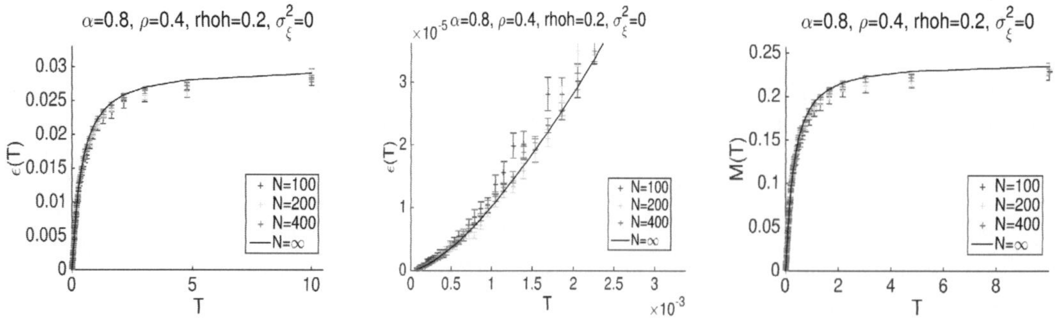

Figure 1. Distortion (left, middle) and MSE (right) plotted against temperature $T = 1/\mu$ for $\alpha = 0.8, \rho = 0.4$, and $\hat{\rho} = 0.2$. The middle panel is a magnified view of the left panel for a small region of T. The solid black curve shows the analytical values, and the color plots give the numerical results. Our numerical results clearly reproduce the correct average values over eq. (8).

This means our SA algorithm follows the equilibrium state up to the zero-temperature limit reasonably well, even though the annealing defined by eq. (12) is very rapid. The MSE goes to zero as T decreases, and so the planted solution is correctly reproduced.

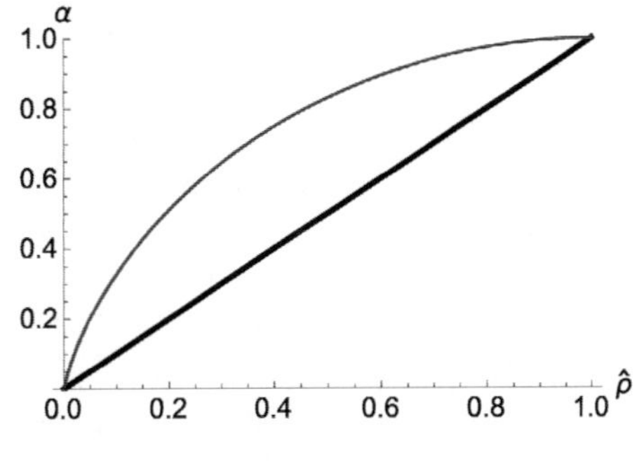

Figure 2. Phase diagram describing typical reconstruction limits in the noiseless case [7]. The straight line $\hat{\rho} = \alpha$ is the limit attained with eq. (3), whereas the curve is the limit for the relaxed problem in which the ℓ_0 norm in eq. (3) is replaced with the ℓ_1 norm. Above these boundaries, the solutions of corresponding optimization problems reconstruct the planted solution $\hat{\boldsymbol{x}}$.

Next, we consider harder cases. It is known that the properties of the problem drastically change as α gets closer to $\hat{\rho}$. The phase diagram in Fig. 2 demonstrates this. The curve is the reconstruction limit of the ℓ_1-relaxed version of (3), which is employed in many realistic cases and has considerable importance. Hence, we first examine the behavior of SA below this ℓ_1 reconstruction limit. Fig. 3 plots ϵ and $\mathcal{M}$ against T for $\alpha = 0.8, \rho = 0.55$, and $\hat{\rho} = 0.5$, where we are below the ℓ_1 reconstruction limit (the boundary is located at $\alpha_c(\hat{\rho} = 0.5) \approx 0.831$). The MSE vanishes as T decreases, meaning that $\hat{\boldsymbol{x}}$ is perfectly reconstructed. Hence, SA can

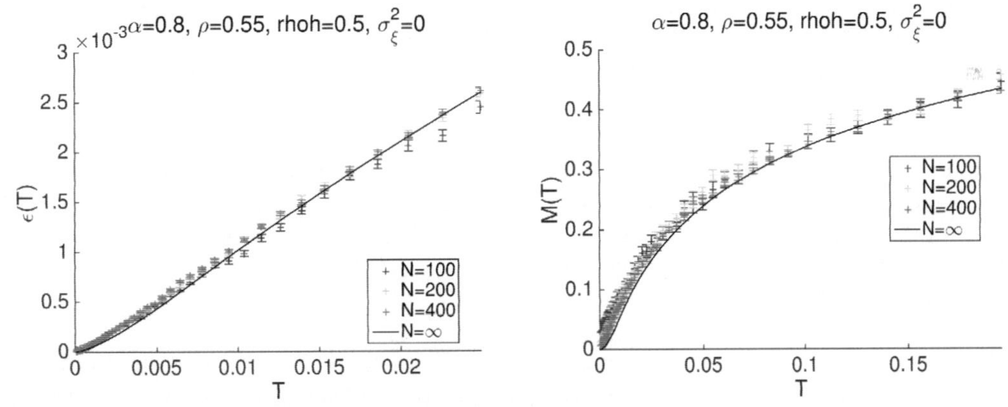

Figure 3. Distortion (left) and MSE (right) plotted against temperature for $\alpha = 0.8, \rho = 0.55$, and $\hat{\rho} = 0.5$ (below the ℓ_1 boundary). Our numerical results accord with the analytical result (black line) and achieve a perfect reconstruction of $\hat{\boldsymbol{x}}$.

outperform the ℓ_1 method of reconstruction, even under the present rapid schedule.

In harder situations, SA cannot always give a perfect reconstruction, even though it exists. Fig. 4 shows plots of T-ϵ and T-$\mathcal{M}$ for $\alpha = 0.75, \rho = 0.65$, and $\hat{\rho} = 0.5$. In this case, there are two stable thermodynamic states at low temperatures: one is connected to the planted solution and is the true equilibrium state in the zero temperature limit, whereas the other is metastable at low temperatures, but is the dominant equilibrium state at higher temperatures [13]. There is a first-order phase transition at a critical temperature $T_c \approx 3 \times 10^{-4}$. Through the SA process,

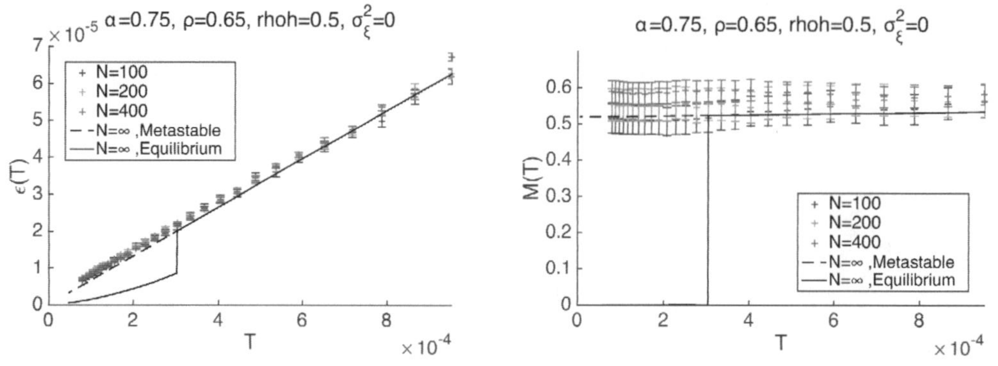

Figure 4. Distortion ϵ (left) and MSE $\mathcal{M}$ (right) plotted against temperature for $\alpha = 0.75, \rho = 0.65$, and $\hat{\rho} = 0.5$. The analytical result shows a first-order phase transition around $T_c \approx 3\times10^{-4}$, and the equilibrium state (black solid line) suddenly decreases for both ϵ and $\mathcal{M}$. However, a metastable state, which continues analytically to the equilibrium state for $T > T_c$, survives below T_c, and the numerical results follow this; hence, the planted solution is not reached.

the system state follows the equilibrium up to $T > T_c$. After the transition, the equilibrium state changes drastically, but the system cannot follow such a jump. Instead, the system remains in the same metastable state in $T < T_c$. Hence, SA cannot find the planted solution in this case, as clearly seen in the non-vanishing MSE $\mathcal{M}$ of Fig. 4. Of course, if τ is large enough, SA can eventually find the planted solution as proved in [21]. In the presence of the first-order phase transition, however, the required τ to do this is scaled with the system size N and rapidly grows as N increases, which prevents reaching the optimal solution in practical times.

Although the system is in the metastable state, the distortion ϵ seems to go to zero in Fig. 4, but unfortunately this is not the case. If we go lower temperatures, the value of ϵ will get stuck at a certain critical temperature and remains a constant below it. This is a freezing transition of this metastable state: the entropy of it becomes zero at the transition temperature. With the present parameters, our analytical solution provides the transition temperature as $T \approx 1.3\times10^{-7}$ and the constant value of distortion is $\epsilon = 8.6 \times 10^{-9}$ which is the achievable limit by our SA algorithm in practical times. This limiting value is not exactly zero although it is still negligibly small. For reference, we have plotted the "free energy" $g = (1/N)\log G$ and the entropy in Fig. 5 for the present parameter values.

Finally, we examined the effect of the parameter ρ, which can be set to any value for a given $\boldsymbol{y}$ and A. Fig. 6 is the counterpart of Fig. 4 with common values of $(\alpha, \hat{\rho}) = (0.75, 0.5)$ and a different value of $\rho = 0.55$. In contrast to Fig. 4, the metastable state is absent. We commonly observed the suppression of this metastable state as ρ decreased [13]. Thus, SA was naively expected to yield the planted solution in the zero temperature limit. The upper panels of Fig. 6 represent the rapid schedule $\tau = 5$, and exhibit a clear deviation from the equilibrium state at low temperatures. A slower schedule, corresponding to $\tau = 100$, is represented in the lower panels, and these numerical data are consistent with the analytical curve of the equilibrium state. The present choice of parameters is in the harder region, so it is unsurprising that the rapid schedule $\tau = 5$ is not sufficiently slow, although it was for Figs. 1-3. Note that this increase in τ is qualitatively different from that caused by the emergence of the metastable state in Fig. 4. The latter is macroscopic, namely, the required waiting time τ increases significantly as N grows to reach the planted solution.

In summary, SA outperforms the ℓ_1 result and can correctly find the planted solution in a

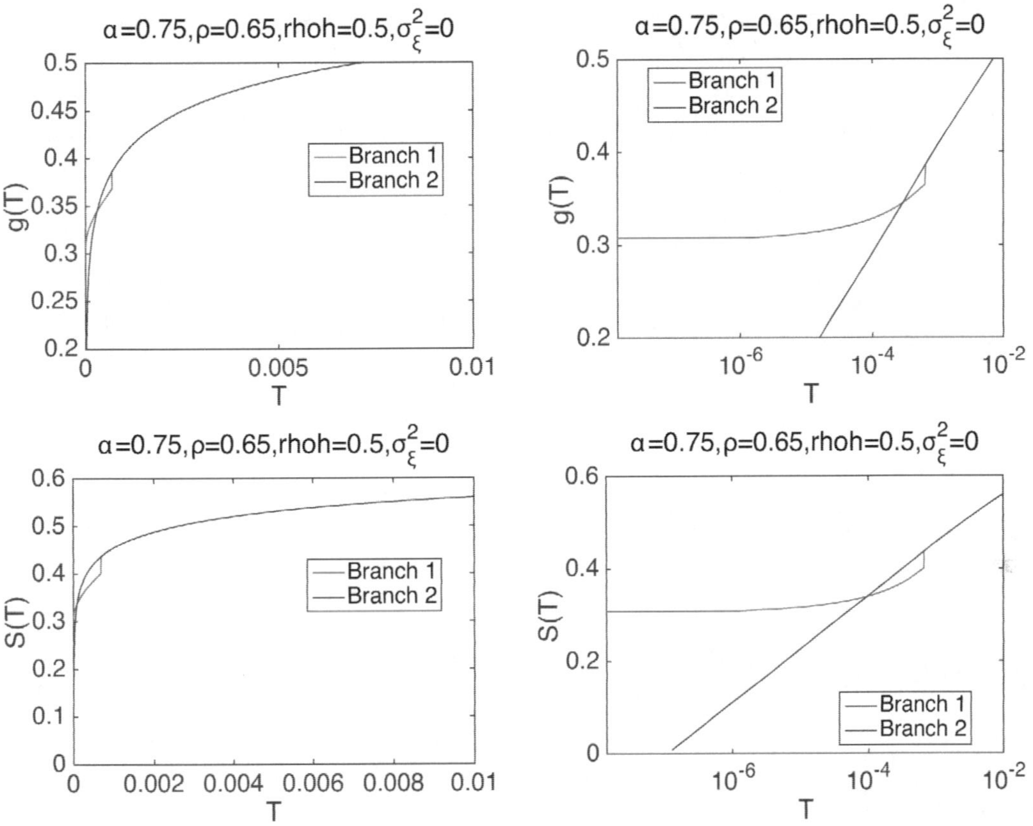

Figure 5. Free energy $g = (1/N)\log G$ (upper) and entropy (lower) plotted against temperature for $\alpha = 0.75, \rho = 0.65$, and $\hat{\rho} = 0.5$. Right panels are the same plots as the left panels but with a semi-logarithmic scale. Two different branches are shown, and the same colors denote the same branches in all panels. A phase transition (where two branches of g cross) can be observed at $T_c \approx 3 \times 10^{-3}$. The larger branch yields the equilibrium state. The entropy of the Branch 2 vanishes at a finite value of T, as clearly shown in the lower-right panel. Branch 1, the red solid curve, is connected to the planted solution.

wider region of parameter space, even under our rather rapid rate of decrease of T. However, for harder regions where α is close to $\hat{\rho}$, the phase space is separated into two distinctive states. The state that is not connected to the planted solution becomes the equilibrium state in the high-temperature region. This prevents SA from correctly finding the planted solution, as the system becomes trapped in the wrong state. Tuning ρ may be a crucial factor in overcoming this. As long as $\rho > \hat{\rho}$, smaller values of ρ are better. This is because the emergence of the metastable state tends to be suppressed as ρ becomes smaller, but such fine tuning requires *a priori* knowledge about the value of $\hat{\rho}$, which is not available in most situations. This problem is of considerable importance, and requires further consideration.

If we do not insist on finding the planted solution, even the metastable state may be desirable, as it shows small distortion. In this case, larger values of ρ are better, because they yield smaller values of distortion, though the compression ratio decreases as ρ increases. Hence, the value of ρ should be chosen according to the requirements of the information processing application.

Journal of Physics: Conference Series **699** (2016) 012017 doi:10.1088/1742-6596/699/1/012017

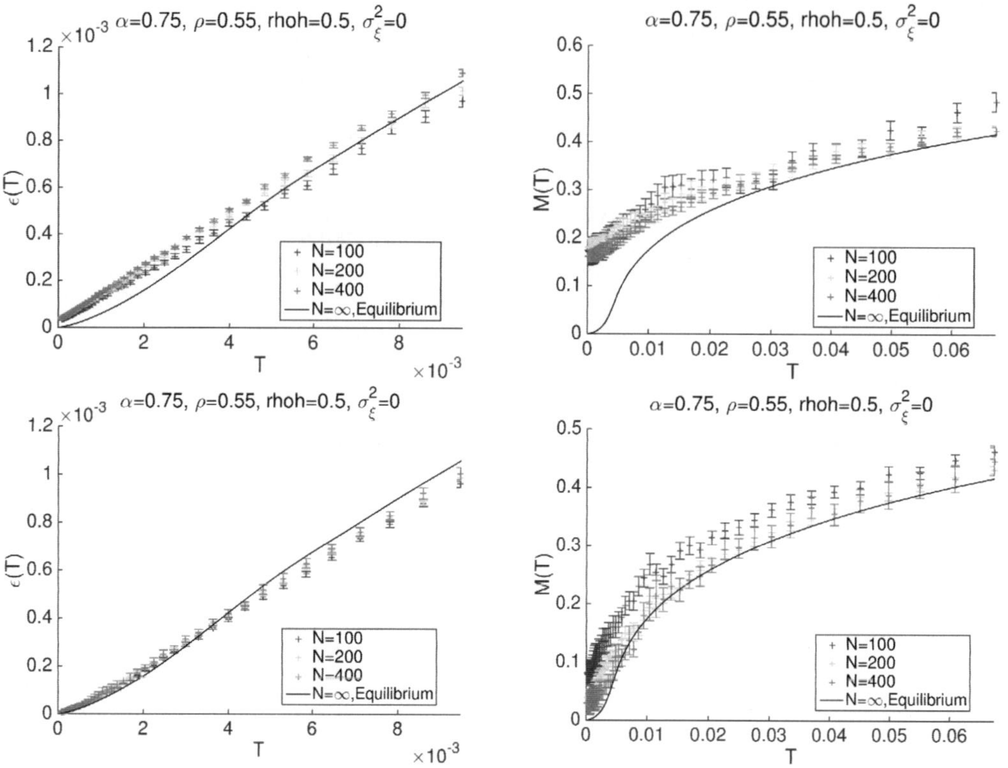

Figure 6. Distortion ϵ (left) and MSE $\mathcal{M}$ (right) plotted against temperature for $\alpha = 0.75, \rho = 0.55$, and $\hat{\rho} = 0.5$. In contrast to Fig. 4 with $\rho = 0.65$, there is no phase transition. The upper panels correspond to a rapid schedule with $\tau = 5$, whereas the lower ones correspond to $\tau = 100$. At low temperatures, the rapid case does not follow the equilibrium state, whereas the slower one is well matched.

3.2. Case without planted solutions

Next, we examine the opposite case with $\sigma_\xi^2 = 1$ and $\sigma_x^2 = 0$. There is now no planted solution, and we focus solely on how small ϵ becomes. For comparison, we present a number of values of ϵ achieved by different methods. The symbol ell$_1$ denotes the ℓ_1 method in [12], in which the ℓ_1-relaxed version of eq. (3) is solved and the resultant solution of $\boldsymbol{x}$ is inserted in eq. (4). Similarly, the symbol ell$_1$+LS corresponds to the method in [12], which gives ϵ obtained by eq. (6), but with the substituted support $\boldsymbol{c}$ determined by solving the ℓ_1-relaxed version of eq. (3). The symbol OMP denotes the results given by orthogonal matching pursuit [22, 23]. The OMP result is not obtained by analytical methods, but by numerical experiments with the same parameters as SA for $N = 400$.

Fig. 7 plots ϵ against temperature T. Note that the thermal average of the distortion stops decreasing at a certain value of T. Below this, ϵ remains constant, giving the achievable limit of the value of distortion in the present case [12], as seen in the analytical result (black solid curve). Over a longer temperature range (left panel), the numerical results seem to agree well with the analytical curve, but over a tighter range of low temperatures (right panel), the numerical data exhibit a systematic deviation from the analytical curve. This is likely to be a result of the rapid nature of our annealing schedule. A slower schedule will improve the achievable value of distortion, as in Fig. 6. Regardless, we can see that the SA result is already better than the

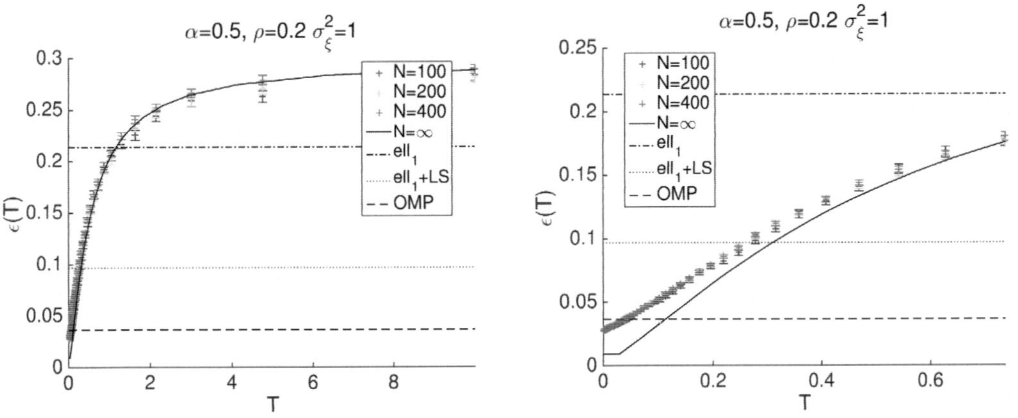

Figure 7. Distortion plotted against temperature for $\alpha = 0.5, \rho = 0.2$ for the case without a planted solution. The right panel is a magnified view of the left one at low temperatures. The black solid curve represents the analytical value of ϵ. Values of ϵ given by different algorithms are also displayed as horizontal lines with different symbols. See the main text for details.

values achieved by the other methods, demonstrating the effectiveness of SA. For reference, the distortion values obtained by ℓ_1, $\ell_1 + \mathrm{LS}$, OMP, and SA at $\mu = \mu_{100} \approx 1.3 \times 10^4$ and $N = 400$ were $\epsilon_{\ell_1} = 0.214, \epsilon_{\ell_1+\mathrm{LS}} = 0.0966, \epsilon_{\mathrm{OMP}} = 0.0365 \pm 6.7 \times 10^{-4}$, $\epsilon_{\mathrm{SA}} = 0.0272 \pm 6.2 \times 10^{-4}$, respectively. The achievable limit was $\epsilon_0 = 0.00919$.

We also examined other parameter values, but the qualitative behavior was the same as that of Fig. 7, and so the results are not shown here.

4. Discussion

4.1. Computation time and its order

For reference, we give the actual computation times for one run of SA under the schedule in eq. (12): for $\alpha = 0.5$ and $\rho = 0.2$, approximately 6, 15, and 38 seconds were required for $N = 100, 200$, and 400, respectively. This experiment was performed on a 1.7 GHz Intel Core i7 with two CPUs using MATLAB®.

As well as these practical results, we can estimate the order of the computation time. Formally, this can be written as $O(N_\mu \tau N N_{\mathrm{MC}})$, the last factor of which is the computational cost of each MC update. The most expensive part is the matrix multiplication and inversion required to calculate the energy. If we use simple multiplication and Gauss elimination in the inversion process for each step, $N_{\mathrm{MC}} = O(M(N\rho)^2 + (N\rho)^3)$. However, we have employed pair flipping in each update of the sparse weights, meaning that the change in the relevant matrices in each move is small and successive. This implies that we can reduce the computation time by successively updating those matrices while employing the matrix inversion formula.

Provided that G_{t+1} is decomposed as

$$G_{t+1} = \begin{pmatrix} G_t & \boldsymbol{g}_{t+1} \\ \boldsymbol{g}_{t+1}^{\mathrm{T}} & g_{t+1} \end{pmatrix}, \tag{15}$$

the matrix inversion formula gives

$$G_{t+1}^{-1} = \begin{pmatrix} G_t^{-1} + \gamma_{t+1} G_t^{-1} \boldsymbol{g}_{t+1} \boldsymbol{g}_{t+1}^{\mathrm{T}} G_t^{-1} & -\gamma_{t+1} G_t^{-1} \boldsymbol{g}_{t+1} \\ (-\gamma_{t+1} G_t^{-1} \boldsymbol{g}_{t+1})^{\mathrm{T}} & \gamma_{t+1} \end{pmatrix} \equiv \begin{pmatrix} U_t & \boldsymbol{u}_{t+1} \\ \boldsymbol{u}_{t+1}^{\mathrm{T}} & u_{t+1} \end{pmatrix} \equiv U_{t+1}, \tag{16}$$

where $\gamma_{t+1} = g_{t+1} - \boldsymbol{g}_{t+1}^{\mathrm{T}} G_t^{-1} \boldsymbol{g}_{t+1}$. We use this as follows. Write G_{t+1} as $(\tilde{A}^{\mathrm{T}}(\boldsymbol{c})\tilde{A}(\boldsymbol{c}))$, and assume that we are given both G_{t+1} and $G_{t+1}^{-1} = U_{t+1}$. First, we treat the deletion part of the MC move, $c_i = 1 \to c_i' = 0$. We want to calculate G_t^{-1} from G_{t+1}^{-1}. This is given by

$$G_t^{-1} = U_t - \boldsymbol{u}_{t+1}\boldsymbol{u}_{t+1}^{\mathrm{T}}/u_{t+1}, \tag{17}$$

which has a computational cost of $O((N\rho)^2)$. The corresponding G_t is obtained by deleting the ith column and row from G_{t+1}. Next, we move to the addition part $c_j = 0 \to c_j' = 1$. We now have G_t and G_t^{-1}. Extending the matrix, $G_t \to G_{t+1}$, involves adding an appropriate column and row to G_t. The kth component of the added column vector can be calculated as $\boldsymbol{g}_{t+1}(k) = \sum_{l=1}^{M} A(l,k)A(l,j)$, where k runs over the indices of ONES, and thus the computational cost is $O(MN\rho)$. Similarly, we can calculate $g_{t+1} = \sum_{l=1}^{M} A(l,j)A(l,j)$, and hence the computational cost of calculating G_{t+1} is $O(MN\rho)$. Now, we can easily calculate G_{t+1}^{-1} by eq. (16) from G_t^{-1}, $\boldsymbol{g}_{t+1}$, and g_{t+1}. The computational cost of this operation is $O((N\rho)^2)$. This completes the successive update of $G = \tilde{A}^{\mathrm{T}}\tilde{A}$ and G^{-1}.

In summary, the factor N_{MC} can be reduced to

$$N_{\mathrm{MC}} = O((N\rho)^2 + MN\rho) = O(N^2\alpha\rho), \ (M = N\alpha \geq N\rho). \tag{18}$$

Thus, the total computational cost of our SA algorithm is $O(\alpha\rho\tau N_\mu N^3)$. As long as N_μ does not scale with the size of the system, we have only third-order dependence on system-size, which is comparable to that of versatile convex optimization solvers used in the ℓ_1-relaxed version of the sparse approximation problem. Hence, our SA algorithm can solve the sparse approximation problem at a computational cost that is of the same order as the ℓ_1-relaxed version, without any need to relax the problem.

4.2. Advantages, disadvantages, and possible extensions

Our study indicates that SA reliably determines a solution with small distortion both in the presence and absence of noise, and has a reasonable computational cost. As long as noise and the metastable state are absent, the solution identified by SA is approximately equal to the planted solution. These findings encourage the use of SA in practical applications of the sparse approximation problem.

To conclude, we summarize the advantages and disadvantages of the present SA algorithm.

Advantages:

- Easy to implement for any $\boldsymbol{y}$ and A.
- Necessarily stops (message passing can be unstable and sometimes does not converge, especially in the presence of noise).
- Solutions at finite temperatures ($\{\boldsymbol{c}_t\}_t$) can be obtained by one iteration of SA. These may be more useful than the optimal solution with minimum distortion, especially in the presence of noise.

Disadvantages:

- Emergence of metastable states in the hard parameter region.
- Presence of ρ. Tuning this parameter is not trivial: larger values of ρ are better for determining the planted solution and decreasing the level of distortion, but are more likely to yield the metastable state.

Journal of Physics: Conference Series **699** (2016) 012017 doi:10.1088/1742-6596/699/1/012017

- Annealing schedule is arbitrary, and it is not *a priori* clear how long the algorithm requires to reach a desired solution.

These disadvantages may be overcome using a range of techniques. Seeding [10] is a good candidate for avoiding trapping in the metastable state. What we should do is to change the matrix A to a structured one as described in [10]. Beyond the pair flipping process, we may flip more sparse weights to generate trial moves, which should shorten the time required for efficient sampling. Extended ensemble may also be beneficial. For example, simulating different values of ρ simultaneously enables a wider region of the phase space to be sampled, which may help escaping from the metastable state. In this way, the second disadvantage will be also diminished. Developing extensions for the present formulation of SA will be helpful to control the sparse approximation problem.

Acknowledgments

This work was supported by JSPS KAKENHI Grant Numbers 26870185 (TO) and 25120013 (YK).

[1] Donoho D L 2006 *IEEE Trans. Inf. Theory* **52** 1289
[2] Candès E J and Tao T 2005 *IEEE Trans. Inf. Theory* **51** 4203
[3] Candès E J, Romberg J and Tao T 2006 *IEEE Trans. Inf. Theory* **52** 489
[4] Candès E J and Tao T 2006 *IEEE Trans. Inf. Theory* **52** 5406
[5] Donoho D L and Tanner J 2009 *J. Am. Math. Soc.* **22** 1
[6] Donoho D L, Malekib A and Montanari A 2009 *Proc. Natl. Acad. Sci.* **106** 18914
[7] Kabashima Y, Wadayama T and Tanaka T 2009 *J. Stat. Mech.* L09003
[8] Ganguli S and Sompolinsky H 2010 *Phys. Rev. Lett.* **104** 188701
[9] Rangan S 2010 *arXiv:1010.5141*
[10] Krzakala F, Mézard M, Sausset F, Sun Y and Zdeborová L 2012 *J. Stat. Mech.* P08009
[11] Sakata A and Kabashima K 2013 *Europhys. Lett.* **103**, 28008
[12] Nakanishi Y, Obuchi T, Kabashima Y and Okada M 2015 *arXiv:1510.02189*
[13] Obuchi T, Nakanishi Y, Kabashima Y and Okada M *in preparation*
[14] Natarajan B K 1995 *SIAM J. Comput.* **24** 227
[15] Temlyakov V N 1998 *Adv. Comput. Math.* **8** 249
[16] Temlyakov V N 1999 *J. Approx. Theory* **98** 117
[17] Tropp J A 2004 *IEEE Trans. Inf. Theory* **50** 2231
[18] Donoho D L, Elad M and Temlyakov V N 2006 *IEEE Trans. Inf. Theory* **51** 6
[19] Monasson R 1996 *Phys. Rev. Lett.* **76** 3881
[20] Krzakala F, Montanari A, Ricci-Tersenghi F, Semerjian G and Zdeborová L 2007 *Proc. Natl. Acad. Sci.* **104** 10318
[21] Geman S and Geman D 1984 *IEEE Trans. Pattern Anal. Mach. Intell.* **6** 721
[22] Pati Y C, Rezaiifar R and Krishnaprasad P S 1993 *Conference Record of the Twenty-Seventh Asilomar Conference on Signals, Systems and Computers* 40
[23] Davis G M, Mallat S G and Zhang Z 1994 *Opt. Eng.* **33** 2183

Asymmetric biclustering with constrained von Mises-Fisher models

Kazuho Watanabe[1], Hsiang-Yun Wu[2], Shigeo Takahashi[3] and Issei Fujishiro[2]

[1] Department of Computer Science and Engineering, Toyohashi University of Technology, 1-1 Hibarigaoka, Tempaku-cho, Toyohashi, Aichi 441-8580, Japan
[2] Department of Information and Computer Science, Keio University, 3-14-1 Hiyoshi, Kohoku-ku, Yokohama, Kanagawa 223-8522, Japan
[3] Department of Computer Science and Engineering, University of Aizu, Tsuruga, Ikki-machi, Aizu-Wakamatsu, Fukushima 965-8580, Japan

E-mail: wkazuho@cs.tut.ac.jp

Abstract. As a probability distribution on the high-dimensional sphere, the von Mises-Fisher (vMF) distribution is widely used for directional statistics and data analysis methods based on correlation. We consider a constrained vMF distribution for block modeling, which provides a probabilistic model of an asymmetric biclustering method that uses correlation as the similarity measure of data features. We derive the variational Bayesian inference algorithm for the mixture of the constrained vMF distributions. It is applied to a multivariate data visualization method implemented with enhanced parallel coordinate plots.

1. Introduction

Finding correlated subspaces, sets of features, is one of promising approaches to understanding high-dimensional data. Visual analysis methods for such a purpose have been developed in the visualization community [1, 2]. Although these methods are effective, the results of the analysis depend heavily on knowledge and observation skill of the users and thus may fail to illuminate important subspaces. To support such data exploration, we developed a multivariate data visualization method, which combines a biclustering technique and parallel coordinate plots [3]. Biclustering, also known as co-clustering or two-mode clustering, performs a simultaneous clustering of the rows (samples or instances) and columns (features or variables) of a data matrix consisting of multivariate data samples [4, 5]. Conventional popular models of biclustering are based on the so-called block models [5]. Although there are extensions of block models using probabilistic inference [6, 7], they treat the clusterings of rows and columns of the data matrix basically symmetrically. This is not necessarily appropriate if a block model is combined with data visualization methods such as parallel coordinate plots, where correlation is considered as a suitable measure of similarity between features (axes), and is in fact used for reordering or contracting axes [8, 9]. Therefore, in our previous work [3], we developed an asymmetric biclustering method, which uses correlation as the similarity measure of features, based on spherical k-means [10, 11]. However, the proposed asymmetric biclustering method assumes that the numbers of clusters are predefined both for rows and columns, and the model selection mechanism has yet to be implemented.

In this paper, we propose a probabilistic model for the asymmetric biclustering method by the von Mises-Fisher (vMF) distribution, which is a probability distribution on a high-dimensional unit sphere. Providing a probabilistic model offers advantages such as dealing with missing data and automatic control of model's complexity by Bayesian methods. However, combining the vMF model with block modeling requires additional constraints on the vMF model that the elements of its mean direction parameter vector are divided into groups and are common within each group. Although the maximum likelihood estimation algorithm was derived for the mixture of (unconstrained) vMF distributions [11] and the variational Bayesian inference algorithms were derived as a deterministic approximation of Bayesian inference [12, 13], their extensions to the constrained vMF distributions are not straightforward due to the constraints. Thus in this paper, we formulate the variational Bayesian inference algorithm for the mixture of the constrained vMF distributions. Since in a small variance limit of the probabilistic model, the algorithm reduces to the asymmetric biclustering method proposed in [3], the derived algorithm is its extension that naturally provides a mechanism for controlling model's complexity, namely, the numbers of data sample clusters and data feature clusters. Variational Bayesian inference offers a criterion for model selection by its objective functional called variational free energy. Furthermore, the variational Bayesian inference algorithm automatically eliminates redundant clusters inheriting the nature of Bayesian inference [14]. Although such a Bayesian framework has been proposed for the usual (symmetric) biclustering model [6], we consider its application to the asymmetric biclustering model based on the mixture of vMF models. The derived inference algorithm is incorporated into the visual analysis framework proposed in [3], which is implemented with enhanced parallel coordinate plots. We demonstrate the model selection ability of the variational Bayesian inference through its applications to synthetic and real datasets.

2. Constrained von Mises-Fisher model for asymmetric biclustering

2.1. Block model

Suppose we are given an $n \times d$ data matrix $\mathbf{X}$ consisting of n samples and d features. Biclustering methods based on the block model [4, 5] divide the data matrix into $K \times L$ submatrices (blocks), each of which has the size $n_k \times d_l$ ($k = 1, \cdots, K$ and $l = 1, \cdots, L$), where n_k is the number of data samples assigned to the kth cluster of samples, and d_l is the number of features assigned to the lth cluster of dimensions. Hence, $\sum_{k=1}^{K} n_k = n$ and $\sum_{l=1}^{L} d_l = d$ hold.

Let $\mathbf{y} = \{y_i\}_{i=1}^{n}$ and $\mathbf{z} = \{z_j\}_{j=1}^{d}$ be latent variables, where $y_i \in \{1, 2, ..., K\}$ indicates the sample cluster assignment of the ith data sample and $z_j \in \{1, 2, ..., L\}$ indicates the feature cluster assignment of the jth feature. Then it holds that $n_k = \sum_{i=1}^{n} \delta_{y_i,k}$ and $d_l = \sum_{j=1}^{d} \delta_{z_j,l}$, where $\delta_{k,l} = 1$ if $k = l$, and $\delta_{k,l} = 0$ otherwise. The 1-of-K and 1-of-L representations of sample and feature cluster assignments are given by $(\delta_{y_i,1}, \cdots, \delta_{y_i,K})$ and $(\delta_{z_j,1}, \cdots, \delta_{z_j,L})$ for the ith sample and jth feature, respectively.

Each block has an associated parameter. Let $\nu_{k,l} \in \mathbb{R}$ be the parameter of the klth block and $\boldsymbol{\nu} = \{\nu_{k,l} : k = 1, ..., K, l = 1, ..., L\}$. Then, the block model estimates $\mathbf{y}$, $\mathbf{z}$ and $\boldsymbol{\nu}$ from the given data matrix $\mathbf{X}$.

2.2. Constrained von Mises-Fisher model

Normalizing each dimension of the data matrix so that each dimension has average zero and unit norm, we denote the ijth element of the normalized data matrix by x_{ij} and the jth column vector by $\mathbf{x}_j = (x_{1j}, \cdots, x_{nj})$, that is, $\frac{1}{n}\sum_{i=1}^{n} x_{ij} = 0$ and $\|\mathbf{x}_j\|^2 = \sum_{i=1}^{n} x_{ij}^2 = 1$ hold for $j = 1, \cdots, d$. This means that each feature lies on the $(n-1)$-dimensional unit sphere, $\mathbf{x}_j \in \mathbb{S}^{n-1}$.

Given the latent variables, we assume the following model on the generation of the data matrix $\mathbf{X}$,

$$p(\mathbf{X}|\boldsymbol{\nu},\mathbf{s},\mathbf{y},\mathbf{z}) = \prod_{j=1}^{d} p_{\text{vMF}}(\mathbf{x}_j|s_j\boldsymbol{\mu}_{z_j}), \tag{1}$$

where

$$p_{\text{vMF}}(\mathbf{x}|\boldsymbol{\mu}) = \frac{\exp(\kappa\mathbf{x}\cdot\boldsymbol{\mu})}{C_{n-1}(\kappa)} \tag{2}$$

is the von Mises-Fisher (vMF) distribution for $\mathbf{x} = (x_1, ..., x_n) \in \mathbb{S}^{n-1}$ with the mean direction parameter $\boldsymbol{\mu} \in \mathbb{S}^{n-1}$ and the concentration parameter $\kappa > 0$. The normalizaton constant is defined by the function,

$$C_n(\kappa) = \int_{\|\mathbf{x}\|=1} \exp(\kappa\mathbf{x}\cdot\boldsymbol{\mu})d\mathbf{x} = \frac{(2\pi)^{n/2}}{\kappa^{\frac{n}{2}-1}} I_{\frac{n}{2}-1}(\kappa),$$

where I_v is the modified Bessel function of the first kind of order v.[1]

For block modeling, we assume that the mean parameter vectors of features in (1) are defined for $l = 1, ..., L$ by

$$\boldsymbol{\mu}_l = (\nu_{y_1,l}, ..., \nu_{y_n,l}),$$

where $\nu_{k,l} \in \mathbb{R}(k = 1, ..., K)$ is the mean parameter of the klth block. This means that the elements of the mean parameter vector $\boldsymbol{\mu}_l$ are constrained to have common values if the cluster assignments of data samples are the same.

Then the constraint of the vMF distribution in (2) entails the constraints,

$$\|\boldsymbol{\mu}_l\|^2 = \sum_{k=1}^{K} n_k\nu_{k,l}^2 = 1, \tag{3}$$

$$\boldsymbol{\mu}_l\cdot\mathbf{1} = \sum_{k=1}^{K} n_k\nu_{k,l} = 0, \tag{4}$$

for the block mean parameter $\boldsymbol{\nu} = \{\nu_{k,l}\}_{k=1,l=1}^{K,L}$.

Additionally, in the model (1), the variable $\mathbf{s} = \{s_j \in \{-1,+1\}\}_{j=1}^{d}$ is introduced to take into account the sign of correlation between $\mathbf{x}_j$ and $\boldsymbol{\mu}_{z_j}$. That is, the cluster centroid vector of features $\boldsymbol{\mu}_{z_j}$ is positively (negatively, resp.) correlated with the jth feature if $s_j = +1$ ($s_j = -1$, resp.).

Hence, given the data matrix $\mathbf{X}$, we estimate latent variables $\mathbf{y}$, $\mathbf{z}$ and $\mathbf{s}$, and the block mean parameters $\boldsymbol{\nu}$ under the above constraints. Note here that the model (1) assumes the independence over data features instead of data samples. Probabilistic models with such an independence assumption have been proposed, for example, in [15] and [16]. In particular, spectral dimensionality reduction methods such as isomap, locally linear embeddings, and Laplacian eigenmaps can be interpreted through such probabilistic models [15].

Since the constraints (3) and (4) depend on the latent variable $\mathbf{y}$ through $\{n_k\}$, we express the block mean parameter $\nu_{k,l}$ as[2]

$$\nu_{k,l} = \frac{\xi_{k,l}}{n_k R_l}, \quad \text{for } \xi_{k,l} \in \mathbb{R},$$

[1] The usual vMF distribution on the $(n-1)$-dimensional sphere has the normalization constant $C_n(\kappa)$ whereas in (2), we have $C_{n-1}(\kappa)$ because of the additional constraint, $\mathbf{x}\cdot\mathbf{1} = \sum_{i=1}^{n} x_i = 0$, which implies that $\mathbf{x}$ lies on $(n-2)$-dimensional unit sphere.

[2] If $n_k = 0$, let $\nu_{k,l} = \xi_{k,l} = 0$.

where

$$R_l = \sqrt{\sum_{k=1}^{K} \frac{\xi_{k,l}^2}{n_k}}.$$

If $\boldsymbol{\xi} = \{\xi_{k,l}\}_{k=1,l=1}^{K,L}$ satisfies $\sum_{k=1}^{K} \xi_{k,l} = 0$ for $l = 1, \cdots, L$ then the constraints (3) and (4) are satisfied. Hence, we consider $\boldsymbol{\xi}$ as a parameter instead of $\boldsymbol{\nu}$ henceforth, and express the model (1) as $p(\mathbf{X}|\boldsymbol{\xi}, \mathrm{s}, \mathbf{y}, \mathbf{z})$.

2.3. The overall generative model

As in the usual probabilistic models for biclustering [6], we assume the following categorical models for latent variables,

$$p_{\text{cat}}(\mathbf{y}|\boldsymbol{\rho}) = \prod_{i=1}^{n} \rho_{y_i} = \prod_{i=1}^{n}\prod_{k=1}^{K} \rho_k^{\delta_{y_i,k}}, \tag{5}$$

$$p_{\text{cat}}(\mathbf{z}|\boldsymbol{\pi}) = \prod_{j=1}^{d} \pi_{z_j} = \prod_{j=1}^{d}\prod_{l=1}^{L} \pi_l^{\delta_{z_j,l}}, \tag{6}$$

where $\boldsymbol{\rho} = \{\rho_k\}_{k=1}^{K}$ and $\boldsymbol{\pi} = \{\pi_l\}_{l=1}^{L}$ are mixing proportions satisfying $\rho_k \geq 0$ $(k = 1, ..., K)$ and $\sum_{k=1}^{K} \rho_k = 1$, and $\pi_l \geq 0$ $(l = 1, ..., L)$ and $\sum_{l=1}^{L} \pi_l = 1$.

Thus the overall generative model of the data matrix $\mathbf{X}$ is given by marginalizing out the latent variables,

$$\begin{aligned} p(\mathbf{X}|\boldsymbol{\xi}, \mathrm{s}, \boldsymbol{\rho}, \boldsymbol{\pi}) &= \sum_{\mathbf{z}}\sum_{\mathbf{y}} p(\mathbf{X}, \mathbf{y}, \mathbf{z}|\boldsymbol{\xi}, \mathrm{s}, \boldsymbol{\rho}, \boldsymbol{\pi}) \\ &= \sum_{\mathbf{z}}\sum_{\mathbf{y}} p_{\text{vMF}}(\mathbf{X}|\boldsymbol{\xi}, \mathrm{s}, \mathbf{y}, \mathbf{z}) p_{\text{cat}}(\mathbf{y}|\boldsymbol{\rho}) p_{\text{cat}}(\mathbf{z}|\boldsymbol{\pi}), \end{aligned}$$

where $\sum_{\mathbf{y}}$ and $\sum_{\mathbf{z}}$ denote the summations over all possible configurations of latent variables $\mathbf{y}$ and $\mathbf{z}$, respectively.

3. Inference and parameter estimation

In this section, we formulate Bayesian inference for the biclustering model in the previous section. We assume the Dirichlet distributions as prior distributions for the mixing proportions of the data sample and feature clusters, respectively,

$$p_{\text{Dir}}(\boldsymbol{\rho}) = \frac{\Gamma(K\alpha)}{\Gamma(\alpha^K)} \prod_{k=1}^{K} \rho_k^{\alpha-1}, \tag{7}$$

$$p_{\text{Dir}}(\boldsymbol{\pi}) = \frac{\Gamma(L\beta)}{\Gamma(\beta^L)} \prod_{l=1}^{L} \pi_l^{\beta-1}, \tag{8}$$

where $\alpha > 0$ and $\beta > 0$ are hyperparameters, and $\Gamma(x) = \int_0^\infty t^{x-1}e^{-t}dt$ is the gamma function.

The posterior distribution of the latent variables conditioned on the data matrix, $\boldsymbol{\xi}$, and $\mathbf{s}$ is

$$p(\mathbf{y}, \mathbf{z}|\mathbf{X}, \boldsymbol{\xi}, \mathrm{s}) = \int p(\mathbf{y}, \mathbf{z}, \boldsymbol{\rho}, \boldsymbol{\pi}|\mathbf{X}, \boldsymbol{\xi}, \mathrm{s}) d\boldsymbol{\rho} d\boldsymbol{\pi},$$

where $p(\mathbf{y}, \mathbf{z}, \boldsymbol{\rho}, \boldsymbol{\pi}|\mathbf{X}, \boldsymbol{\xi}, \mathrm{s}) = p(\mathbf{X}, \mathbf{y}, \mathbf{z}|\boldsymbol{\xi}, \mathrm{s}, \boldsymbol{\rho}, \boldsymbol{\pi}) p_{\text{Dir}}(\boldsymbol{\rho}) p_{\text{Dir}}(\boldsymbol{\pi}) / p(\mathbf{X}|\boldsymbol{\xi}, \mathrm{s})$, which is intractable because the likelihood function $p(\mathbf{X}|\boldsymbol{\xi}, \mathrm{s})$ requires the summations $\sum_{\mathbf{y}}$ and $\sum_{\mathbf{z}}$ of K^n and L^d terms, respectively.

3.1. Variational Bayesian inference

To obtain a tractable upper bound for the negative log-likelihood, $-\log p(\mathbf{X}|\boldsymbol{\xi}, \mathbf{s})$, we introduce an approximating posterior $q(\mathbf{y}, \mathbf{z}, \boldsymbol{\rho}, \boldsymbol{\pi})$, that factorizes as

$$q(\mathbf{y}, \mathbf{z}, \boldsymbol{\rho}, \boldsymbol{\pi}) = q(\mathbf{y})q(\mathbf{z})q(\boldsymbol{\rho})q(\boldsymbol{\pi}),$$

where the distribution $q(\mathbf{y})$ further factorizes as $q(\mathbf{y}) = \prod_{i=1}^{n} q(y_i)$. Then we obtain an upper bound for $F(\boldsymbol{\xi}, \mathbf{s}) = -\log p(\mathbf{X}|\boldsymbol{\xi}, \mathbf{s})$ as follows,

$$\begin{aligned} F(\boldsymbol{\xi}, \mathbf{s}) &\leq F(\boldsymbol{\xi}, \mathbf{s}) + \mathrm{KL}[q(\mathbf{y}, \mathbf{z}, \boldsymbol{\rho}, \boldsymbol{\pi})||p(\mathbf{y}, \mathbf{z}|\mathbf{X}, \boldsymbol{\xi}, \mathbf{s})] && (9) \\ &= \left\langle \log \frac{q(\mathbf{y}, \mathbf{z}, \boldsymbol{\rho}, \boldsymbol{\pi})}{p(\mathbf{X}, \mathbf{y}, \mathbf{z}, \boldsymbol{\rho}, \boldsymbol{\pi}|\boldsymbol{\xi}, \mathbf{s})} \right\rangle_{q(\mathbf{y}, \mathbf{z}, \boldsymbol{\rho}, \boldsymbol{\pi})} \equiv \overline{F}[q(\mathbf{y}), q(\mathbf{z}), q(\boldsymbol{\rho}), q(\boldsymbol{\pi}), \boldsymbol{\xi}, \mathbf{s}], && (10) \end{aligned}$$

where $\mathrm{KL}[q||p]$ denotes the Kullback information from q to p, and $\langle\cdot\rangle_q$ the expectation with respect to q. The inequality (9) follows from the non-negativity of the Kullback information or equivalently from Jensen's inequality. The upper bound $\overline{F}$, called the variational free energy, is minimized alternately with respect to one of $q(\mathbf{y})$, $q(\mathbf{z})$, $q(\boldsymbol{\rho})$, $q(\boldsymbol{\pi})$, $\boldsymbol{\xi}$, and $\mathbf{s}$ while the others are fixed. The variational free energy provides an estimate of model's negative log-likelihood. It is used as a model selection criterion by choosing the numbers of clusters, K and L, that minimize the variational free energy among candidate models.

Variational posteriors of mixing proportions: Minimizing $\overline{F}$ with respect to $q(\boldsymbol{\rho})$ and $q(\boldsymbol{\pi})$ while the other variational posteriors and parameters fixed, we have

$$q(\boldsymbol{\rho}) \propto \exp \langle \log(p_{\mathrm{cat}}(\mathbf{y}|\boldsymbol{\rho})p_{\mathrm{Dir}}(\boldsymbol{\rho}))\rangle_{q(\mathbf{y})},$$

and

$$q(\boldsymbol{\pi}) \propto \exp \langle \log(p_{\mathrm{cat}}(\mathbf{z}|\boldsymbol{\pi})p_{\mathrm{Dir}}(\boldsymbol{\pi}))\rangle_{q(\mathbf{z})}.$$

It follows from (5), (7), (6), and (8) that

$$\begin{aligned} q(\boldsymbol{\rho}) &\propto \prod_{k=1}^{K} \rho_k^{\overline{n}_k+\alpha-1}, && (11) \\ q(\boldsymbol{\pi}) &\propto \prod_{l=1}^{L} \pi_l^{\overline{d}_l+\beta-1}. && (12) \end{aligned}$$

This means that $q(\boldsymbol{\rho})$ and $q(\boldsymbol{\pi})$ are the Dirichlet distributions with hyperparameters $\{\overline{n}_k + \alpha\}$ and $\{\overline{d}_l + \beta\}$, respectively, where

$$\begin{aligned} \overline{n}_k &= \langle n_k\rangle_{q(\mathbf{y})} = \sum_{i=1}^{n} q(y_i = k), && (13) \\ \overline{d}_l &= \langle d_l\rangle_{q(\mathbf{z})} = \sum_{j=1}^{d} q(z_j = l), && (14) \end{aligned}$$

are the expected number of data samples assigned to the kth sample cluster and that of data features assigned to the lth feature cluster, respectively.

Variational posterior of feature cluster assignments and sign of correlation: As a function of only $q(\mathbf{z})$, $\overline{F}$ is minimized when

$$q(\mathbf{z}) \propto \exp \langle \log(p_{\mathrm{vMF}}(\mathbf{X}|\mathbf{z}, \mathbf{y}, \boldsymbol{\xi}) p_{\mathrm{cat}}(\mathbf{z}|\boldsymbol{\pi})) \rangle_{q(\mathbf{y})q(\boldsymbol{\pi})} .$$

Combined with (2), (6) and (12), this yields that

$$q(\mathbf{z}) = \prod_{j=1}^{d} q(z_j),$$

and

$$q(z_j = l) \propto \exp\left(\kappa s_j \eta_l^{(j)} + \Psi(\overline{d}_l + \beta) - \Psi(d + L\beta)\right), \tag{15}$$

where $\Psi(x) = d \log \Gamma(x)/dx$ is the digamma function, $\eta_l^{(j)} = \mathbf{x}_j \cdot \overline{\boldsymbol{\mu}}_l$, $\overline{\boldsymbol{\mu}}_l$ is the n-dimensional vector whose ith element is $\overline{\nu}_{y_i,l}$, which can be interpreted as the expected cluster centroid vector of the lth feature cluster, and

$$\overline{\nu}_{k,l} = \langle \nu_{k,l} \rangle_{q(\mathbf{y})} \simeq \frac{\xi_{k,l}}{\overline{n}_k \overline{R}_l}, \tag{16}$$

$$\overline{R}_l = \sqrt{\sum_{k=1}^{K} \frac{\xi_{k,l}^2}{\overline{n}_k}}. \tag{17}$$

Here, we have used the approximation that the expectation of $\nu_{k,l}$ with respect to $q(\mathbf{y})$ is computed by the expectation of n_k. Since $\overline{F}$ depends on the sign s_j through $-s_j \sum_{l=1}^{L} q(z_j = l)\eta_l^{(i)}$, we compute it for the two candidates of $q(z_j)$ with $s_j = +1$ and -1, and adopt s_j and $q(z_j)$ yielding smaller $\overline{F}$.

Variational posterior of sample cluster assignments: Similarly, $\overline{F}$ as a function of $q(y_i)$ is minimized by

$$q(y_i) \propto \exp \langle \log(p_{\mathrm{vMF}}(\mathbf{X}|\mathbf{z}, \mathbf{y}, \boldsymbol{\xi}) p_{\mathrm{cat}}(\mathbf{y}|\boldsymbol{\rho})) \rangle_{q(\mathbf{y}^{\neg i})q(\mathbf{z})q(\boldsymbol{\rho})} ,$$

where $q(\mathbf{y}^{\neg i}) = \prod_{i' \neq i} q(y_{i'})$. This yields that

$$q(y_i = k) \propto \exp\left(\kappa \gamma_k^{(i)} + \Psi(\overline{n}_k + \alpha) - \Psi(n + K\alpha)\right), \tag{18}$$

where

$$\gamma_k^{(i)} = \sum_{j=1}^{d} s_j x_{ij} \left(\sum_{l=1}^{L} q(z_j = l) \left\langle \nu_{k,l}^{(y_i = k)} \right\rangle_{q(\mathbf{y}^{\neg i})} \right) + \sum_{i' \neq i} \sum_{k'=1}^{K} \sum_{j=1}^{d} s_j x_{i'j} \left(\sum_{l=1}^{L} q(z_j = l) \left\langle \nu_{k',l}^{(y_i = k)} \right\rangle_{q(\mathbf{y}^{\neg i})} \right), \tag{19}$$

$$\begin{aligned}
\left\langle \nu_{k',l}^{(y_i = k)} \right\rangle_{q(\mathbf{y}^{\neg i})} &= \frac{\xi_{k',l}}{(\overline{n}_{k'}^{\neg i} + \delta_{k',k}) \overline{R}_l^{(y_i = k)}}, \\
\overline{R}_l^{(y_i = k)} &= \sqrt{\sum_{m=1}^{K} \frac{\xi_{m,l}^2}{\overline{n}_m^{\neg i} + \delta_{m,k}}}, \\
\text{and } \overline{n}_k^{\neg i} &= \sum_{i' \neq i} q(y_{i'} = k) = \overline{n}_k - q(y_i = k).
\end{aligned}$$

Note that the computation of $\gamma_k^{(i)}$ for each i requires the complexity of $O(n)$. If we approximate it by ignoring the dependency of $\nu_{k,l}$ on $\mathbf{y}$, $\gamma_k^{(i)}$ in (19) is replaced by

$$\gamma_k^{(i)} = -\frac{\|\tilde{\mathbf{x}}_i - \overline{\mathbf{m}}_k\|^2}{2}, \tag{20}$$

where $\tilde{\mathbf{x}}_i$ is the ith row vector of the data matrix and $\overline{\mathbf{m}}_k$ is the d-dimensional vector consisting of $s_j \langle \nu_{k,z_j} \rangle_{q(z_j)}$, which can be interpreted as the expected cluster centroid vector of the kth sample cluster.

Parameter estimation: The variational free energy depends on the block mean parameter $\xi_{k,l}$ through

$$-\kappa \sum_{j=1}^{d} \sum_{i=1}^{n} \sum_{l=1}^{L} \sum_{k=1}^{K} q(z_j = l) q(y_i = k) \langle \nu_{k,l} \rangle_{q(\mathbf{y})} s_j x_{ij},$$

if we approximate the expectation with respect to $q(\mathbf{y})$ by separately applying it for $\delta_{y_i,k}$ and $\nu_{k,l}$. If $\langle \nu_{k,l} \rangle_{q(\mathbf{y})}$ in the above sum is further approximated by (16), the Cauchy-Schwarz inequality implies that $\overline{F}$ is minimized by

$$\xi_{k,l} = \sum_{j=1}^{d} \sum_{i=1}^{n} q(z_j = l) q(y_i = k) s_j x_{ij}, \tag{21}$$

up to multiplication of a constant independent of k.

The update rule of the concentration parameter κ is obtained by differentiating $\overline{F}$ with respect to κ and equating it to zero, which, combined with (17) and (21), yields,

$$A_{n-1}(\kappa) \equiv \frac{C'_{n-1}(\kappa)}{C_{n-1}(\kappa)} = \frac{I_{\frac{n-1}{2}}(\kappa)}{I_{\frac{n-1}{2}-1}(\kappa)} = \frac{1}{d} \sum_{l=1}^{L} \overline{R}_l.$$

We use the method proposed in [17] to solve this equation for κ. Moreover, if we introduce a separate concentration parameter κ_l to the lth feature cluster, the above equation is replaced by

$$A_{n-1}(\kappa_l) = \frac{\overline{R}_l}{\overline{d}_l},$$

for $l = 1, ..., L$.

The hyperparameters α and β of the Dirichlet priors can also be estimated by using the Newton-Raphson updates as shown in [6].

In the limit, $\kappa \to \infty$, the probabilistic (soft) cluster assignments given in (15) and (18) become deterministic (hard), and the algorithm reduces to the asymmetric biclustering method proposed in [3], which maximizes the objective function, $\sum_{j=1}^{d} s_j \mathbf{x}_j \cdot \boldsymbol{\mu}_{z_j}$ where $\boldsymbol{\mu}_l = (\nu_{y_1,l}, ..., \nu_{y_n,l})$.

4. Experiments

In this section, we present experimental results of the proposed approach with synthetic and real-world datasets. We used the hyperparameters $\alpha = \beta = 1$ so that the prior distributions (7) and (8) are uniform. We ran five trials of the variational Bayesian algorithm and took the result with the smallest variational free energy. We observed that the variational free energy monotonically decreases in spite of the approximations introduced in Section 3.1 except for the rough approximation given by (20), which we did not use in the experiments below.

4.1. IRIS data

As a low-dimensional real dataset for demonstration, we applied the variational Bayesian inference algorithm to the *Iris* dataset, which has often been used in pattern recognition [18]. The data matrix consists of $150(= n)$ samples and $4(= d)$ features. We observed that the concentration parameter grows unboundedly, if we estimate it for fixed K and L that we considered. Although this means that the deterministic (non-probabilistic) asymmetric biclustering is favored, the model selection mechanism is lost in this limit, while if we set $L = 4$, one of the feature clusters is automatically eliminated during the estimation as is shown in Figure 1 for a different setting of κ.

We fixed $\kappa \in \{100, 150, 200, 250, 300\}$ and examined if the model selection mechanism works. Fixing $L = 4$, we compared the variational free energy for $K \in \{3, 4, \cdots, 8\}$ (Table 1). We can see that as κ grows, the more complex model is favored. This means that κ works as a regularization parameter. If its value is fixed, the numbers of clusters K and L can be determined by the minimization of the vatiational free energy. In addition, because of the prior distributions (7) and (8), redundant components are automatically eliminated if $\overline{n}_k/n$ or $\overline{d}_l/d$ is estimated near zero, or two cluster centers are estimated to be overlapping [14]. In fact, for all the results in Table 1, at least one of the four feature clusters was eliminated and the feature cluster consisting of the two features, *petal length* and *petal width* was obtained as illustrated in Figure 1 for $\kappa = 200$ and $K = 3$ with the clustered parallel coordinate plot proposed in [3], where sample cluster assignments are indicated by colors of polylines and feature clusters are divided by thick separators.

Table 1. Variational free energy (negated) for different number K of sample clusters and the concentration parameter κ. The number of sample clusters with the smallest variational free energy is highlighted for each column.

	$\kappa = 100$	$\kappa = 150$	$\kappa = 200$	$\kappa = 250$	$\kappa = 300$
$K = 3$	**719.028**	**766.249**	**801.694**	827.433	845.829
$K = 4$	717.768	764.923	800.353	831.528	853.828
$K = 5$	717.887	763.457	799.726	**832.243**	856.360
$K = 6$	716.763	757.156	798.302	830.458	**857.001**
$K = 7$	715.442	755.101	796.768	829.424	853.641
$K = 8$	714.556	755.072	795.784	828.766	852.105

In this example, the growth of κ was observed by the minimization of the variational free energy, which is equivalent to the maximum likelihood estimation of κ. A finite κ can be estimated by introducing a prior distribution and incorporating Bayesian inference for it [13]. Also in such Bayesian inference of κ, the model selection results by variational free energy minimization or automatic elimination of redundant components depend on the choice of the prior distribution of κ.

4.2. Synthetic data

We demonstrate our experimental study on a 750 records of 12-dimensional synthetic data employed in [1, 3]. This multivariate data includes four 3-dimensional clusters with 10% noise and two 6-dimensional clusters without noise, while the data samples are uniformly distributed in other dimensions.

The estimation of the concentration parameter chose the deterministic biclustering ($\kappa \to \infty$) also for this dataset. Hence, we fixed $\kappa \in \{200, 225, 250, 275, 300\}$, and compared the variational free energy for $K \in \{3, \cdots, 8\}$ and $L \in \{4, \cdots, 9\}$. Table 2 shows the selected pairs of the numbers of clusters, (K, L), that attained the minimum variational free energy for different κ.

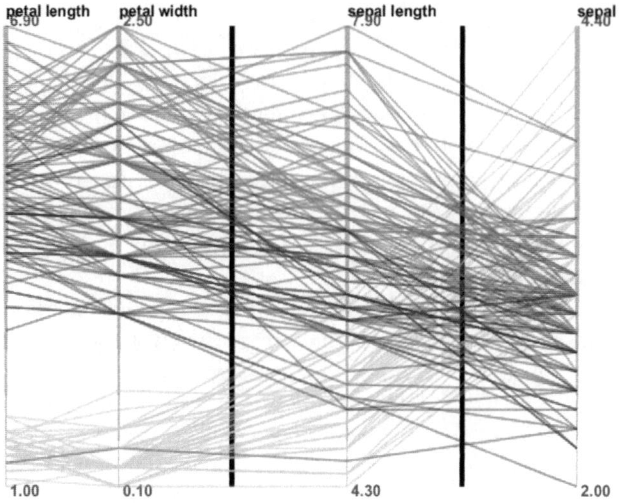

Figure 1. Clustered parallel coordinate plot for $K = 3$ and $L = 4$.

Table 2. Selected numbers of sample and feature clusters

	$\kappa = 200$	$\kappa = 225$	$\kappa = 250$	$\kappa = 275$	$\kappa = 300$
(K, L)	$(4, 3)$	$(4, 5)$	$(6, 8)$	$(5, 8)$	$(6, 7)$

We see the tendency that the greater κ is, the more complex model is selected if the variational free energy is used as the model selection criterion. Figure 2 (left) shows the variational free energy for pairs (K, L), with its sign flipped for presentation purpose, when $\kappa = 250$. Its minimum is attained at $(K = 6, L = 8)$ and $(K = 6, L = 9)$. The result of biclustering for $K = 6$ and $L = 8$ is demonstrated in Figure 2 (right) with the clustered parallel coordinate plot. Although we set $L = 8$, one of the feature clusters is eliminated, and there are seven feature clusters, one cluster with six features and six clusters with each of the other features.

In the clustered parallel coordinate plot (Figure 2 (right)), we can find the three feature clusters with each of the three uniformly distributed dimensions, which can be detected as less correlated clusters by looking at the error of each block defined by

$$\frac{1}{\overline{n}_k \overline{d}_l} \sum_{i=1}^{n} \sum_{j=1}^{d} q(y_i = k) q(z_j = l) \left(x_{ij} - s_j \nu_{k,l} \right)^2 ,$$

for the klth block. Feature or sample clusters with high block errors can be considered mismatched with the block model. The interactive visual data analysis framework developed in [3] enables to keep such clusters aside and continue the analysis for the remaining data matrix.

4.3. Supernova data

We employed the Berkeley supernova dataset [19] as another real-world example, to which the deterministic asymmetric biclustering has been applied in a previous work [20]. The dataset consists of 132 data samples corresponding to supernovae of a subtype, called "Type Ia", and 14 variables including both photometric and spectroscopic features.

We observed the tendency that if the concentration parameter κ is fixed, the more complex model is favored as κ grows also for this dataset. We also found that κ does not grow

Journal of Physics: Conference Series **699** (2016) 012018 doi:10.1088/1742-6596/699/1/012018

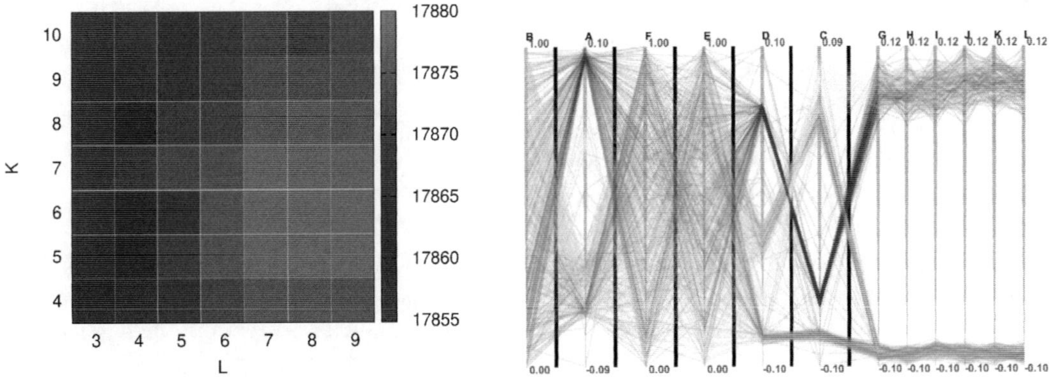

Figure 2. Variational free energy (negated) for different numbers of sample and feature clusters (left). Clustered parallel coordinate plot for $K = 6$ and $L = 8$ (right).

unboundedly if it is estimated for moderate numbers of clusters. Thus, we introduced separate concentration parameters to respective feature clusters and estimated them by the update rule given in Section 3.1. We set the numbers of clusters to $K = L = 6$, from which a detailed interactive analysis of this dataset was conducted in the previous work [20]. Figure 3(left) shows the result with the contracted parallel coordinate plot where axes of each feature cluster are contracted to a single axis by the linear discriminant analysis using sample cluster assignments as class labels [3]. We focus on the feature clusters, {Si6V},{Si6EW, Si6A, Si6FWHM}, {Si5EW, Si5DEW, x1, MBc}, which are consistent with the previous analysis except for the fact that MBc is included in the last cluster. We also observe that the scatter plot between the two feature clusters, {Si6EW, Si6A, Si6FWHM}, {Si5EW, Si5DEW, x1, MBc} appears similar to that discussed in the previous work (Figure 3(right)). Additionally, the feature x1 is negatively correlated with the other three features, Si5EW, Si5DEW, MBc, and included in the cluster with its sign flipped. This suggests that the variables $\mathbf{s}$ in the model (1) are also estimated appropriately.

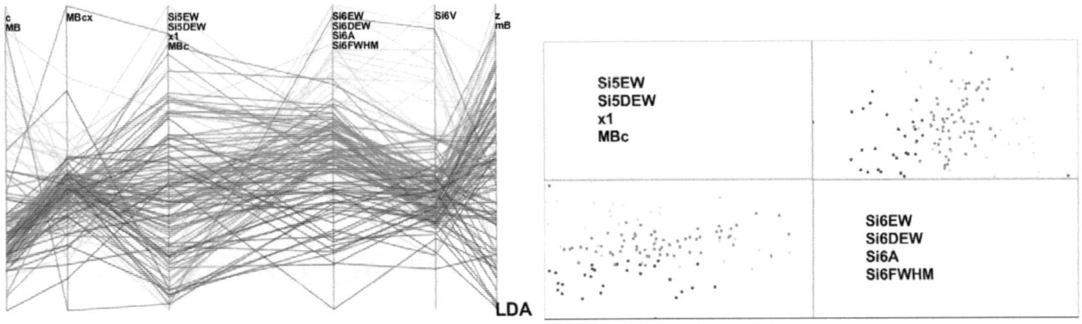

Figure 3. Contracted parallel coordinate plot of the supernova dataset (left). Scatter plot of the third and fourth feature clusters on the left, {Si5EW, Si5DEW, x1, MBc} and {Si6EW, Si6A, Si6FWHM} (right).

5. Discussion and conclusion

In this paper, we derived the variational Bayesian inference algorithm for the mixture of the constrained vMF distributions, which provides a probabilistic model of asymmetric biclustering.

The derived algorithm was applied to the multivariate data visualization tool with enhanced parallel coordinate plots. Through numerical experiments, we demonstrated that the Bayesian formulation offers a selection criterion of the number of feature and sample clusters.

The assumption of the block structure may be too restrictive for some datasets. Our current framework enables to interactively select clustered rows and columns of the data matrix and set aside them from the main target of the analysis. It would be important to model more complex structures directly, such as the nested partitioning proposed for the ordinary biclustering method in [7].

Another type of parallel coordinate plots chooses the ordering, locations and scales of axes so that data samples are aligned as horizontally as possible [21]. It is an important undertaking to explore an extension of our visualization framework in such a direction, in particular, by incorporating probabilistic cluster assignments of data samples and features.

Acknowledgments

This work has been partially supported by MEXT KAKENHI under Grants-in-Aid for Scientific Research on Innovative Areas No. 25120014. We would like to thank Professor Makoto Uemura, Hiroshima Astrophysical Science Center, Hiroshima University for his valuable comments on the analysis of the supernova dataset.

References

[1] Tatu A, Maaß F, Farber I, Bertini E, Schreck T, Seidl T and Keim D 2012 *Proc. IEEE Conference on Visual Analytics Science and Technology* pp 63–72
[2] Yuan X, Ren D, Wang Z and Guo C 2013 *IEEE Transactions on Visualization and Computer Graphics* **19** 2625–2633
[3] Watanabe K, Wu H-Y, Niibe Y, Takahashi S and Fujishiro I 2015 *Proc. 2015 IEEE Pacific Visualization Symp.* pp 287–294
[4] Hartigan J A 1972 *Journal of the American Statistical Association* **67** 123–129
[5] Mechelen I V, Bock H H and Boeck P D 2004 *Statistical methods in medical research* **13** 363–394
[6] Shan H and Banerjee A 2008 *Proc. 2008 IEEE International Conference on Data Mining* pp 530–539
[7] Roy D M and Teh Y W 2009 *Advances in Neural Information Processing Systems*
[8] Peng W, Ward M O and Rundensteiner E A 2004 *Proc. IEEE Conference on Information Visualization* pp 89–96
[9] Nohno K, Wu H-Y, Watanabe K, Takahashi S and Fujishiro I 2014 *Proc. International Conference on Information Visualisation* pp 7–12
[10] Dhillon I S and Modha D S 2001 *Machine Learning* **42** 143–175
[11] Banerjee A, Dhillon I S, Ghosh J and Sra S 2005 *Journal of Machine Learning Research* **6** 1345–1382
[12] Tanabe A, Fukumizu K, Oba S and Ishii S 2004 *Proc. 2004 Workshop on Information-Based Induction Sciences* pp 46–51 (in Japanese)
[13] Gopal S and Yang Y 2014 *Proc. International Conference on Machine Learning* pp 154–162
[14] Attias H 1999 *Proc. 15th Conference on Uncertainty in Artificial Intelligence* pp 21–30
[15] Lawrence N D 2012 *Journal of Machine Learning Research* **13** 1609–1638
[16] Kemp C and Tenenbaum J B 2008 *Proc. National Academy of Sciences* **105** 10687–10692
[17] Sra S 2012 *Computational Statistics* **27** 177–190
[18] Asuncion A and Newman D J 2007 UCI Machine Learning Repository
[19] Silverman J M, Foley R J, Filippenko A V, Ganeshalingam M *et al* 2012 *Monthly Notices of the Royal Astronomical Society* **425** 1789–1818
[20] Uemura M, Kawabata K S, Ikeda S, Maeda K, Wu H-Y, Watanabe K, Takahashi S and Fujishiro I 2016 Data-driven approach to Type Ia supernovae: variable selection on the peak luminosity and clustering in visual analytics *Journal of Physics: Conference Series* in press
[21] Kumasaka N and Shibata R 2008 *Computational Statistics & Data Analysis* **52** 3616–3644

Stochastic gradient method with accelerated stochastic dynamics

Masayuki Ohzeki

Department of Systems Science, Graduate School of Informatics, Kyoto University, 36-1 Yoshida Hon-machi, Sakyo-ku, Kyoto, 606-8501, Japan

E-mail: mohzeki@i.kyoto-u.ac.jp

Abstract. We implement the simple method to accelerate the convergence speed to the steady state and enhance the mixing rate to the stochastic gradient Langevin method. The ordinary stochastic gradient method is based on mini-batch learning for reducing the computational cost when the amount of data is extraordinary large. The stochasticity of the gradient can be mitigated by the injection of Gaussian noise, which yields the stochastic Langevin gradient method; this method can be used for Bayesian posterior sampling. However, the performance of the stochastic Langevin gradient method depends on the mixing rate of the stochastic dynamics. In this study, we propose violating the detailed balance condition to enhance the mixing rate. Recent studies have revealed that violating the detailed balance condition accelerates the convergence to a stationary state and reduces the correlation time between the samplings. We implement this violation of the detailed balance condition in the stochastic gradient Langevin method and test our method for a simple model to demonstrate its performance.

1. Introduction

Since massive amounts of data can be acquired from various sources, the importance of the so-called big-data analysis is rapidly increasing. In particular, extracting relevant parameters that facilitate describing the acquired data is critical for analyzing high-dimensional data. For high-dimensional data analysis, Boltzmann machine learning can be used to extract parameters that describe the structure of given data [1]. Boltzmann machine learning has proven to be effective and has stimulated increasing interest in deep learning [2, 3, 4, 5]. We can here implement the L_1-norm regularization to determine the sparse parameters for characterizing given data [5] and and a type of the relaxation method to the L_0 norm [6, 7]. Although Boltzmann machine learning has been successfully implemented to extract features from high-dimensional data, the computational cost is usually expensive for high-precision estimations. The crucial reason for the high computational cost is the expectation value in the Gibbs-Boltzmann distribution. In order to mitigate the computational cost, some approximations such as the mean-field analysis, belief propagation, and their improvements can be adopted [8, 9, 10, 11, 12, 13, 14]. In addition, increase of the amount of the data also increases the computational cost for evaluating the update rules in the learning algorithms.

In this study, we develop a technique to mitigate the computational cost of learning from large amount of high-dimensional data. In the middle of the 20th century, Robbins and Monro proposed a stochastic gradient method, which utilized a mini-batch dataset for gradient computation to update the tentative estimations at each iteration in the learning process [15].

The stochasticity of the gradient will be averaged out because the entire dataset is used by the end of the learning process. Then, the learning rate is gradually decreased to ensure convergence to a local maximum of the likelihood function. In order to mitigate the stochasticity of the gradient, Welling and Teh proposed an improved version of the stochastic gradient method by introducing Gaussian noise at each iteration such as by using the Langevin dynamics [16]. At the first stage of the method, the stochastic gradient noise dominates the injected noise in the learning process. On the other hand, at the last stage, the injected noise equilibrates the system to converge to a stationary state.

For enhancing the precision of the learning, it is beneficial to accelerate the convergence to a stationary state in the stochastic gradient Langevin method. Considering the recent advancement in stochastic dynamics in the field of nonequilibrium statistical physics, a novel technique to accelerate the convergence to the desired stationary distribution has been proposed, which performs the violation of the detailed balance condition (vDBC). The detailed balance condition (DBC) is typically satisfied in stochastic dynamics simulated by the Langevin equation, the master equation, and the Fokker-Planck equation. One such representative method is the Markov-chain Monte-Carlo method. The condition assures convergence to a stationary distribution and facilitates the construction of the stochastic rule as in the transition matrix. However, the DBC is just a sufficient condition for the convergence and thus is not necessarily to be satisfied. Hence, we may violate the DBC to attain faster convergence to the stationary distribution. One of the techniques that do not hold the DBC is the Suwa-Todo method [17]. In this method, the elements of the transfer matrix are optimized in the master equation simulated in the Markov-Chain Monte-Carlo method, while ignoring the DBC. Another method is the skewed DBC, which utilizes double stochastic rules for acceleration while violating the DBC [18]. The vDBC is characterized by the inherent asymmetry of the transfer matrix at the master equation or the Fokker-Planck equation level. This asymmetry leads to an eigenvalue shift as compared to the case when the DBC is satisfied and results in the acceleration of the convergence to a stationary state [19]. From a different point of view, the vDBC can be captured as rare-event sampling, which produces an alternative pathway to the stationary state [20]. A simple implementation of the vDBC has been performed in the Langevin dynamics especially for a system with continuous-valued degrees of freedom [21, 22]. The vDBC can be recast as the resultant feature of the optimization of stochastic dynamics from a point of view of the variational principle [23]. Although several innovative studies to enhance the sampling efficacy and accelerate the convergence to a predetermined distribution have been extensively reported [24, 25, 26, 27, 28, 29, 30, 31], the concept of vDBC is much easier to implement. In this study, we aim to implement the concept of the vDBC in the stochastic gradient Langevin method and verify its performance.

The remainder of the paper is organized as follows: First, we briefly review the stochastic gradient Langevin method. In the next section, we introduce the scheme of the accelerated stochastic dynamics through the vDBC. In the fourth section, we formulate the application of the accelerated stochastic dynamics to the stochastic gradient Langevin method. We test our scheme, in the following section, for the simplest learning model, which is a mixture of the Gaussian distribution, to confirm the efficacy of our method. The last section is devoted for discussion on the future direction of our study.

2. Stochastic gradient Langevin method

In this section, we briefly review the stochastic gradient Langevin method. The stochastic gradient Langevin method has been invented for Bayesian learning, which captures the uncertainty in the learned parameters and avoids overfitting. This method combines the Robbins-Monro method, which optimizes a likelihood function in a stochastic manner, with overdamped Langevin dynamics, which injects noise into the parameter updates. Each stochasticity has a different property. The former stochasticity is a resultant property to reduce the computational cost for large-scale data. In the latter, the noise makes the trajectory of the parameters converge to full posterior distribution rather than just the maximum a posteriori mode.

We consider the learning of the parameters $\boldsymbol{\theta}$ with a prior distribution $p(\boldsymbol{\theta})$ and the probability of the datum $\mathbf{x}$ given the parameter $\boldsymbol{\theta}$, $P(\mathbf{x}|\boldsymbol{\theta})$; namely the posterior distribution $P(\mathbf{x}|\boldsymbol{\theta})p(\boldsymbol{\theta})$. The basic task is to find the maximum a posteriori (MAP) parameters $\boldsymbol{\theta}$ by maximizing the following logarithm of the posterior distribution:

$$\boldsymbol{\theta}^* = \arg\max_{\boldsymbol{\theta}} \left\{ \sum_{k=1}^{D} \log P(\mathbf{x}^{(k)}|\boldsymbol{\theta}) + \log p(\boldsymbol{\theta}) \right\}. \quad (1)$$

In the context of the optimization, the prior distribution regularizes the parameters while the likelihood terms constitute the cost function to be optimized. In the basic framework of the maximum a posteriori, by taking the derivative of the parameters $\boldsymbol{\theta}$, we obtain the following iteration for seeking the maximum of the posterior distribution.

$$\boldsymbol{\theta}' = \boldsymbol{\theta} + dt\frac{\partial \mathcal{L}(\boldsymbol{\theta})}{\partial \boldsymbol{\theta}}, \quad (2)$$

where

$$\frac{\partial \mathcal{L}(\boldsymbol{\theta})}{\partial \boldsymbol{\theta}} = \frac{\partial}{\partial \boldsymbol{\theta}} \log p(\boldsymbol{\theta}) + \sum_{k=1}^{D} \frac{\partial}{\partial \boldsymbol{\theta}} \log P(\mathbf{x}^{(k)}|\boldsymbol{\theta}). \quad (3)$$

The logarithm of the posterior distribution is defined as $\mathcal{L}(\boldsymbol{\theta})$. Increasing the amount of given data results in the computational cost to sum over the gradient of the log-likelihood in the second term. To mitigate the problem, a stochastic gradient method has been proposed. In this method, the gradient of the log-likelihood function for a randomly chosen subset of the given data is used:

$$\frac{\partial \mathcal{L}(\boldsymbol{\theta})}{\partial \boldsymbol{\theta}} = \frac{\partial}{\partial \boldsymbol{\theta}} \log p(\boldsymbol{\theta}) + \frac{D}{d}\sum_{k=1}^{d} \frac{\partial}{\partial \boldsymbol{\theta}} \log P(\mathbf{x}^{(k)}|\boldsymbol{\theta}). \quad (4)$$

where $d < D$. The subset of the given data is randomly chosen. Over a number of iterations, all the data is used and the noise in the gradient caused by using subsets is averaged out. To ensure convergence to a local maximum, the step size is gradually decreased. The step size dt depends on the step as dt_i and satisfies $\sum_i dt_i = \infty$ and $\sum_t dt_i^2 < \infty$ [15]. Typically, step size is set as $dt_i = a(b+i)^{-\gamma}$ where $\gamma \in (0.5, 1]$. This is the stochastic gradient method and just an approximation of the MAP estimation (or also applicable to the maximum likelihood estimation).

To avoid the problem that it does not capture parameter uncertainty and can potentially overfit data in the MAP estimation, stochastic noise as shown below can be introduced.

$$\boldsymbol{\theta}' = \boldsymbol{\theta} + dt\frac{\partial \mathcal{L}(\boldsymbol{\theta})}{\partial \boldsymbol{\theta}} + \sqrt{2T}dW, \quad (5)$$

where T is the temperature representing the strength of the injected noise (usually set as unity) and $d\mathbf{W}$ denotes the Wiener process, whose order is $O(\sqrt{dt})$, i. e., the Gaussian noise with a vanishing mean and a variance of dt. This simply represents the overdamped Langevin equation in the context of statistical physics. The overdamped Langevin equation has a corresponding

Fokker-Planck equation, which describes the time-evolution of the distribution function. It can be confirmed that the equilibrium distribution can be the posterior distribution. Welling and Teh have proposed the combination of Langevin dynamics with the stochastic gradient method, i.e., the stochastic gradient Langevin method, to generate the posterior distribution for learning from large-scale data [16]. The step size is then decreased similar to the stochastic gradient method introduced above. By decreasing the step size gradually, the injected noise will become dominant and the effective dynamics will converge to the Langevin equation with the exact gradient. A number of the iterations will thus generate the posterior distribution. In other words, the performance of the stochastic gradient Langevin method strongly depends on the convergence rate of the Langevin equation to the equilibrium distribution, i.e., the posterior distribution. The increase in the convergence speed improves the performance and reduces the computational cost of the method. This fact motivates the study of the stochastic gradient Langevin method from a point of view of nonequilibrium statistical physics.

3. Accelerated stochastic dynamics

The stochastic gradient Langevin method aims at the sampling following the posterior distribution while mitigating the computational cost for evaluation of the gradient. Hereafter let us consider how to perform sampling from the desired distribution function. The distribution function can be characterized by the Gibbs-Boltzmann distribution associated with many degrees of freedom. Therefore it is difficult to straightforwardly perform sampling following the distribution function. We then assume that a fictitious system consisting of several degrees of freedom is governed by the Langevin equation. After sufficient long equilibration, the degrees of freedom follows the Gibbs-Boltzmann distribution. However the relaxation to the Gibbs-Boltzmann distribution takes extremely long time. Here, we employ the recent advancement in stochastic dynamics, i.e., the vDBC. For the case of the standard Langevin equation, the DBC holds to ensure the convergence to a stationary state. However, this is just a sufficient condition and can be violated. We modify the standard form of the Langevin equation to accelerate the convergence speed towards a stationary state. This is referred as the accelerated stochastic dynamics.

3.1. Langevin equation and its corresponding Fokker-Planck equation

We begin with the N-dimensional overdamped Langevin dynamics defined as

$$d\boldsymbol{\theta} = \mathbf{A}(\boldsymbol{\theta})dt + \sqrt{2T}d\mathbf{W}, \tag{6}$$

where $\boldsymbol{\theta}$ is the vector representing the N-dimensional degrees of freedom (parameters to be estimated for machine learning) and $d\boldsymbol{\theta}$ denotes the infinitesimal change in $\boldsymbol{\theta}$ during dt. Here we assume that the step size dt is homogeneous for the sake of simplicity. In addition, $\mathbf{A}(\boldsymbol{\theta})$ is a time-independent force vector, T is the temperature, and $d\mathbf{W}$ denotes the Wiener process for the N-dimensional degrees of freedom as stated earlier. The force usually takes the gradient of the energy as $-\partial E(\boldsymbol{\theta})/\partial\boldsymbol{\theta}$ in the context of the statistical physics. or our problem, the force is denoted as $\partial \log p(\boldsymbol{\theta})/\partial\boldsymbol{\theta} + (1/D)\sum_{k=1}^{D} \partial \log P(\mathbf{x}^{(k)}|\boldsymbol{\theta})/\partial\boldsymbol{\theta}$. In other words, the energy function is defined as

$$E(\boldsymbol{\theta}) = -\log p(\boldsymbol{\theta}) - \sum_{k=1}^{D} \log P(\mathbf{x}^{(k)}|\boldsymbol{\theta}) \tag{7}$$

For simplicity, we use the notion of the energy function hereafter. In accelerated stochastic dynamics, we do not stick with the standard case. In order to accelerate the convergence to a stationary state, we modify the force from its usual form. The probability distribution for the degrees of freedom evolved in the Langevin dynamics can be formulated as the Fokker-Planck equation as follows:

$$\frac{\partial P(\boldsymbol{\theta},t)}{\partial t} = -\sum_{k=1}^{N} \frac{\partial}{\partial \theta_k} J_k(\boldsymbol{\theta},t), \tag{8}$$

where $\mathbf{J}(\boldsymbol{\theta},t) = (J_1(\boldsymbol{\theta},t), J_2(\boldsymbol{\theta},t), \cdots, J_N(\boldsymbol{\theta},t))$ denotes the probabilistic flow defined as

$$\mathbf{J}(\boldsymbol{\theta},t) = \left(\mathbf{A}(\boldsymbol{\theta}) - T\frac{\partial}{\partial \boldsymbol{\theta}}\right) P(\boldsymbol{\theta},t). \tag{9}$$

When the system is in any stationary state, the divergence of the probabilistic flow vanishes. Therefore, the following equality should be satisfied in the steady state:

$$0 = -\sum_{k=1}^{N} \frac{\partial}{\partial \theta_k} J_k^{(\mathrm{ss})}(\boldsymbol{\theta}), \tag{10}$$

where

$$\mathbf{J}^{(\mathrm{ss})}(\boldsymbol{\theta}) = \left(\mathbf{A}(\boldsymbol{\theta}) - T\frac{\partial}{\partial \boldsymbol{\theta}}\right) P^{(\mathrm{ss})}(\boldsymbol{\theta}). \tag{11}$$

We refer to the above equality as the divergence-free condition. This is a balanced condition in the context of the master equation. The stationary state should be set for the accomplishment of our aim as follows:

$$P^{(\mathrm{ss})}(\boldsymbol{\theta}) = \prod_{k=1}^{D} P(\mathbf{x}^{(k)}|\boldsymbol{\theta})p(\boldsymbol{\theta}), \tag{12}$$

where we set $T = 1$ (in machine learning, the inverse of temperature is absorbed into the energy definition or the likelihood function and prior distribution).

We can immediately obtain a trivial solution of the divergence-free condition. When the probabilistic flow in the steady state does not exist, the divergence-free condition is satisfied.

$$\mathbf{J}^{(\mathrm{eq})}(\boldsymbol{\theta}) = \mathbf{0}, \tag{13}$$

where the superscript has been changed from "ss" to "eq". This stationary state is in particular called the equilibrium state because current does not exist here. Then, the force can be defined as

$$\mathbf{A}(\boldsymbol{\theta}) = -\frac{\partial}{\partial \boldsymbol{\theta}} E(\boldsymbol{\theta}). \tag{14}$$

If we simulate the Langevin dynamics with the trivial force as shown in Equation (14), the long-time relaxation yields the equilibrium distribution.

3.2. Nontrivial solution

We impose the probabilistic flow in a steady state to obtain the following form:

$$\mathbf{J}^{(\mathrm{ss})}(\boldsymbol{\theta}) = \gamma \mathbf{B}(\boldsymbol{\theta}) P^{(\mathrm{ss})}(\boldsymbol{\theta}), \tag{15}$$

where γ is a degree of vDBC. Following the divergence-free condition, the vector field $\mathbf{B}(\boldsymbol{\theta}) = (B_1(\boldsymbol{\theta}), B_2(\boldsymbol{\theta}), \cdots, B_N(\boldsymbol{\theta}))$ must satisfy

$$0 = \gamma \sum_{k=1}^{N} \left(\frac{\partial B_k(\boldsymbol{\theta})}{\partial \theta_k} - B_k \frac{\partial E(\boldsymbol{\theta})}{\partial \theta_k}\right) P^{(\mathrm{ss})}(\boldsymbol{\theta}), \tag{16}$$

where we have used the fact that the distribution function is set as $P^{(\mathrm{ss})}(\boldsymbol{\theta}) \propto \exp(-E(\boldsymbol{\theta}))$ and the subscript for the bracket denotes the element of the vector. A trivial solution can be given by

$$B_k(\boldsymbol{\theta}) \propto \exp(E(\boldsymbol{\theta})). \tag{17}$$

However, we have to be careful about the instability when using the exponential term, which often leads to instability in the integration of the Langevin equality

Let us find an alternative to the force obtained above to remove the instability of the exponential term. We set the vector field as

$$B_k(\boldsymbol{\theta}) = \left(\frac{\partial E(\boldsymbol{\theta})}{\partial \theta_{k-1}} - \frac{\partial E(\boldsymbol{\theta})}{\partial \theta_{k+1}}\right), \tag{18}$$

where $\theta_{N+1} = \theta_1$ and $\theta_0 = \theta_N$. Then the divergence-free condition holds. The resulting force is given as

$$A_k(\boldsymbol{\theta}) = -\frac{\partial E(\boldsymbol{\theta})}{\partial \theta_k} + \gamma\left(\frac{\partial E(\boldsymbol{\theta})}{\partial \theta_{k-1}} - \frac{\partial E(\boldsymbol{\theta})}{\partial \theta_{k+1}}\right). \tag{19}$$

This force drives the system while satisfying the divergence-free condition but not the DBC. The violation of the DBC can be confirmed through the formulation of the path probability during infinitesimal time. For a two-dimensional case, the above nontrivial force can be implemented as

$$A_1(\boldsymbol{\theta}) = -\frac{\partial E(\boldsymbol{\theta})}{\partial \theta_1} - \gamma\frac{\partial E(\boldsymbol{\theta})}{\partial \theta_2} \tag{20}$$

$$A_2(\boldsymbol{\theta}) = -\frac{\partial E(\boldsymbol{\theta})}{\partial \theta_2} + \gamma\frac{\partial E(\boldsymbol{\theta})}{\partial \theta_1}. \tag{21}$$

Notice that the nontrivial force can be simply denoted as the combination of the gradient of the energy.

The message of the above calculation is as follows. Even if we use the nontrivial force, we can obtain the same distribution function in the steady state as that of the equilibrium case. We may use the nontrivial force to evolve the system to reach the steady state. The question is which is faster. The answer is the case with the nontrivial force. The nontrivial force violates the DBC and the resultant effect found even in the Fokker-Planck equation. The Fokker-Planck equation can be reformulated by the operator form with the partial derivative. The ordinary Fokker-Planck equation only has the Hermitian operators. On the other hand, the nontrivial force yields the anti-Hermitian operator as in Reference [22]. The part of the Hermitian operators does not change even after induction of the nontrivial force. This is a remarkable property of the nontrivial force. In the same sense as in the case of the master equation [19], the anti-Hermitian (asymmetric) part of the operators shift the eigenvalues characterizing the convergence speed to the steady state. The previous study has proved that the asymmetric part of the transition matrix in the master equation (finite-dimensional case) affects the eigenvalues so that the convergence speed can be faster than that in the case without asymmetric part. In a strict sense, the case for the infinite-dimensional system as in the Fokker-Planck equation has not been yet established but the same discussion can be applied to this case. Therefore the nontrivial force accelerates the convergence to the steady state, which is characterized by the desired distribution function.

3.3. Replication of the system

In addition to the above nontrivial force, we further find a nontrivial solution satisfying the divergence-free condition by considering the replication of the system. Note that this is slightly different from the replica exchange Monte-Carlo simulation [25]. The replication stands for that we use the replicas of the system governed by the Langevin dynamics. The energy function and the temperature are common. We then consider the evolution of the system following the Langevin equation. In order to accelerate the stochastic dynamics, we introduce additional force in the replicate system. As in the previous subsection, we find the nontrivial force to hold the same distribution function in the steady state as that of the equilibrium case. Let us look at the replicate Fokker-Planck equation

$$\frac{\partial P(\Theta,t)}{\partial t} = -\sum_{r=1}^{R}\sum_{k=1}^{N}\frac{\partial}{\partial \theta_{rk}}J_{rk}(\Theta,t), \tag{22}$$

where Θ is a matrix $\Theta = (\boldsymbol{\theta}_1, \boldsymbol{\theta}_2, \cdots, \boldsymbol{\theta}_R)$ containing the N-dimensional vector for each system denoted by $r = 1, 2, \cdots, R$, and $\mathbf{J}_r(\Theta,t) = (J_{r1}(\Theta,t), J_{r2}(\Theta,t), \cdots, J_{rN}(\Theta,t))$ is the probabilistic flow defined as

$$\mathbf{J}_r(\Theta,t) = \left(\mathbf{A}_r(\Theta) - T\frac{\partial}{\partial \boldsymbol{\theta}_r}\right)P(\Theta,t). \tag{23}$$

The gradient indexed by r is considered for each system. Let us impose the divergence-free condition. The divergence-free condition for the replicate system can be recast as

$$0 = -\sum_{r=1}^{R}\sum_{k=1}^{N}\frac{\partial}{\partial \theta_{rk}}J_{rk}^{(\mathrm{ss})}(\Theta), \tag{24}$$

where

$$\mathbf{J}_r^{(\mathrm{ss})}(\Theta) = \left(\mathbf{A}_r(\Theta) - T\frac{\partial}{\partial \boldsymbol{\theta}_r}\right)\prod_{r=1}^{R}P^{(\mathrm{ss})}(\boldsymbol{\theta}_r). \tag{25}$$

We impose the stationary state of the replicate system as the the product of the identical independent distribution. We can then determine a nontrivial solution by considering a combination of the forces on the replicate system as

$$\mathbf{A}_r(\Theta) = -\frac{\partial}{\partial \boldsymbol{\theta}_r}E(\boldsymbol{\theta}_r) + \gamma\left(\frac{\partial}{\partial \boldsymbol{\theta}_{r+1}}E(\boldsymbol{\theta}_{r+1}) - \frac{\partial}{\partial \boldsymbol{\theta}_{r-1}}E(\boldsymbol{\theta}_{r-1})\right), \tag{26}$$

where the boundary index is taken periodically. Then the probabilistic flow in the steady state is given as

$$\mathbf{J}_r^{(\mathrm{ss})}(\Theta) = \gamma\left(\frac{\partial}{\partial \boldsymbol{\theta}_{r+1}}E(\boldsymbol{\theta}_{r+1}) - \frac{\partial}{\partial \boldsymbol{\theta}_{r-1}}E(\boldsymbol{\theta}_{r-1})\right)\prod_{r=1}^{R}P(\boldsymbol{\theta}_r). \tag{27}$$

We can immediately confirm that the divergence of the above probabilistic flow will vanish because the result of the summation becomes zero. However, the probabilistic flow exists even in the stationary state. This is quite different from an equilibrium system, where there is no divergence of flow. We can confirm that the DBC does not hold in the formulation of the path probability [21, 22]. In other words, the remainder of the probabilistic flow is obtained from the vDBC. This is the Ohzeki-Ichiki method. In the original formulation in the literature, a duplicate system was considered, where $R = 2$. Here, we generalize the original formulation into the replicate system for it to be suitable in the application of the stochastic gradient Langevin method.

A series of the previous studies reveal that faster convergence can be assured by a mathematical argument [19] and is understood as the rare-event sampling [20]. In addition, the Ohzeki-Ichiki method may avoid the critical slowing down that occurs during the phase transition [21]. The method is powerful in this sense and very simple to implement because only the force modification results in remarkable performance. The previous study confirms the reduction in the correlation time as well as faster convergence to the stationary state [22].

Recall the replica exchange Monte Carlo simulation, in which multiple systems with "different" temperatures are driven simultaneously, while each realization is exchanged several times. The technique exhibits outstanding performance with fast convergence to the equilibrium state. However, in the above formulation, the replicate system has a common temperature. We can find a nontrivial solution with the inhomogeneous temperature in the replicate system. In this study, we restrict ourselves to the case with homogeneous temperature, by considering inhomogeneous temperature to be a part of future work.

4. Application to the stochastic gradient method

We are now in a position to apply the Ohzeki-Ichiki method to the stochastic gradient Langevin method. We propose two ways to apply this method to the stochastic gradient Langevin method. Whichever we choose, the Ohzeki-Ichiki method is very simple. In order to perform sampling following the posterior distribution, we employ the stochastic gradient Langevin method. We only evaluate the numerical integration over the Langevin equation, in which the energy function is given by the logarithm of the posterior distribution. In addition to the ordinary force, namely the gradient of the energy function, we employ the nontrivial force as described in the previous section. The mathematical proof ensures the faster convergence to the steady state [19]. Therefore the nontrivial force always accelerate the stochastic dynamics to make the system converge to the steady state. The convergence speed can be increased depending on the strength of the nontrivial force, namely the value of γ. However the upper limit of the value of γ depends on the numerical recipe to integrate the stochastic differential equation. In the following calculation, we employ the Heun scheme not Euler-Maruyama scheme to mitigate the errors in the numerical computation errors. The remaining arbitrariness is choice of the nontrivial force. One is the nontrivial force between the different components in the single system. The other is the nontrivial force in the replicate system. The latter one is advantage for the parallel computation.

4.1. Ohzeki-Ichiki method

In the stochastic gradient Langevin method, we compute the gradient of the subset of the given data ($k = 1, 2, \cdots, d$). The indices of the data are randomly chosen. In the stochastic gradient method, we approximate the force term, i.e., gradient, as follows:

$$-\frac{\partial}{\partial \boldsymbol{\theta}} E(\boldsymbol{\theta}) \approx \frac{\partial}{\partial \boldsymbol{\theta}} \log p(\boldsymbol{\theta}) + \frac{D}{d} \sum_{k=1}^{d} \frac{\partial}{\partial \boldsymbol{\theta}} \log P(\mathbf{x}^{(k)}|\boldsymbol{\theta}). \tag{28}$$

A simple way to apply the Ohzeki-Ichiki method is the direct insertion of this approximate gradient into Equation (19). In other words, we only change the force term for the kth element in the Langevin equation as

$$-\frac{\partial}{\partial \theta_k} E(\boldsymbol{\theta}) \to -\frac{\partial}{\partial \theta_k} E(\boldsymbol{\theta}) + \gamma \left(\frac{\partial}{\partial \theta_{k-1}} E(\boldsymbol{\theta}) - \frac{\partial}{\partial \theta_{k+1}} E(\boldsymbol{\theta}) \right). \tag{29}$$

Therefore no more computational cost is needed for implementation of the Ohzeki-Ichiki method. This is a remarkable property of the Ohzeki-Ichiki method. To enhance precision and improve the computation speed, additional efforts for its performance are usually needed. However this is not the case. In addition, we propose an alternative application of the Ohzeki-Ichiki method by considering the replicate system as shown below.

4.2. Ohzeki-Ichiki method for replicate system

In order to obtain more sampling points following the posterior distribution, we may perform parallel computation of the stochastic gradient Langevin method with a smaller batch size. This motivation is consistent with the concept of the Ohzeki-Ichiki method, in which we may consider that the replication of the system will generate a nontrivial force. We then distribute the approximate gradient into the replicated system. In other words, we apply the gradient determined by a different dataset, i.e., the further divided subset of a given data where the number of subsets is set to be d_r such that $d = \sum_{r=1} d_r$. For simplicity, we use $d_r = d/R$ in the numerical test.

We first divide the given subset of d into the replicate system as d_r for each r system. Then we evaluate the gradient for each replicate system. The computational cost for evaluation of the gradient can be reduced to a factor d_r/d. We have the value of the gradient for each replicate system as follows:

$$-\frac{\partial}{\partial \boldsymbol{\theta}_r} E(\boldsymbol{\theta}_r) \approx \frac{\partial}{\partial \boldsymbol{\theta}} \log p(\boldsymbol{\theta}) + \frac{D}{d_r} \sum_{k_r=1}^{d_r} \frac{\partial}{\partial \boldsymbol{\theta}} \log P(\mathbf{x}^{(k_r)}|\boldsymbol{\theta}). \tag{30}$$

Then we make the combination of the gradient conforming the nontrivial force as in Equation (27). In other words, we only change the force term in the Langevin equation as

$$-\frac{\partial}{\partial \boldsymbol{\theta}_r} E(\boldsymbol{\theta}_r) = -\frac{\partial}{\partial \boldsymbol{\theta}_r} E(\boldsymbol{\theta}_r) + \gamma \left(\frac{\partial}{\partial \boldsymbol{\theta}_{r+1}} E(\boldsymbol{\theta}_{r+1}) - \frac{\partial}{\partial \boldsymbol{\theta}_{r-1}} E(\boldsymbol{\theta}_{r-1}) \right) \tag{31}$$

In this sense, the nontrivial force is a collection of the gradient computed in different replica. The different replica has the individual data as the subset for evaluation of the gradient. To make the nontrivial force, all we need to know is the value of the gradient for the different data set distributed to each replica. This is done by communication between different replicas. The communication cost to send the data flow between different replicas is demanded but not significant. The gain by the replicate system overcomes the communication cost. The computational cost can be basically reduced to a factor d_r/d due to the parallel computation. In addition, because we replicate the system, we generate more sampling points from the given datasets. In this sense, the proposed scheme is very important for sampling following the posterior distribution.

The convergence speed compared to the ordinary evolution by the Langevin dynamics is definitely faster since the Ohzeki-Ichiki method holds mathematical assurance as in the previous study [19]. We should emphasize that its implementation is extremely simple. Additional computation does not need to perform the Ohzeki-Ichiki method. We only reuse the value of the gradient to make the nontrivial force. In the present paper, we propose the application of the Ohzeki-Ichiki method to the stochastic gradient Langevin method. In the next section, we test our method in the simple model although we push up the quantitative evaluation of performance in the future study.

5. Numerical tests

Let us test the stochastic gradient Langevin method with the Ohzeki-Ichiki method. Similarly to the first study on the stochastic gradient Langevin method [16], we test our algorithm for the

Gaussian mixture distribution with tied mean, in which the probability distribution of the data is defined as

$$P(x|\boldsymbol{\theta}) = \frac{1}{2}\sqrt{\frac{b}{2\pi}}\exp\left(-\frac{1}{2}b\left(x-\theta_1\right)^2\right) + \frac{1}{2}\sqrt{\frac{b}{2\pi}}\exp\left(-\frac{1}{2}b\left(x-\theta_1-\theta_2\right)^2\right) \tag{32}$$

where b denotes the precision (the inverse of the variance) and the parameters $\boldsymbol{\theta} = (\theta_1, \theta_2)$ are related to the mean of the Gaussian distribution. The prior distribution is assumed to be a Gaussian distribution as

$$P(\boldsymbol{\theta}) = \left(\frac{a_1 a_2}{2\pi}\right)\exp\left(-\frac{1}{2}a_1\theta_1^2 - \frac{1}{2}a_2\theta_2^2\right) \tag{33}$$

where a_1 and a_2 denote the precisions for the parameters. This model consists of a bimodal distribution. Thus the MAP estimation cannot capture the uncertainty of the parameters. The posterior distribution can be computed exactly as

$$P(\boldsymbol{\theta}|x) = \frac{1}{Z(x)}P(x|\boldsymbol{\theta})P(\boldsymbol{\theta}), \tag{34}$$

where

$$Z(x) = \sqrt{\frac{\alpha_1}{2\pi}}\exp\left(-\frac{\alpha_1}{2}x^2\right) + \sqrt{\frac{\alpha_2}{2\pi}}\exp\left(-\frac{\alpha_2}{2}x^2\right) \tag{35}$$

and

$$\alpha_1 = \frac{a_1 b}{a_1 + b} \tag{36}$$

$$\alpha_2 = \frac{a_1 a_2 b}{a_1 a_2 + (a_1 + a_2)b}. \tag{37}$$

We generate $D = 100$ samples of $x^{(k)}$ from one of the Gaussian distribution at $(\theta_1, \theta_2) = (2, -2)$. The model has another-mode solution as $(\theta_1, \theta_2) = (0, 2)$, which has strong negative correlation between the parameters.

Let us perform the stochastic gradient Langevin method with the Ohzeki-Ichiki method. The gradient of the log-likelihood function is evaluated as

$$\begin{aligned}
\frac{\partial}{\partial\theta_1}\log P(x|\boldsymbol{\theta}) &= \frac{b(x-\theta_1)\exp(-b(x-\theta_1)^2/2) + b(x-\theta_1-\theta_2)\exp(-b(x-\theta_1-\theta_2)^2/2)}{\exp(-b(x-\theta_1)^2/2) + \exp(-b(x-\theta_1-\theta_2)^2/2)} \\
\frac{\partial}{\partial\theta_2}\log P(x|\boldsymbol{\theta}) &= \frac{b(x-\theta_1-\theta_2)\exp(-b(x-\theta_1-\theta_2)^2/2)}{\exp(-b(x-\theta_1)^2/2) + \exp(-b(x-\theta_1-\theta_2)^2/2)}.
\end{aligned} \tag{38}$$

We set $a_1 = a_2 = 0.1$ and $b = 10$, which yield a large barrier between the two modes at $(\theta_1, \theta_2) = (0, 2)$ and $(\theta_1, \theta_2) = (2, -2)$. We iterate the update process for 10^6 steps. The step size decreases as $\beta(t + \delta) - \epsilon$, where $\epsilon = 0.55$ and δ and β are determined such that step sizes changes from $dt = 0.01$ to $dt = 0.0001$. First let us compare the results with the batch size for the stochastic gradient $d = 1$. Under this setting, the original stochastic gradient Langevin method usually fails to generate the sampling points following the posterior distribution in short time. Here, we demonstrate a more difficult case for the generation of the sampling points following the posterior distribution than the previous study [16]. We show the obtained sampling points through a density plot. We compare the stochastic gradient Langevin method with and without the application of the Ohzeki-Ichiki method with $\gamma = 2$ and 5 using the nontrivial force in Equations (20) and (21) as shown in Fig. 1. We can confirm that a wide range of the sampling

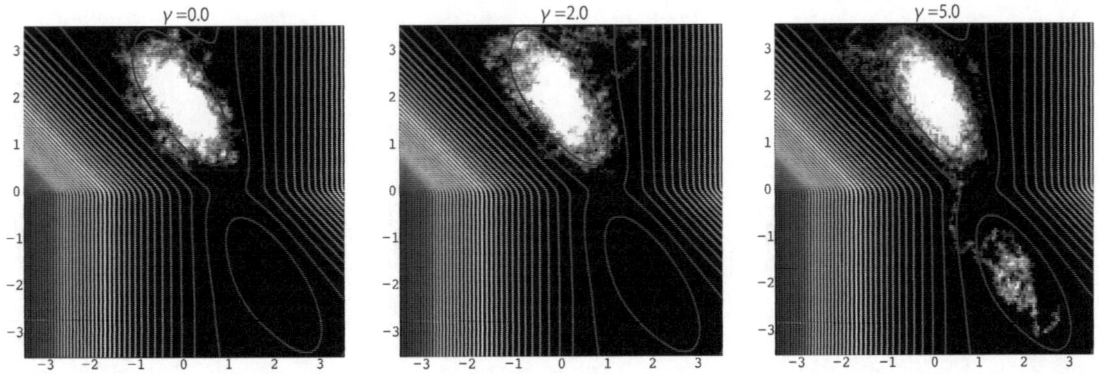

Figure 1. (Colour online) Density plot and the contour plot of the exact posterior distribution of the batch size $d = 1$. The number of iterations is $\Delta t = 10^6$. The left panel shows the stochastic gradient Langevin method, the centre one describes that with the Ohzeki-Ichiki method with $\gamma = 2$, and the right one represents that with $\gamma = 5$. The posterior distribution is plotted by the colour-gradation curves from red (higher values) to blue (lower values).

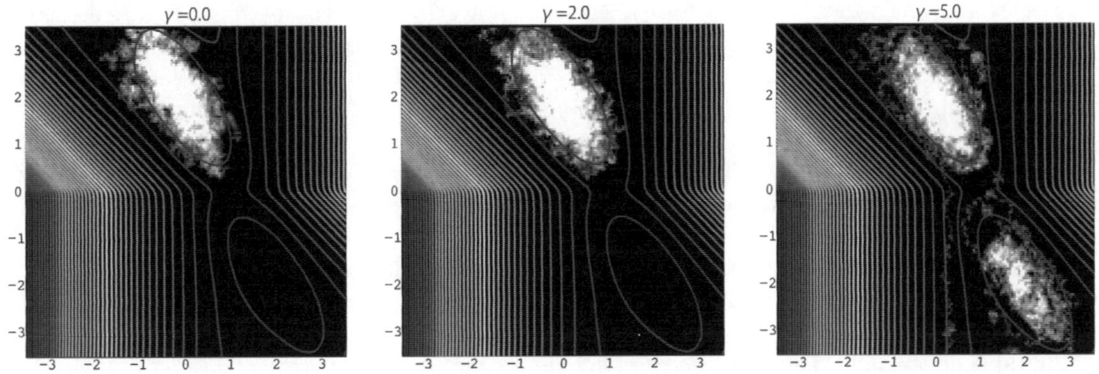

Figure 2. (Colour online) Density plot and the contour plot of the exact posterior distribution of the batch size $d = 10$. The number of iterations is $\Delta t = 10^6$. The same symbols are used in Figure 1.

points are obtained by the Ohzeki-Ichiki method. In particular, the case where $\gamma = 5$ exhibits a leap from one mode to the other mode. In other words, for the case with the application of the Ohzeki-Ichiki method with a larger value of the degree of vDBC estimates more accurate the posterior distribution.

Figure 2 shows the results of the stochastic gradient Langevin method and its improved version by the application of the Ohzeki-Ichiki method with the batch size $d = 10$. Similar to the previous case, for $d = 1$, we again confirm the efficient sampling by the application of the Ohzeki-Ichiki method.

Let us consider another application of the Ohzeki-Ichiki method dealing with the replicate system. We employ the Ohzeki-Ichiki method for the parallel update of the $R = d = 10$ replicate system. For each system, we use one datum randomly chosen for the gradient to update the tentative estimation value. We perform parallel computation of the update equation over $R = d = 10$ replicate system. Then we introduce the nontrivial force. We only need to

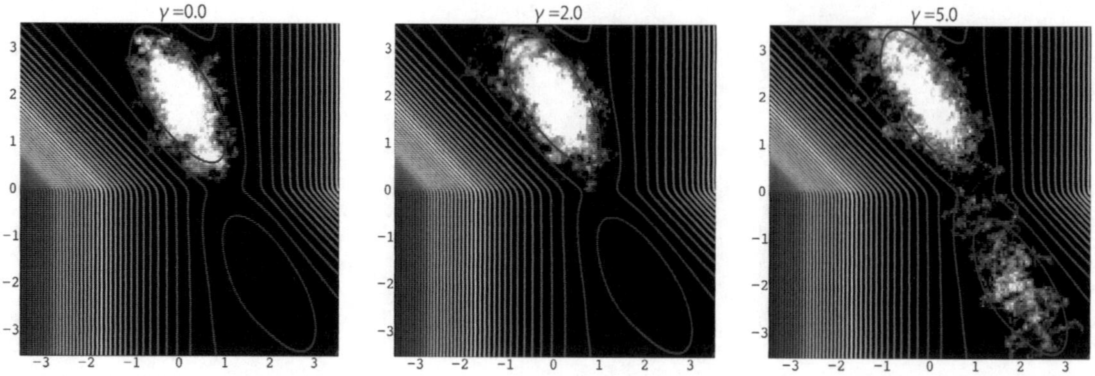

Figure 3. (Colour online) Density plot and the contour plot of the exact posterior distribution by the replicate system $R = 10$ and the batch size $d = 10$. We divide the mini batch d into d/R. The number of iterations is $\Delta t = 10^5$. The same symbols are used in Figure 1.

communicate the value of the gradient for the given subset $d_r = d/R = 1$ with different replicas. In other words, we "reuse" the data for efficient sampling as the nontrivial force. Figure 3 shows the results of the parallel computation of the stochastic gradient Langevin method and those of the case induced by the nontrivial force with $\gamma = 2$ and $\gamma = 5$. In this computation, we reduce the number of steps to $\Delta t = 10^5$. Efficient sampling can be performed by using the replicate system.

We have not yet compared the quantitative performance of the methods. However one can confirm their qualitative performance from the results shown in the figures. We find that a wide range of the sampling can be obtained by use of the larger value of γ. Similarly to the previous case, the sampling points without the nontrivial force ($\gamma = 0$) can not go over the ridge to the other mode of the posterior distribution. On the other hand, the nontrivial force of $\gamma = 5$ in the replicate system brings the sampling point in the other mode.

In addition, we plot the results of the long-time computation up to $\Delta t = 10^6$ by using the replicate system in Figure 4. Even for the case without the nontrivial force, the sampling points are found in a relatively wide range. The sampling from the bimodal distribution seems to be performed. However increase in the γ yields the more efficient sampling more correctly following the posterior distribution, as shown in Figure 4.

It may be assumed that the infinitely large nontrivial force causes the system to be in a stationary state immediately. This may or may not be true. Indeed increase in γ makes the system mixed rapidly and accelerates the convergence to a stationary state. However, depending on the value of γ, the numerical integration of the Langevin dynamics may become worse. To stabilize the computation of the Langevin dynamics, more sophisticated integration methods or the Metropolis-Hasting method may be utilized as in the original proposal of the stochastic gradient Langevin method. From mathematical and physical point of view, the Ohzeki-Ichiki method can ensure that a stationary state is reached faster than the case under the DBC. Thus, the increase in the degree of vDBC has some tradeoff between performance and instability. The optimal value of γ will be part of future work. The answer may be obtained from the variational principle discussed in a previous study [23].

6. Summary

In this study, we proposed the application of the Ohzeki-Ichiki method in the stochastic gradient Langevin method. The Ohzeki-Ichiki method violates the DBC, which is a sufficient condition

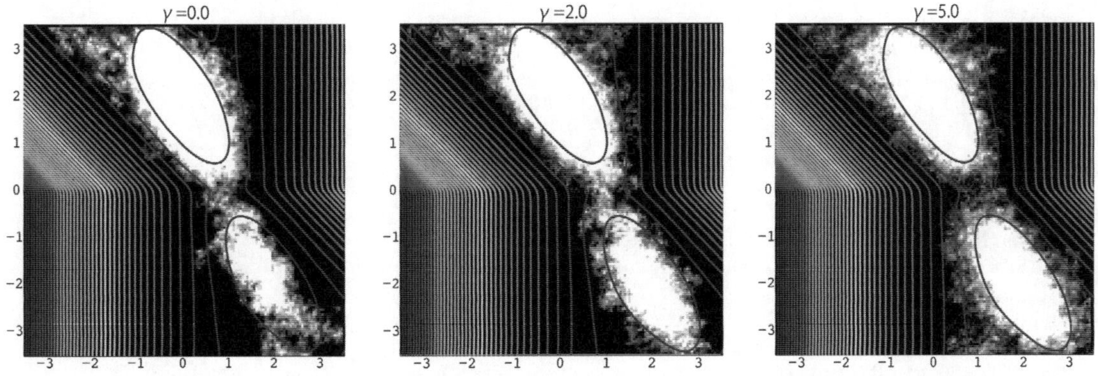

Figure 4. (Colour online) Density plot and the contour plot of the exact posterior distribution by the replicate system $R = 10$ and the batch size $d = 10$. We divide the mini batch d into d/R. The number of iterations is $\Delta t = 10^6$. The same symbols are used in Figure 1.

to ensure convergence to a stationary state, and shows remarkable performance for reaching a stationary state compared to the standard equilibrium case under the DBC. We applied this method to the stochastic gradient Langevin method, which approximates the gradient of the log-likelihood function in a stochastic manner and utilizes prior distribution to capture the uncertainty of the leaned parameters. The acceleration of the convergence to a stationary state improved the performance of the stochastic gradient Langevin method as expected. The range for the sampling points was wider as compared to the original method. In addition, we developed a new technique to implement the Ohzeki-Ichiki method based on the idea of the stochastic gradient method. We introduced the nontrivial force in the replicated system, which is the gradient determined by the distributed data from the given subset. This replication generated more sampling points and the nontrivial force made the replicate system converge to the stationary state faster than the ordinary approach. These are just the preliminary results of the Ohzeki-Ichiki method for machine learning problems. Future study will reveal its significance in this field and confirms the quantitative assessment of the Ohzeki-Ichiki method. We again emphasize that the implementation of the Ohzeki-Ichiki method is relatively simpler than other methods to accelerate the convergence to a stationary state. This method is simply the addition of the nontrivial force. Hence, the proposed method makes advancement in the machine learning process.

Acknowledgement

The author would like to thank M. Yasuda, and A. Ichiki for the fruitful discussions. This work is financially supported by the JST-CREST, JSPS KAKENHI Grants No. 25120008 and No. 15H03699, and the Kayamori Foundation of Informational Science Advancement.

References

[1] Ackley D H, Hinton G E and Sejnowski T J 1985 *Cognitive Science* **9** 147–169
[2] Hinton G E, Osindero S and Teh Y W 2006 *Neural Comput.* **18** 1527–1554
[3] Hinton G E and Salakhutdinov R R 2006 *Science* **313** 504–507
[4] Pankaj M and David J S 2014 **stat.ML/1410.3831**
[5] Ohzeki M 2015 *Journal of the Physical Society of Japan* **84** 054801
[6] Decelle A and Ricci-Tersenghi F 2014 *Phys. Rev. Lett.* **112**(7) 070603
[7] Yamanaka S, Ohzeki M and Decelle A 2015 *Journal of the Physical Society of Japan* **84** 024801
[8] Sessak V and Monasson R 2009 *Journal of Physics A: Mathematical and Theoretical* **42** 055001

[9] Cocco S and Monasson R 2011 *Phys. Rev. Lett.* **106**(9) 090601
[10] Cocco S and Monasson R 2012 *Journal of Statistical Physics* **147** 252–314
[11] Ricci-Tersenghi F 2012 *Journal of Statistical Mechanics: Theory and Experiment* **2012** P08015
[12] Yasuda M and Tanaka K 2013 *Phys. Rev.* E **87**(1) 012134
[13] Raymond J and Ricci-Tersenghi F 2013 *Phys. Rev.* E **87**(5) 052111
[14] Ohzeki M 2013 *Journal of Physics: Conference Series* **473** 012005
[15] Robbins H and Monro S 1951 *Ann. Math. Statist.* **22** 400–407
[16] Welling M and Teh Y W 2011 *Proceedings of the International Conference on Machine Learning*
[17] Suwa H and Todo S 2010 *Phys. Rev. Lett.* **105**(12) 120603
[18] Turitsyn K S, Chertkov M and Vucelja M 2011 *Physica D: Nonlinear Phenomena* **240** 410 – 414
[19] Ichiki A and Ohzeki M 2013 *Phys. Rev.* E **88**(2) 020101
[20] Ichiki A and Ohzeki M 2015 *Phys. Rev.* E **91**(6) 062105
[21] Ohzeki M and Ichiki A 2015 *Phys. Rev.* E **92**(1) 012105
[22] Ohzeki M and Ichiki A 2015 *Journal of Physics: Conference Series* **638** 012003
[23] Takahashi K and Ohzeki M 2016 *Phys. Rev.* E **93**(1) 012129
[24] Swendsen R H and Wang J S 1987 *Phys. Rev. Lett.* **58**(2) 86–88
[25] Hukushima K and Nemoto K 1996 *Journal of the Physical Society of Japan* **65** 1604–1608
[26] Neal R 2001 *Statistics and Computing* **11** 125–139
[27] Ohzeki M 2010 *Phys. Rev. Lett.* **105**(5) 050401
[28] Ohzeki M and Nishimori H 2010 *J. Phys. Soc. Jpn.* **79**(8) 084003
[29] Ohzeki M and Nishimori H 2011 *Journal of Physics: Conference Series* **302** 012047
[30] Ohzeki M and Nishimori H 2011 *Physica E: Low-dimensional Systems and Nanostructures* **43** 782–785
[31] Ohzeki M 2012 *Phys. Rev.* E **86**(6) 061110

Bayes method for low rank tensor estimation

†Taiji Suzuki and ‡Heishiro Kanagawa

†‡Graduate School of Information Science and Engineering, Tokyo Institute of Technology, O-okayama 2-12-1, Meguro-ku, Tokyo 152-8552, JAPAN
† PRESTO, JST

E-mail: †s-taiji@is.titech.ac.jp, ‡kanagawah.ab@m.titech.ac.jp

Abstract. We investigate the statistical convergence rate of a Bayesian low-rank tensor estimator, and construct a Bayesian nonlinear tensor estimator. The problem setting is the regression problem where the regression coefficient forms a tensor structure. This problem setting occurs in many practical applications, such as collaborative filtering, multi-task learning, and spatio-temporal data analysis. The convergence rate of the Bayes tensor estimator is analyzed in terms of both in-sample and out-of-sample predictive accuracies. It is shown that a fast learning rate is achieved without any strong convexity of the observation. Moreover, we extend the tensor estimator to a nonlinear function estimator so that we estimate a function that is a tensor product of several functions.

1. Introduction

Tensor modeling is a powerful tool for representing higher order relations between several data sources. The second order correlation has been a main tool in data analysis for a long time. However, because of an increase in the variety of data types, we frequently encounter a situation where higher order correlations are important for transferring information between more than two data sources. In this situation, a tensor structure is required. For example, a recommendation system with a three-mode table, such as user $\times$ movie $\times$ context, is regarded as comprising the tensor data analysis [1]. The noteworthy success of tensor data analysis is based on the notion of the *low rank* property of a tensor, which is analogous to that of a matrix. The rank of a tensor is defined by a generalized version of the singular value decomposition for matrices. This enables us to decompose a tensor into a few factors and find higher order relations between several data sources.

A naive approach to computing tensor decomposition requires non-convex optimization [2]. Several authors have proposed convex relaxation methods to overcome the computational difficulty caused by non-convexity [3–7]. The main idea of convex relaxations is to unfold a tensor into a matrix, and apply trace norm regularization to the matrix thus obtained. This technique connects low rank tensor estimation to the well-investigated convex low rank matrix estimation. Thus, we can apply the techniques developed in low rank matrix estimation in terms of optimization and statistical theories. To address the theoretical aspects, [6] gave the statistical convergence rate of a convex tensor estimator that utilizes the so-called *overlapped Schatten 1-norm* defined by the sum of the trace norms of all unfolded matricizations. [8] showed that the bound given by [6] is tight, but can be improved by a modified technique called *square deal.* [7] proposed another approach called *latent Schatten 1-norm* regularization that is defined by the infimum convolution of trace norms of all unfolded matricizations, and analyzed its convergence rate. These theoretical studies revealed the qualitative dependence of learning rates on the rank and size of the underlying tensor. However, one problem of convex methods is that, reducing the problem to a matrix estimation, one may lose statistical efficiency in exchange for computational efficiency.

One important question is whether we can construct a tractable estimator that possesses a (near) optimal learning rate for tensor estimation. The Bayes method is a promising way to satisfy this requirement [9–13]. In particular, [13] showed that a Bayes estimator can achieve the near-minimax optimal rate for the tensor estimation problem. Roughly speaking, it is shown that the mean squared error of the estimator $\hat{A}$ is evaluated as

$$\|\hat{A} - A^*\|_n^2 = O_p\left(\frac{d^*(\sum_{k=1}^K M_k)\log(K\sqrt{n(\sum_k M_k)^K})}{n}\right),$$

where n is the sample size, $\hat{A}$ is the Bayes estimator, A^* is the true tensor, d^* is the *CP-rank* of the true tensor (its definition will be given in Section 2), and $(M_1, \ldots, M_K)$ is the size ($A^* \in \mathbb{R}^{M_1 \times \cdots \times M_K}$). A remarkable point of the analysis of [13] is that the rate is proven *without* assuming any strong convexity on the empirical L_2 norm. A kind of restricted strong convexity [14, 15] has been usually assumed in the existing regularization approaches. However, for the analysis of the predictive accuracy of a Bayes estimator, we don't need that assumption.

Another issue we deal with in this article is the nonlinear extension of the tensor model. All the works introduced above are about a linear model where the function we want to estimate is a linear function having a tensor form. This model can be extended to a nonlinear model [16]. So far, an alternating minimization with an RKHS regularization has been investigated numerically [16]. However, its theoretical properties are not well investigated. In this article, we present a Gaussian process approach to estimate the nonlinear tensor model. The Gaussian process approach is derived as a natural extension of the linear one. We will show its effectiveness in numerical experiments.

2. Problem Settings

In this section, the problem setting of this article is shown. Suppose that there exists the true tensor $A^* \in \mathbb{R}^{M_1 \times \cdots \times M_K}$ of order K, and we observe n samples $D_n = \{(Y_i, X_i)\}_{i=1}^n$ from the following linear model:

$$Y_i = \langle A^*, X_i \rangle + \epsilon_i. \tag{1}$$

Here, X_i is a tensor in $\mathbb{R}^{M_1 \times \cdots \times M_K}$ and the inner product $\langle \cdot, \cdot \rangle$ between two tensors $A, X \in \mathbb{R}^{M_1 \times \cdots \times M_K}$ is defined by $\langle A, X \rangle = \sum_{j_1,\ldots,j_K=1}^{M_1,\ldots,M_K} A_{j_1,\ldots,j_K} X_{j_1,\ldots,j_K}$. ϵ_i is i.i.d. noise from a normal distribution $N(0, \sigma^2)$ with mean 0 and variance σ^2.

Now, we assume the true tensor A^* is "low-rank." The notion of rank considered in this article is *CP-rank* (Canonical Polyadic rank) [17, 18]. We say a tensor $A \in \mathbb{R}^{M_1 \times \cdots \times M_K}$ has CP-rank d' if there exist matrices $U^{(k)} \in \mathbb{R}^{d' \times M_k}$ $(k = 1, \ldots, K)$ such that $A_{j_1,\ldots,j_K} = \sum_{r=1}^{d'} U_{r,j_1}^{(1)} U_{r,j_2}^{(2)} \ldots U_{r,j_K}^{(K)}$, and d' is the minimum number to yield this decomposition (we do not require the orthogonality of $U^{(k)}$). This is called *CP-decomposition*. When A satisfies this relation for $\boldsymbol{U} = (U^{(1)}, U^{(2)}, \ldots, U^{(K)})$, we write

$$A = A_{\boldsymbol{U}} = [[U^{(1)}, U^{(2)}, \ldots, U^{(K)}]] := \left(\sum_{r=1}^{d'} U_{r,j_1}^{(1)} U_{r,j_2}^{(2)} \ldots U_{r,j_K}^{(K)}\right)_{j_1,\ldots,j_K}. \tag{2}$$

We denote by d^* the CP-rank of the true tensor A^*. Notice that, for the special case of matrices ($K = 2$), the CP-rank coincides with the usual rank of a matrix.

In this article, we are interested in the predictive accuracy of the linear model with the assumption that A^* has low CP-rank. Because of the low CP-rank assumption, the learning problem becomes more structured than an ordinary linear regression problem on a vector. This problem setting includes the well-known low rank matrix estimation as a special case $K = 2$. There are two types of predictive accuracy: *in-sample* and *out-of-sample* ones. The in-sample predictive accuracy of an estimator $\hat{A}$ is defined by

$$\|\hat{A} - A^*\|_n^2 := \frac{1}{n}\sum_{i=1}^{n}\langle X_i, \hat{A} - A^*\rangle^2, \tag{3}$$

where $\{X_i\}_{i=1}^n$ is the observed input samples. The out-of-sample one is defined by

$$\|\hat{A} - A^*\|_{L_2(P(X))}^2 := \mathrm{E}_{X\sim P(X)}[\langle X, \hat{A} - A^*\rangle^2], \tag{4}$$

where $P(X)$ is the distribution of X that generates the observed samples $\{X_i\}_{i=1}^n$ and the expectation is taken over independent realization X from the observed ones.

Example 1. Tensor completion under random sampling. *Suppose that we have partial observations of a tensor. A tensor completion problem consists of denoising the observational noise and completing the unobserved elements. In this problem, X_i is independently identically distributed from a set $\{e_{j_1,\dots,j_K} \mid 1 \le j_k \le M_k\ (k = 1,\dots,K)\}$, where $e_{j_1,\dots,j_K}$ is an indicator tensor that has 1 at its $(j_1,\dots,j_K)$-element and 0 elsewhere, and thus, Y_i is an observation of one element of A^* contaminated with noise ϵ_i.*

The out-of-sample accuracy measures how accurately we can recover the underlying tensor A^* from the partial observation. If X_i is uniformly distributed, $\|\hat{A} - A^*\|_{L_2(P(X))} = \frac{1}{\sqrt{M_1 \dots M_K}}\|\hat{A} - A^*\|_2$, where $\|\cdot\|_2$ is the ℓ_2-norm obtained by summing the squares of all the elements. If $K = 2$, this problem is reduced to the standard matrix completion problem. In that sense, our problem setting is a wide generalization of the low rank matrix completion problem.

Example 2. Multi-task learning. *Suppose that several tasks are aligned across a 2-dimensional space. For each task $(s,t) \in \{1,\dots,M_1\} \times \{1,\dots,M_2\}$ (indexed by two numbers), there is a true weight vector $a^*_{(s,t)} \in \mathbb{R}^{M_3}$. The tensor A^* is an array of the weight vectors $a^*_{(s,t)}$, that is, $A^*_{s,t,j} = a^*_{(s,t),j}$.*

The input vector X_i is a vector of predictor variables for one specific task, say (s,t), and takes a form such that $X_{i,(s',t',:)} = \begin{cases} x_i^{(s,t)} \in \mathbb{R}^{M_3}, & ((s',t') = (s,t)), \\ \mathbf{0}, & (otherwise). \end{cases}$

By assuming A^* is low-rank in the sense of CP-rank, the problem becomes a multi-task feature learning with a two dimensional structure in the task space [19].

As shown in the examples, the estimation problem of low-rank tensor A^* is a natural extension of low-rank matrix estimation. However, it has a much richer structure than matrix estimation. Thus far, some convex regularized learning problems have been proposed analogously to spectrum regularization on a matrix, and their theoretical analysis has also been provided. However, no method have been proved to be statistically optimal. There is a huge gap between a matrix and higher order array. One reason for this gap is the computational complexity of the convex envelope of CP-rank. It is well known that the trace norm of a matrix is a convex envelope of the matrix rank on a set of matrices with a restricted operator norm [20]. However, as for tensors, computing CP-rank and CP-decomposition themselves is NP-hard [21]. A nonlinear model as an extension of the model (1) will be considered in Section 6.

3. Bayesian tensor estimator of linear model

Here, we provide the prior distribution of the Bayes estimator for the linear tensor model (1) (as for the nonlinear model we present that in Section 6.1). On a decomposition of a rank d' tensor $A = [[U^{(1)}, U^{(2)}, \dots, U^{(K)}]]$ $(U^{(k)} \in \mathbb{R}^{d' \times M_k})$, we place a Gaussian prior:

$$\pi(U^{(1)}, \dots, U^{(K)}|d') \propto \exp\left\{-\frac{d'}{2\sigma_{\mathrm{p}}^2}\sum_{k=1}^{K}\mathrm{Tr}[U^{(k)\top}U^{(k)}]\right\},$$

where $\sigma_{\rm p} > 0$. Moreover, we placed a prior distribution on the rank $1 \le d' \le d_{\max}$ as

$$\pi(d') = \frac{1}{N_\xi}\xi^{d'(M_1+\cdots+M_K)},$$

where $0 < \xi < 1$ is some positive real number, $d_{\max}$ is a sufficiently large number that is supposed to be larger than d^*, and N_ξ is the normalizing constant, $N_\xi = \frac{1-\xi^{M_1+\cdots+M_K}}{\xi-\xi^{d_{\max}(M_1+\cdots+M_K)}}$.

Now, since the noise is Gaussian, the likelihood of a tensor A is given by

$$p(D_n|A) =: p_{n,A} \propto \exp\left\{-\frac{1}{2\sigma^2}\sum_{i=1}^{n}(Y_i - \langle A, X_i\rangle)^2\right\}.$$

The posterior distribution is given by

$$\Pi(A \in \mathcal{C}|D_n) = \frac{\sum_{d=1}^{d_{\max}}\int_{A_U\in\mathcal{C}} p_{n,A_U}\pi((U^{(k)})_{k=1}^K|d)\pi(d)\mathrm{d}U^{(1)}\ldots\mathrm{d}U^{(K)}}{\sum_{d=1}^{d_{\max}}\int p_{n,A_U}\pi((U^{(k)})_{k=1}^K|d)\pi(d)\mathrm{d}U^{(1)}\ldots\mathrm{d}U^{(K)}},$$

where $\mathcal{C} \subseteq \mathbb{R}^{M_1\times\ldots M_K}$. It is noteworthy that the posterior distribution of $U^{(k)}$ conditioned by d and $U^{(k')}$ $(k' \neq k)$ is a Gaussian distribution, because the prior is conjugate to Gaussian distributions. Therefore, the posterior mean $\int_A f(A)\Pi(\mathrm{d}A|D_n)$ of a function $f : \mathbb{R}^{M_1\times\ldots M_K} \to \mathbb{R}$ can be computed by an MCMC method, such as Gibbs sampling (as for the Bayes tensor estimator, see [10–12]). In this article, we consider the posterior mean estimator $\hat{A} = \int A\Pi(\mathrm{d}A|D_n)$ which is the Bayes estimator corresponding to the square loss.

4. Convergence rate analysis of linear tensor estimator

In this section, we present the statistical convergence rate of the Bayes estimator of the linear tensor model (1). Before we give the convergence rate, we define some quantities and provide the assumptions. We define the *max-norm* of A^* as

$$\|A^*\|_{\max,2} := \min_{\{U^{(k)}\}}\{\max_{i,k}\|U^{(k)}_{:,i}\|_2 \mid A^* = [[U^{(1)},\ldots,U^{(K)}]],\ U^{(k)} \in \mathbb{R}^{d^*\times M_k}\}.$$

The ℓ_p-norm of a tensor A is given by $\|A\|_p := (\sum_{j_1,\ldots,j_K}|A_{j_1,\ldots,j_K}|^p)^{\frac{1}{p}}$. The prior mass around the true tensor A^* is a key quantity for characterizing the convergence rate, which is denoted by Ξ:

$$\Xi(\delta) := -\log(\Pi(A : \|A - A^*\|_n < \delta)),$$

where $\delta > 0$. Small $\Xi(\delta)$ means that the prior is well concentrated around the truth. Thus, it is natural to consider that, if Ξ is large, the Bayes estimator could be close to the truth. However, clearly, we do not know beforehand the location of the truth. Thus, it is not beneficial to place too much prior mass around one specific point. Instead, the prior mass should cover a wide range of possibilities of A^*. The balance between concentration and dispersion has a similar meaning to that of the bias-variance trade-off.

To normalize the scale, we assume that the ℓ_1-norm of X_i is bounded[1].

Assumption 1. *We assume that the ℓ_1-norm of X_i is bounded by 1: $\|X_i\|_1 \le 1$, a.s..*

Finally, we define the technical quantities $C_{n,K} := 3K\sqrt{n}\left(\frac{4\sigma_{\rm p}^2\Xi(\frac{1}{\sqrt{n}})}{d^*}\right)^{\frac{K}{2}}$ and $c_\xi := \min\{|\log(\xi)|/\log(C_{n,K}), 1\}/4$.

[1] ℓ_1-norm could be replaced by another norm such as ℓ_2. This difference affects the analysis of out-of-sample accuracies, but rejecting samples with $\max_X|\langle X, A\rangle| \le R$ gives an analogous result for other norms.

4.1. *In-sample predictive accuracy*

We now give the convergence rate of the in-sample predictive accuracy. We suppose the inputs $\{X_i\}_{i=1}^n$ are fixed (not random). The in-sample predictive accuracy conditioned by $\{X_i\}_{i=1}^n$ is given as follows.

Theorem 1. *Under Assumption 1, there exists a universal constant C such that the posterior mean of the in-sample accuracy is upper bounded by*

$$\mathrm{E}\left[\int \|A - A^*\|_n^2 \mathrm{d}\Pi(A|Y_{1:n})\right]$$
$$\leq \frac{C}{n}\left[d^*\left(\sum\nolimits_k M_k + \frac{1}{|\log(\xi)|}\right)\log(C_{n,K}) + \frac{\Xi(\sqrt{\frac{c_\xi}{n}})}{c_\xi} + \log(d_{\max}) + K + 8^K(K+1)!\right]. \quad (5)$$

The proof is given by [13]. This theorem provides the speed at which the posterior mass concentrates around the true A^*. It should be noted that the integral in the LHS of Eq. (5) is taken *outside* $\|A - A^*\|_n^2$. This gives not only information about the posterior concentration but also the convergence rate of the posterior mean estimator. This can be shown as follows. By Jensen's inequality, we have $\mathrm{E}\left[\|\int A\mathrm{d}\Pi(A|Y_{1:n}) - A^*\|_n^2\right] \leq \mathrm{E}\left[\int \|A - A^*\|_n^2 \mathrm{d}\Pi(A|Y_{1:n})\right]$. Therefore, Theorem 1 gives a much stronger claim on the posterior than just stating the convergence rate of the posterior mean estimator.

Since the rate (5) is rather complicated, we give a simplified bound. By assuming $\log(d_{\max})$ and $K!$ are smaller than $d^*(\sum_k M_k)$, we rearrange it as

$$\mathrm{E}\left[\int \|A - A^*\|_n^2 \mathrm{d}\Pi(A|Y_{1:n})\right] = O\left(\frac{d^*(M_1 + \cdots + M_K)}{n}\log\left(K\sqrt{n(\textstyle\sum_{k=1}^K M_k)^K \frac{\sigma_{\mathrm{p}}^K}{\xi}}\right)\right).$$

Inside the $O(\cdot)$ symbol, a constant factor depending on $K, \|A^*\|_{\max,2}, \sigma_{\mathrm{p}}, \xi$ is hidden. This bound means that the convergence rate is characterized by the actual degree of freedom up to a log term. That is, since the true tensor has rank d^* and thus has a decomposition (2), the number of unknown parameters is bounded by $d^*(M_1 + \cdots + M_K)$. Thus, the rate is basically $O(\frac{\text{degree of freedom}}{n})$ (up to log order), which is optimal (see Section 5 for more precise argument). Here, we would like to emphasize that the true rank d^* is unknown, but by placing a prior distribution on a rank the Bayes estimator can appropriately estimate the rank and gives an almost optimal rate depending on the true rank. In this sense, the Bayes estimator has adaptivity to the true rank.

More importantly, we do not assume *any strong convexity* on the design. Usually, to derive a fast convergence rate of sparse estimators, such as Lasso and the trace norm regularization estimator, we assume a variant of strong convexity, such as a restricted eigenvalue condition [14] and restricted strong convexity [15]. It is difficult to check the strong convexity condition in practice. However, our convergence rate does not require such conditions. One reason why this is possible is that we are interested in the predictive accuracies rather than the actual distance between the tensors $\|A - A^*\|_2^2$ (parameter estimation accuracy) . It is known that this phenomenon occurs also in high dimensional regression of vectors, see [22] for example.

4.2. *Out-of-sample predictive accuracy*

Next, we turn to the convergence rate of the out-of-sample predictive accuracy. In this setting, the input sequence $\{X_i\}_{i=1}^n$ is not fixed, but an i.i.d. random variable generated by a distribution $P(X)$.

To obtain fast convergence of the out-of-sample accuracy, we need to bound the difference between the empirical and population L_2-errors: $\|A - A^*\|_n^2 - \|A - A^*\|_{L_2(P(X))}^2$. To ensure that this quantity is small using Bernstein's inequality, the sup-norm $\max_X |\langle X, A\rangle|$ should be bounded. However, the sup-norm of the posterior mean could be large in tensor estimation. This difficulty can be avoided by rejecting posterior sample A with a large sup-norm, say R. That is, we accept only a sample A_U that satisfies $\boldsymbol{U} \in \{(U^{(1)}, \ldots, U^{(K)}) \mid \|U_{:,j}^{(k)}\| \leq R \ (1 \leq k \leq K, 1 \leq j \leq M_k)\} =: \mathcal{U}_R$. If $\|A\|_{\max,2} \leq R$, then under Assumption 1, we have $|\langle X, A\rangle| \leq R^K$. Therefore, by restricting the max-norm of the posterior sample, we can control the sup-norm. The estimation with this posterior can be implemented merely by rejecting the posterior samples with a sup-norm larger than a threshold R during the sampling scheme.

Here, we assume that the max-norm $\|A^*\|_{\max,2}$ of the true tensor is approximately known, that is, we know $R > 0$ such that $2\|A^*\|_{\max,2} < R$. Otherwise, we may apply cross validation. We employ this R for the posterior rejection threshold.

The resultant posterior distribution is expressed as the conditional posterior distribution $\Pi(\cdot|\|A\|_{\max,2} \leq R, D_n)$.

Theorem 2. *Under Assumption 1 and $\|A^*\|_{\max,2} + \sigma_{\rm p} < R$, we have*

$$\begin{aligned}&\mathrm{E}_{D_n}\left[\int \|A_{\boldsymbol{U}} - A^*\|^2_{L_2(P(X))} \mathrm{d}\Pi(A_{\boldsymbol{U}}|\boldsymbol{U} \in \mathcal{U}_R, D_n)\right]\\ \leq & C\left(\frac{d^*(\sum_{k=1}^K M_k)}{n}(1 \vee R^{2K})\log\left(K\sqrt{n}R^{\frac{K}{2}}\frac{\sigma_{\rm p}^K}{\xi}\right)\right),\end{aligned}$$

where C is a constant depending on $K, \log(d_{\max}), \sigma_{\rm p}$.

It should be noted that the optimal rate is achieved, although we do not assume any strong convexity on the distribution $L_2(\Pi)$. This can be so because we are not analyzing the actual L_2-norm $\|A - A^*\|_2$. If we do not assume strong convexity like $\|A - A^*\|_2 \leq C\|A - A^*\|_{L_2(P(X))}$, it is impossible to derive fast convergence of $\|A - A^*\|_2$. The trick is that we focus on the "weighted" L_2-norm $\|A - A^*\|_{L_2(P(X))}$ instead of $\|A - A^*\|_2$.

Finally, it is remarked that, if X_i is the uniform at random observation in the tensor completion problem, then $\|A - A^*\|^2_{L_2(P(X))} = \frac{1}{\prod_k M_k}\|A - A^*\|_2^2$ (note that in this setting $\|X_i\|_1 = 1$). Thus, our analysis yields fast convergence of the tensor recovery. If $K = 2$, the analysis recovers the well known rate of matrix completion problems up to a $\log(nM_1M_2)$ term [15, 23, 24]:

$$\frac{1}{M_1M_2}\|\hat{A} - A^*\|_2^2 = O_p\left(\frac{d^*(M_1 + M_2)}{n}\log(nM_1M_2)\right).$$

5. Minimax optimality

In this section, the learning rates derived above are actually minimax optimal up to log terms. To prove this, we specify the $L_2(P(X))$ norm. We take $L_2(P(X))$ as a uniform observation of the entries. That is, $\langle X, A\rangle = A_{i_1,i_2,\dots,i_K}$ for some $i_k \in \{1, \dots, M_k\}$ $(k = 1, \dots, K)$, and the choice of $(i_1, i_2, \dots, i_K)$ is uniform. The hypothesis space is given by

$$\mathcal{T}_R = \{[[U^{(1)}, \dots, U^{(K)}]] \mid U^{(k)} \in \mathbb{R}^{d^* \times M_k}, \|U^{(k)}_{:,j}\| \leq R \ (1 \leq k \leq K, 1 \leq j \leq M_k)\},$$

the set of tensors with rank d^* and the max norm not more than R. The minimax optimal rate is the convergence rate that can not be improved by *any* estimator: for any estimator, there is a tensor $A^* \in \mathcal{T}_R$ such that the predictive accuracy corresponding to the true tensor A^* is lower bounded by the minimax optimal risk.

Theorem 3. *The minimax learning rate of the tensor estimation is lower bounded as follows. Suppose that $R \geq 1$, $M_k > 4$ $(\forall k < K)$, $M_K/d^* > 4$ and M_K/d^* is integer. Then there exists a constant C such that*

$$\inf_{\hat{A}} \sup_{A^* \in \mathcal{T}_R} \mathrm{E}[\|\hat{A} - A^*\|^2_{L_2(P(X))}] \geq C\min\left\{\sigma^2\left(\frac{d^*(\sum_{k=1}^K M_k)}{n}\right), (R^2/d^*)^K\right\}$$

where $\inf_{\hat{A}}$ is taken over all estimator and the expectation is taken for the training samples.

The proof is given by [13]. The theorem is proven by using the information theoretic argument developed by [25] to derive the minimax optimal rate. The assumption, M_K/d^* is integer, is just a technical assumption and is not essential. We can see that the convergence rates given in Theorems 2 are minimax optimal upto log and R^{2K} terms.

6. Extension to nonlinear tensor estimator

So far, we have considered a linear relations. However, the model can be easily extended to capture nonlinear relations.

We consider the input X_i is a concatenation of K variables, i.e., $X_i = (x_i^{(1)}, \cdots, x_i^{(K)})$ where each $x_i^{(k)}$ is an element of a set $\mathcal{X}_k$. Here, we don't assume that $x^{(k)}$ is a real vector. That can be of any data type such as graph as long as we can define a positive definite kernel function on it. We consider the regression problem where these samples are generated according to the nonparametric model [16]:

$$Y_i = \sum_{r=1}^{d^*} \prod_{k=1}^{K} f_r^{*(k)}(X_i^{(k)}) + \epsilon_i, \tag{6}$$

where $\{\epsilon_i\}_{i=1}^n$ are i.i.d. zero-mean random variables. In this regression problem, our objective is to estimate the true function $f^*(x^{(1)}, \ldots, x^{(K)}) = \sum_{r=1}^{d^*} \prod_{k=1}^{K} f_r^{*(k)}(x_i^{(k)})$ from the data $D_n = \{(X_i, Y_i)\}_{i=1}^n$.

This model captures the effect of non-linear higher order interactions among the input components $\{x^{(k)}\}_{k=1}^K$ to the output y, thus is useful for a regression problem where the output is determined by complex relations between the input components. That kind of regression problem appears in several applications such as multi-task learning, recommendation system, spatiotemporal data analysis [19, **?**, **?**].

To understand the model (6), it is helpful to consider the linear model (1) as a special case. If we assume that X_i is rank-1, i.e., $X_i = x_i^{(1)} \circ \cdots \circ x_i^{(K)}$, then the inner product between the tensors X_i and A^* is written as:

$$\langle A^*, X_i \rangle = \sum_{r=1}^{d^*} \prod_{k=1}^{K} \langle U_{r,:}^{*(k)}, x_i^{(k)} \rangle.$$

This is equivalent to the case where we limit $f_r^{*(k)}$ in Eq.(6) to the linear function $\langle U_{r,:}^{*(k)}, x_i^{(k)} \rangle$. Hence, the linear model based on CP-decomposition can be understood as a special case of the nonlinear model (6).

6.1. The prior and posterior distributions for the nonlinear tensor estimation

We put a Gaussian process prior $\mathrm{GP}_{r,k}$ on each component $f_r^{*(k)}$. A (zero-mean) Gaussian process $f = (f(x) : x \in \mathcal{X})$ on some input space $\mathcal{X}$ is a set of random variable $f(x)$ indexed by $\mathcal{X}$ and defined on a common probability space $(\Omega, \mathcal{U}, P)$ such that each finite subset $(f(x_1), \ldots, f(x_j))$ $(j = 1, 2, \ldots)$ obeys a zero-mean multivariate normal distribution. Corresponding to the Gaussian process, we can define the kernel function $k : \mathcal{X} \times \mathcal{X} \to \mathbb{R}$ as the covariance function defined by $k(x, x') := E[f(x)f(x')]$. We denote by $k_{r,k}$ the kernel function corresponding to the Gaussian process $\mathrm{GP}_{r,k}$.

The number d of components (in other word, rank) can be set by cross validation. Otherwise, we put a prior distribution on the rank $1 \leq d \leq d_{\max}$ as

$$\pi(d) = \frac{\xi^d}{\sum_{d'=1}^{d_{\max}} \xi^{d'}},$$

where $0 < \xi < 1$ is some positive real number and $d_{\max}$ is taken so that it is sufficiently larger than the supposed true rank d^*. Given a rank d, let $\mathcal{F} = (f_r^{(k)})_{r=1,\ldots,d,\ k=1,\ldots,K}$ be a concatenation of functions $\{f_r^{(k)}\}_{r=1,\dots,d,\ k=1,\dots,K}$ and we consider the following prior distribution on the product space $\mathrm{d}\mathcal{F} = (\mathrm{d}f_r^{(k)})_{r=1,\dots,d,\ k=1,\dots,K}$:

$$\Pi(\mathrm{d}\mathcal{F}|d) = \Pi_{r=1}^{d}\Pi_{k=1}^{K}\mathrm{GP}_{r,k}(\mathrm{d}f_r^{(k)}).$$

Based on the prior distribution introduced above, we provide the posterior distribution and the corresponding Bayesian estimator. The posterior measure is constructed as:

$$\Pi(\mathrm{d}\mathcal{F}|D_n) \propto \sum_{d=1}^{d_{\max}} \frac{\Pi(D_n|\mathcal{F},d)}{\sum_{d'=1}^{d_{\max}}\int \Pi(D_n|\tilde{\mathcal{F}})\Pi(\mathrm{d}\tilde{\mathcal{F}}|d')\pi(d')}\Pi(\mathrm{d}\mathcal{F}|d)\pi(d),$$

where $\Pi(D_n|\mathcal{F})$ is a quasi likelihood defined by

$$\Pi(D_n|\mathcal{F}) = \exp\left\{-\beta\sum_{i=1}^{n}\left(Y_i - \sum_{r=1}^{d}\prod_{k=1}^{K} f_r^{(k)}(x_i^{(k)})\right)^2\right\}$$

with an inverse temperature parameter $\beta > 0$. Corresponding to the posterior, we have the postierior mean estimator $\hat{f}$: $\hat{f} = \int f\Pi(\mathrm{d}f|D_n)$.

6.2. *Posterior sampling*

Here, we present how to obtain samples from the posterior distribution. Let $\mathbf{f}_r^{(k)}$ be the function values of $f_r^{(k)}$ on the input data points: $(f_r^{(k)}(x_1^{(k)}),\dots,f_r^{(k)}(x_n^{(k)})) \in \mathbb{R}^n$. Since the distribution of $\mathbf{f}_r^{(k)}$ for some (r,k) given d and $\mathbf{f}_{r'}^{(k')}$ $((r',k') \neq (r,k))$ is a Gaussian distribution, we use Gibbs sampling for the computation of the mean estimator. Let $\boldsymbol{K}_r^{(k)}$ be the Gram matrix corresponding to the kernel function $k_{r,k}$; $\boldsymbol{K}_r^{(k)} = (k_{r,k}(x_i,x_j))_{i=1,j=1}^{n,n}$ The conditional distribution of $\mathbf{f}_r^{(k)}$ is

$$\begin{aligned}
&\pi(\mathbf{f}_r^{(k)}|\mathbf{f}_{-r}^{(-k)},D_n) \sim N(\boldsymbol{\mu}_r^{(k)},\boldsymbol{\Sigma}_r^{(k)}),\\
&\boldsymbol{\mu} = \boldsymbol{\Sigma}\big(\boldsymbol{a} * (\boldsymbol{y}-\boldsymbol{b})\big),\\
&\boldsymbol{\Sigma} = \boldsymbol{K}_r^{(k)} - \boldsymbol{K}_r^{(r)}\big(\boldsymbol{K}_r^{(k)} + \mathrm{diag}(a_1^2,\dots,a_n^2)\big)^{-1}\boldsymbol{K}_r^{(k)}.
\end{aligned}$$

where the symbol $*$ is the Hadamamard product, $N(\boldsymbol{\mu},\boldsymbol{\Sigma})$ is the multivariate normal distribution with mean $\boldsymbol{\mu}$ and variance-covariance $\boldsymbol{\Sigma}$, $\mathbf{f}_{-r}^{(-k)} = \left\{\mathbf{f}_{r'}^{(k')}|\,(r',k')\neq(r,k)\right\}$, $\boldsymbol{y} = (y_i)_{i=1}^n$, $\boldsymbol{a} = \left(\prod_{k'\neq k} f_r^{(k')}(x_i^{(k')})\right)_{i=1}^n$, and $\boldsymbol{b} = \left(\sum_{r'\neq r}\left(\prod_{k=1}^{K} f_{r'}^{(k)}(x_i^{(k)})\right)\right)_{i=1}^n$. Therefore, we can sample the function value $\mathbf{f}_r^{(k)}$ for $(r,k)\in[d]\times[K]$ from the posterior using the Gibbs sampling technique. That is, we sample $\mathbf{f}_r^{(k)}$ recursively one after another. We denote the j-th sample by $\mathbf{f}_{r,[j]}^{(k)}$ $((r,k)\in[d]\times[K])$ for $j=1,\dots,J$.

Based on the posterior samples generated from the above formula, we can predict the outcome on the input x through the following distribution:

$$\begin{aligned}
&\pi(f_r^{(k)}(x^{(k)})|\mathbf{f}_{(r)}^{(k)},\mathbf{f}_{(-r)}^{(-k)},D_n) \sim N(\mu,\sigma^2),\\
&\mu = k_{x^{(k)}}^{\top}(\mathrm{K}_r^{(k)})^{-1}\mathbf{f}_r^{(k)},\\
&\sigma^2 = k(x^{(k)},x^{(k)}) - k_{x^{(k)}}^{\top}(\mathrm{K}_r^{(k)})^{-1}k_{x^{(k)}},
\end{aligned}$$

where $k_{x^{(k)}} := (k(x_1^{(k)},x^{(k)}),\dots,k(x_n^{(k)},x^{(k)}))^{\top}$. Along with the posterior sample $\{\mathbf{f}_{r,[j]}^{(k)}\ (r,k)\in[d]\times[K]\}_j$, the prediction on the new input point $x=(x^{(1)},\dots,x^{(K)})$ is given by

$$\hat{f}(x) = \frac{1}{J}\sum_{j=1}^{J}\sum_{r=1}^{d}\prod_{k=1}^{K} k_{x^{(k)}}^{\top}(\mathrm{K}_r^{(k)})^{-1}\mathbf{f}_{r,[j]}^{(k)}.$$

7. Numerical experiments

We now present numerical experiments to show how the Bayes estimator works in practice. The numerical experiments are performed on both linear and nonlinear models.

7.1. Linear tensor estimation

We compare the convex optimization approach and the Bayes method for linear tensor estimation on artificial data. The problem is the tensor completion problem where each observation is a random selection of one element of A^* with observational noise $N(0,1)$ (see Example 1). The true tensor A^* was randomly generated such that each element of $U^{(k)}$ $(k = 1,\ldots,K)$ was uniformly distributed on $[-1,1]$. σ_{p} was set at 5, and the true tensor was estimated by the posterior mean obtained by the rejection sampling scheme with $R = 10$. $d_{\max}$ and ξ were set at 10 and 0.5. The posterior sampling was terminated after 500 iterations. The experiments were executed in five different settings, called settings 1 to 5: $\{(M_1,\ldots,M_K),d^*\} = \{(10,10,10),4\}$, $\{(10,10,40),5\}$, $\{(20,20,30),8\}$, $\{(20,30,40),5\}$, $\{(30,30,40),6\}$. For each setting, we repeated the experiments five times and computed the average of the in-sample predictive accuracy and out-of-sample accuracy over all five repetitions. The number of samples was chosen as $n = n_{\mathrm{s}} \prod_k M_k$, where n_{s} varied from 0.3 to 0.9.

We compare the Bayes estimator with the overlapped Schatten 1-norm regularization approach [6]. The comparison is executed in the settings 2 and 5. As for the regularization parameter of the convex regularized approach, we have chosen the best parameter at each sample size and each problem setting. The dashed lines correspond to the convex approach, and the solid line correspond to the Bayes approach. The accuracies of both methods are improved as the sample size increases. It can be seen that the Bayes approach much outperforms the convex approach in terms of both in-sample and out-of-sample accuracies.

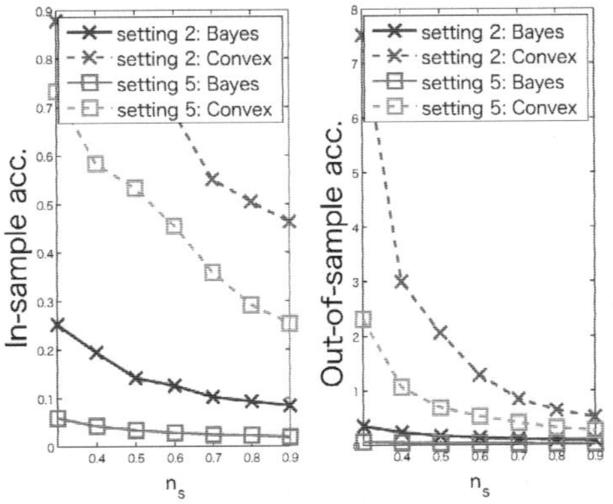

Figure 1: In-sample and out-of-sample accuracy comparison between the convex regularization approach and the Bayes approach, averaged over five repetitions.

7.2. Nonlinear tensor estimation

Here, we apply the nonlinear tensor estimator to the multi-task learning problem on the *school dataset*. The school dataset is from the inner London Education Authority (ILEA) and consists of examination records of 15362 students at 139 secondary schools in years 1985, 1986, and 1987. We modeled this data as a MLMTL problem in which the task is to predict exam scores for students at schools in certain years based on student and school dependent inputs. Following [26] for the preprocessing of the categorical attributes, we obtained 44 features. Therefore, in this setting, $M_1 = 139$ (number of school), $M_2 = 3$ (number of years), and $M_3 = 44$ (number of features).

We put a Gaussian kernels on the 44-dimensional feature space, and put the following kernel on the year and the school (we consider three settings (a), (b) and (c)):

(a) School: $k(p, p') = \delta_{p,p'}$, Year: $k(q, q') = \delta_{q,q'}$,

(b) School: $k(p, p') = \delta_{p,p'} + 0.5 \cdot (1 - \delta_{p,p'})$, Year: $k(q, q') = \delta_{q,q'} + 0.8 \cdot (1 - \delta_{q,q'})$,

(c) School: $k(p, p') = \delta_{p,p'} + 0.8 \cdot (1 - \delta_{p,p'})$, Year: $k(q, q') = \delta_{q,q'} + 0.8 \cdot (1 - \delta_{q,q'})$,

where p, p' are indexes of schools and q, q' are those of years, and $\delta_{i,j}$ is the delta function that is 1 if $i = j$, and 0 otherwise. We fixed $d = 3$.

We compare the Bayes estimator and the regularization method proposed by [27]. The regularization parameter for [27] was chosen from $[100, 900]$ so that it achieves the optimal test MSE. As in [27], we used the percentage of explained variance, $100 \cdot (\text{test MSE})/(\text{varianec of } y)$, as the evaluation metric. The result is shown in Figure 3a. We can see the nonlinear Bayes estimator much improves the accuracy.

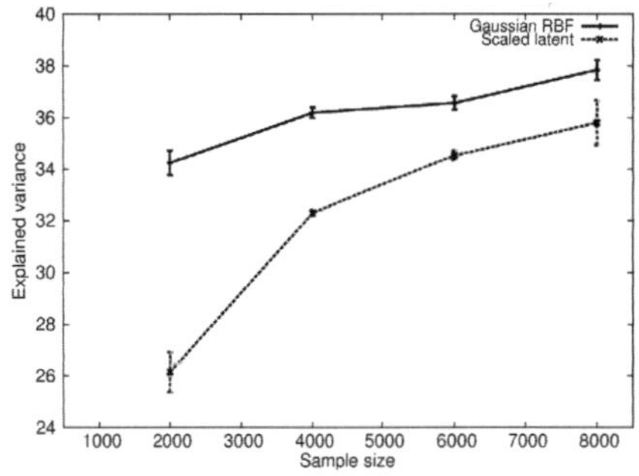

(a) Comparison between the Bayesian nonparametic method and scaled latent norm regularization [27]

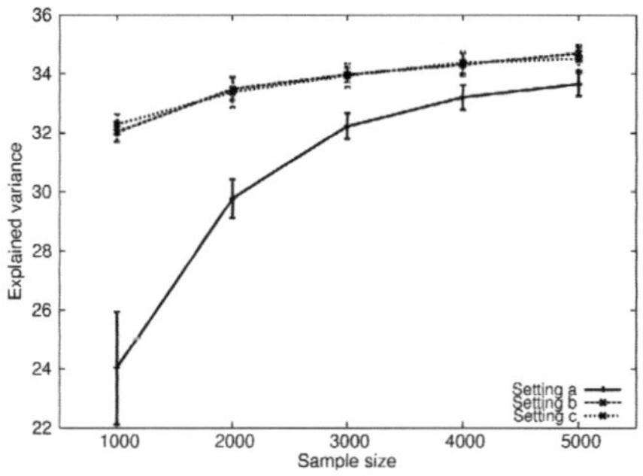

(b) Comparison between different kernels on the tasks

Figure 2: Results for the school data set

We also compared different settings of kernels on tasks (Figure 4b). The two settings (b) and (c) that incorporate similarities between tasks performed better than the setting (a).

8. Conclusion and discussions

We presented the statistical convergence rate of a Bayesian low rank tensor estimator. The notion of a tensor's rank in this article was based on CP-rank. It is noteworthy that the predictive accuracy was derived *without* any strong convexity assumption. Moreover, we presented the minimax optimal rate of the out-of-sample predictive accuracy. Moreover, we presented a nonparametric method to estimate a nonlinear tensor function using a Gaussian process technique. Numerical experiments showed effectiveness of the Bayes estimators.

Acknowledgment

This work was partially supported by MEXT Kakenhi (25730013, 25120012, 26280009, 15H01678 and 15H05707), JST-PRESTO and JST-CREST.

[1] Karatzoglou A, Amatriain X, Baltrunas L and Oliver N 2010 *Proceedings of the 4th ACM Conference on Recommender Systems 2010* pp 79–86

[2] Kolda T G and Bader B W 2009 *SIAM Review* **51** 455–500

[3] Liu J, Musialski P, Wonka P and Ye J 2009 *Proceedings of the 12th International Conference on Computer Vision (ICCV)* pp 2114–2121

[4] Signoretto M, Lathauwer L D and Suykens J 2010 Nuclear norms for tensors and their use for convex multilinear estimation Tech. Rep. 10-186 ESAT-SISTA K.U.Leuven

[5] Gandy S, Recht B and Yamada I 2011 *Inverse Problems* **27** 025010
[6] Tomioka R, Suzuki T, Hayashi K and Kashima H 2011 *Advances in Neural Information Processing Systems 24* pp 972–980 nIPS2011
[7] Tomioka R and Suzuki T 2013 *Advances in Neural Information Processing Systems 26* pp 1331–1339 nIPS2013
[8] Mu C, Huang B, Wright J and Goldfarb D 2014 *Proceedings of the 31th International Conference on Machine Learning* pp 73–81
[9] Chu W and Ghahramani Z 2009 *Proceedings of the 12th International Conference on Artificial Intelligence and Statistics (AISTATS)* (*JMLR Workshop and Conference Proceedings* vol 5)
[10] Xu Z, Yan F and Qi Y 2013 *IEEE Transactions on Pattern Analysis and Machine Intelligence* **99** 1 prePrints
[11] Xiong L, Chen X, Huang T K, Schneider J and Carbonell J G 2010 *Proceedings of SIAM Data Mining* pp 211–222
[12] Rai P, Wang Y, Guo S, Chen G, Dunson D and Carin L 2014 *Proceedings of the 31th International Conference on Machine Learning* (*JMLR Workshop and Conference Proceedings* vol 32) pp 1800–1808
[13] Suzuki T 2015 *Proceedings of the 32nd International Conference on Machine Learning (ICML2015)* ed Blei D and Bach F (JMLR Workshop and Conference Proceedings) pp 1273–1282
[14] Bickel P J, Ritov Y and Tsybakov A B 2009 *The Annals of Statistics* **37** 1705–1732
[15] Negahban S, Ravikumar P, Wainwright M J and Yu B 2012 *Statistical Science* **27** 538–557
[16] Signoretto M, De Lathauwer L and Suykens J A 2013 Learning tensors in reproducing kernel hilbert spaces with multilinear spectral penalties arXiv:1310.4977
[17] Hitchcock F L 1927 *Journal of Mathematics and Physics* **6** 164–189
[18] Hitchcock F L 1927 *Journal of Mathematics and Physics* **7** 39–79
[19] Romera-Paredes B, Aung H, Bianchi-Berthouze N and Pontil M 2013 *Proceedings of the 30th International Conference on Machine Learning* (*JMLR Workshop and Conference Proceedings* vol 28) pp 1444–1452
[20] Srebro N, Rennie J and Jaakkola T 2005 *Advances in Neural Information Processing Systems 17* (MIT Press) pp 1329–1336
[21] Hillar C J and Lim L H 2013 *Journal of the ACM* **60** 45:1–45:39
[22] Dalalyan A S and Tsybakov A B 2008 *Machine Learning* **72** 39–61
[23] Negahban S and Wainwright M J 2012 *Journal of Machine Learning Research* **13** 1665–1697
[24] Rohde A and Tsybakov A B 2011 *The Annals of Statistics* **39** 887–930
[25] Yang Y and Barron A 1999 *The Annals of Statistics* **27** 1564–1599
[26] Bakker B and Heskes T 2003 *Journal of Machine Learning Research* **4** 2003
[27] Wimalawarne K, Sugiyama M and Tomioka R 2014 *Advances in Neural Information Processing Systems 27* (Curran Associates, Inc.) pp 2825–2833

Higher order stationary subspace analysis

Danny Panknin[1,†], Paul von Bünau[†], Motoaki Kawanabe[2,‡], Frank C. Meinecke[†], Klaus-Robert Müller[3,†]

[†] Berlin Institute of Technology (TU Berlin), Machine Learning Group, Computer Science
[‡] Advanced Telecommunications Research Institute International (ATR), 2-2-2 Hikaridai, Seika-cho, Soraku-gun, Kyoto 619-0288, Japan

E-mail: [1] danny.panknin@tu-berlin.de, [2] kawanabe@atr.jp, [3] klaus-robert.mueller@tu-berlin.de
E-mail correspondence to the authors Panknin and Müller.

Abstract. Non-stationarity in data is an ubiquitous problem in signal processing. The recent stationary subspace analysis procedure (SSA) has enabled to decompose such data into a stationary subspace and a non-stationary part respectively. Algorithmically only weak non-stationarities could be tackled by SSA. The present paper takes the *conceptual* step generalizing from the use of first and second moments as in SSA to higher order moments, thus defining the proposed higher order stationary subspace analysis procedure (HOSSA). The paper derives the novel procedure and shows simulations. An obvious trade-off between the necessity of estimating higher moments and the accuracy and robustness with which they can be estimated is observed. In an ideal setting of plenty of data where higher moment information is dominating our novel approach can win against standard SSA. However, with limited data, even though higher moments actually dominate the underlying data, still SSA may arrive on par.

Key words. Stationary subspace analysis, blind source separation, non-stationary data, multivariate time series analysis, dimensionality reduction.

1. Introduction

Non-stationary effects in observed data are a common phenomenon, ranging from the neurosciences (e.g. [1, 2, 3, 4, 5, 6]) to econometrics (e.g. [7]). The causes for temporal distribution changes are manifold. In electroencephalography (EEG) analysis, for instance, non-stationary effects have been attributed to slow drifts of the mental state, the neural response to external stimulation and non-neural (e.g. technical) artefacts (see [8] for discussion).

However, detecting distribution changes is difficult in the case where (a) no auxiliary information regarding the distribution changes is available and (b) the observed multivariate data is generated as a mixture of underlying latent factors. In the absence of auxiliary information, such as timing information or a target variable, regression or classification methods are infeasible (unsupervised or explorative setting). Secondly, when each observed variable is generated as a linear combination of latent variables, changes in the joint distribution may not be easily visible in the original coordinates.

Therefore, stationary subspace analysis (SSA) [9] has been proposed to discern the stationary and non-stationary contributions to each observed variable in a completely unsupervised approach. In the SSA model, the observed D-variate time series is generated as a linear mixture of d_s latent stationary sources $s^{\mathfrak{s}}(t)$ and d_n latent non-stationary sources $s^{\mathfrak{n}}(t)$,

$$x(t) = A \begin{bmatrix} s^{\mathfrak{s}}(t) \\ s^{\mathfrak{n}}(t) \end{bmatrix}, \tag{1}$$

where A is an unknown time-constant square mixing matrix of dimension $d_s + d_n$. In the original SSA algorithm [9], this mixture model is inverted by minimizing the distance in distribution of the estimated stationary sources over epochs of the time series (that are defined e.g. by a sliding window). This SSA algorithm has been applied successfully to brain-computer-interfacing (e.g. [9, 10]), myoelectric control [11], computer vision [12], domain adaptation [13], geophysical data analysis [14] and change detection [15, 16]; there is an open-source implementation of the algorithm [17].

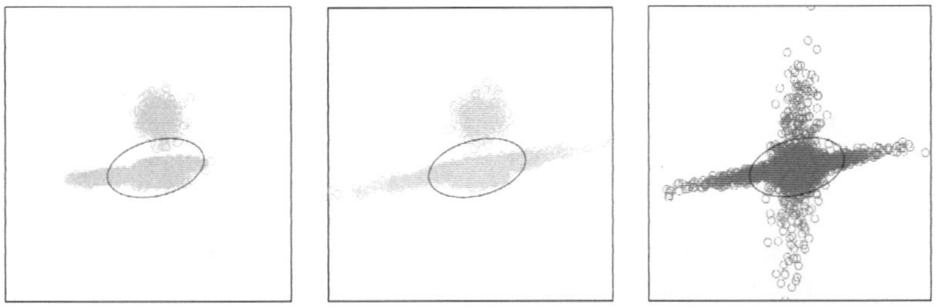

(a) Epoch 1 (b) Epoch 2 (c) Epoch 3

Figure 1: The distribution of this bivariate dataset changes between epochs but the mean and the covariance matrix stays the same; the non-stationarity affects only the higher order moments.

However, the SSA algorithm [9] is based on the restricted notion of *weak stationarity*. That is, a time series is considered stationary when its mean and covariance remains constant over time. This means that information in the higher order moments is ignored; e.g. a component with constant mean and covariance but changing kurtosis would be considered stationary. Figure 1 shows an example. In the presence of such a latent factor, the existing SSA algorithm fails to invert the mixture (1). It has been shown that higher moments contain useful information, e.g. in information theory [18, 19], independent component analysis [20] and vision [21].

In this paper, we present the first SSA algorithm which utilizes information from higher order moments based on an approximation of the Kullback-Leibler divergence. For ease of exposition, we restrict ourselves to the third and fourth order moments. Note that our contribution is primarily conceptual in nature. In controlled simulations, we demonstrate the validity of our contribution for higher order information.

The remainder of this paper is organized as follows. In the next Section 2, we briefly review the SSA model and algorithms based on weak stationarity. In Section 3 we derive the proposed method, making use of some concepts of the SSA approach. We compare the performance of SSA and HOSSA in toy examples in Section 4 to reveal drawbacks and advantages of both methods. In the final section, we recapitulate in which situations using HOSSA instead of SSA may improve results and when to stick with SSA.

2. Stationary Subspace Analysis

The aim of SSA is to invert the mixing model (1) given only samples from $x(t)$. That is, we want to find the estimated demixing matrix,

$$\hat{B} = \begin{bmatrix} \hat{B}^{\mathfrak{s}} \\ \hat{B}^{\mathfrak{n}} \end{bmatrix},$$

that separates the contribution of the stationary and the non-stationary sources in the observations. Let $A = \begin{bmatrix} A^{\mathfrak{s}} & A^{\mathfrak{n}} \end{bmatrix}$ be the unknown mixing matrix where the first d_s columns $A^{\mathfrak{s}}$ and the last d_n columns $A^{\mathfrak{n}}$ span the stationary and the non-stationary subspace respectively. In this paper, we assume that the subspace dimensions d_s and d_n are fixed in advance. By using the maximum likelihood version of SSA [10], it may be possible to construct selection procedures of these dimensions from data. However, this is out of the scope of our paper. By applying the demixing matrix $\hat{B}$ to the observations $x(t)$ we obtain estimates $\hat{s}^{\mathfrak{s}}(t)$ and $\hat{s}^{\mathfrak{n}}(t)$ for the underlying latent sources,

$$\begin{bmatrix} \hat{s}^{\mathfrak{s}}(t) \\ \hat{s}^{\mathfrak{n}}(t) \end{bmatrix} = \hat{B}x(t) = \begin{bmatrix} \hat{B}^{\mathfrak{s}} \\ \hat{B}^{\mathfrak{n}} \end{bmatrix} x(t) = \begin{bmatrix} \hat{B}^{\mathfrak{s}}A^{\mathfrak{s}} & \hat{B}^{\mathfrak{s}}A^{\mathfrak{n}} \\ \hat{B}^{\mathfrak{n}}A^{\mathfrak{s}} & \hat{B}^{\mathfrak{n}}A^{\mathfrak{n}} \end{bmatrix} \begin{bmatrix} s^{\mathfrak{s}}(t) \\ s^{\mathfrak{n}}(t) \end{bmatrix}.$$

The inverse of the mixing matrix A^{-1} is a demixing matrix, but it is not unique: any linear transformation *within* the two groups of estimated sources yields another valid separation, because it leaves the stationary resp. non-stationary nature of the sources unchanged. But also the separation into $\mathfrak{s}$- and $\mathfrak{n}$-sources itself is not unique: adding stationary components to a non-stationary source leaves it non-stationary, whereas the converse is not true. That is, the entire set of solutions to the SSA problem is given by the condition $\hat{B}^{\mathfrak{s}}A^{\mathfrak{n}} = 0$. Hence we can identify the true stationary sources derived by applying $\hat{B}^{\mathfrak{s}}$, up to the linear transformation, and the true non-stationary subspace spanned by the columns of $A^{\mathfrak{n}}$, whereas the non-stationary sources and the stationary subspace are not identifiable in general (see also [22]).

The SSA algorithm [9] finds the projections $\hat{B}^{\mathfrak{s}}$ and $\hat{B}^{\mathfrak{n}}$ that minimize resp. maximize the distance in distribution across epochs of the time series. The design of the epoch structure depends on the particular applications; common choices include a sliding window or a chronological segmentation. The distribution in each of the N epochs is approximated by a multivariate Gaussian distribution where $\hat{\mu}_i$ and $\hat{\Sigma}_i$ are the sample mean and sample covariance matrix in the i-th epoch. After a pre-whitening and centering of the average mean and covariance, such that

$$\frac{1}{N}\sum_{i=1}^{N}\hat{\mu}_i = 0 \quad \text{and} \quad \frac{1}{N}\sum_{i=1}^{N}\hat{\Sigma}_i = I,$$

the distance in distribution across the epochs is measured as the sum of the Kullback-Leibler (KL) divergences D_{KL} between each epoch and the average epoch. KL divergence is the most standard criterion measuring differences of two probability distributions used in statistics, information theory and statistical signal processing. The minimum 0 of this sum is achieved for projections that keep the means and covariances constant. Thus, by minimizing the objective

$$\begin{aligned} \hat{B}^{s} &= \underset{BB^{\top}=I}{\operatorname{argmin}} \sum_{i=1}^{N} D_{\mathrm{KL}}\Big[\mathcal{N}(B\hat{\mu}_i, B\hat{\Sigma}_i B^{\top}) \,\Big|\Big|\, \mathcal{N}(0, I)\Big] \\ &= \underset{BB^{\top}=I}{\operatorname{argmin}} \sum_{i=1}^{N} \Big(-\log\det\Big(B\hat{\Sigma}_i B^{\top}\Big) + \|B\hat{\mu}_i\|^2\Big), \end{aligned}$$

we find a stationary projection due to the weak stationarity assumption. This optimization problem can be solved efficiently by using multiplicative updates with rotation matrices parametrized as matrix exponentials of antisymmetric matrices [9, 23, 17]. Alternatively an algebraic version of SSA has been proposed [22].

3. SSA With Higher Order Moments

Keeping the concepts of second order SSA in mind, we now formulate the task of finding the $\mathfrak{s}$-sources as an optimization problem in an analogous way.

3.1. Establishing an Objective Function

Since we now need to consider differences between the first four moments Gaussian distributions are not adequate, as they are already uniquely determined given the first two moments, such that there exists no degree of freedom to adjust third and fourth moments. Instead we make use of a density approximation, that allows for a direct specification of the first four moments. More specifically, we employ Gram-Charlier expansion which was also applied to ICA for defining non-Gaussianity criteria [24]. Beforehand we will discuss in which form and how to estimate the required moments from data.

3.1.1. Efficient Estimation of Cumulants: Cumulants are a transformed version of simple moments and can therefore be used to represent the corresponding distribution. They are more convenient for our purpose as we can directly find distributions with specified moments setting up a series expansion in terms of the cumulants.

Unlike estimating the first two (multivariate) moments or their variants from data, estimating higher order moments is less commonly known. Given n samples $\{X_1, \ldots, X_n\}$ of a D-dimensional distribution, an unbiased j-th order cumulant estimate $\widehat{\kappa}^j = [\widehat{\kappa}^{i_1,\ldots,i_j}]_{i_1,\ldots,i_j=1}^D$ is in general, component-wise, given by (see e.g. [25], Chapter 4):

$$\widehat{\kappa}^{i_1,\ldots,i_j} = \frac{1}{n} \sum_{k_1,\ldots,k_j=1}^{n} \phi^{k_1\ldots k_j} X_{k_1}^{i_1} \cdot \ldots \cdot X_{k_j}^{i_j}$$

where $\phi^{k_1\ldots k_j} = \frac{(-1)^{\nu-1}}{\binom{n-1}{\nu-1}}$ with $\nu = \#\{k_1, \ldots, k_j\}$, the number of different indices.

Determination of these components in that general form is computationally inefficient, i.e. in $\mathcal{O}(n^j)$. However, rearranging the formulas can reduce the complexity to linear time in n: The j-th order cumulant can be expressed in terms of weighted, already determined lower order cumulants and a linear time sum of j-th order data products. This yields the well known sample mean and covariance for the first and second cumulant. For our purpose we also require the cumulant estimates of third and fourth order, given by

$$\widehat{\kappa}^{r,s,t} = \frac{n}{(n-1)(n-2)} \Big[\sum_{i=1}^{n} X_i^r X_i^s X_i^t - n\widehat{\kappa}^r \widehat{\kappa}^s \widehat{\kappa}^t - (n-1)(\widehat{\kappa}^r \widehat{\kappa}^{s,t} + \widehat{\kappa}^s \widehat{\kappa}^{r,t} + \widehat{\kappa}^t \widehat{\kappa}^{r,s}) \Big]$$

and

$$\begin{aligned}
\widehat{\kappa}^{r,s,t,u} &= \frac{1}{(n-1)(n-2)(n-3)} \Big[n(n+1) \sum_{i=1}^{n} X_i^r X_i^s X_i^t X_i^u - n^2(n+1) \widehat{\kappa}^r \widehat{\kappa}^s \widehat{\kappa}^t \widehat{\kappa}^u \\
&- n(n-1)(n+1)(\widehat{\kappa}^r \widehat{\kappa}^s \widehat{\kappa}^{t,u} + \widehat{\kappa}^r \widehat{\kappa}^t \widehat{\kappa}^{s,u} + \widehat{\kappa}^r \widehat{\kappa}^u \widehat{\kappa}^{s,t} + \widehat{\kappa}^s \widehat{\kappa}^t \widehat{\kappa}^{r,u} + \widehat{\kappa}^s \widehat{\kappa}^u \widehat{\kappa}^{r,t} + \widehat{\kappa}^t \widehat{\kappa}^u \widehat{\kappa}^{r,s}) \\
&- (n-1)^3 (\widehat{\kappa}^{r,s} \widehat{\kappa}^{t,u} + \widehat{\kappa}^{r,t} \widehat{\kappa}^{s,u} + \widehat{\kappa}^{r,u} \widehat{\kappa}^{s,t}) \\
&- (n+1)(n-1)(n-2)(\widehat{\kappa}^r \widehat{\kappa}^{s,t,u} + \widehat{\kappa}^s \widehat{\kappa}^{r,t,u} + \widehat{\kappa}^t \widehat{\kappa}^{r,s,u} + \widehat{\kappa}^u \widehat{\kappa}^{r,s,t}) \Big].
\end{aligned}$$

3.1.2. Difference Measurement with KL Divergence: After estimating the first four cumulants for each epoch $\widehat{\mathcal{K}}_i^1, \dots, \widehat{\mathcal{K}}_i^4$ and over the whole time $\widehat{\mathcal{K}}_0^1, \dots, \widehat{\mathcal{K}}_0^4$ we try, equivalently to second order SSA, to find a transformation of the form $I^{d_s}BW$, which, applied on the time series, gives the d_s estimated stationary components, where I^{d_s} are the upper d_s rows of the unity matrix, W is the already mentioned whitening matrix, and $B \in \mathsf{O}(D)$ is an orthogonal matrix.

Note that cumulants obey the transformation law of contravariant tensors (see e.g. [25], Chapter 2). Therefore it holds that, given a linear transformation $A = [a^{ij}]_{i,j=1}^{d,D} \in \mathbb{R}^{d \times D}$ and an D-dimensional random vector X with j^{th}-order cumulants $\mathcal{K}^j = [\kappa^{i_1,\dots,i_j}]_{i_1,\dots,i_j=1}^{D}$, the cumulants of the transformed random vector AX, denoted as $\widetilde{\mathcal{K}}^j$, are given by

$$\widetilde{\mathcal{K}}^j = [\widetilde{\kappa}^{i_1,\dots,i_j}]_{i_1,\dots,i_j=1}^{d},$$

$$\widetilde{\kappa}^{i_1,\dots,i_j} = \sum_{k_1,\dots,k_j=1}^{D} a^{i_1 k_1} \cdot \ldots \cdot a^{i_j k_j} \kappa^{k_1,\dots,k_j}.$$

Let $\widetilde{\mathcal{K}}_i^1, \dots, \widetilde{\mathcal{K}}_i^4$ be the epoch-wise cumulant estimates transformed by $I^{d_s}BW$, and $\widetilde{\mathcal{K}}_0^1, \dots, \widetilde{\mathcal{K}}_0^4$ the respective estimates over the whole time.

Note that due to centering, the definition of W and the constraint $BB^\top = I$ it is

$$\widetilde{\mathcal{K}}_0^1 = 0 \quad \text{and} \quad \widetilde{\mathcal{K}}_0^2 = I.$$

Now, the matrix parameterized by B transforms the time series to the d_s stationary components, when the epoch wise moments are constant over time, i.e. equalling the moments over the whole time. This can be reformulated as the KL divergence being zero between adequately chosen distributions $p_i := p_i(\,\cdot\,; \widetilde{\mathcal{K}}_i^1, \widetilde{\mathcal{K}}_i^2, \widetilde{\mathcal{K}}_i^3, \widetilde{\mathcal{K}}_i^4)$, and $p_0 := p_0(\,\cdot\,; 0, I, \widetilde{\mathcal{K}}_0^3, \widetilde{\mathcal{K}}_0^4)$ respectively, possessing these moments. Formally we want to minimize

$$L(B) = \sum_{i=1}^{N} D_{\mathrm{KL}}\Big[\, p_0 \,\Big|\Big|\, p_i \,\Big]$$

subject to $BB^\top = I$. How such distributions can be constructed is part of the next subsection.

3.1.3. The Gram Charlier Expansion Assume all moments of a distribution exist and were known. When formulating appropriate coefficients $\eta^{i_1 \dots i_j}$, which are functions of the first j moments we can expand the distribution as an infinite series

$$f(\mathbf{t}) = \Phi(\mathbf{t}; \mathcal{K}^1, \mathcal{K}^2) \left\{ 1 + \sum_{j=3}^{\infty} \sum_{i_1,\dots,i_j=1}^{D} \frac{1}{j!} \eta^{i_1 \dots i_j} h^{i_1 \dots i_j}(\mathbf{t}) \right\} \tag{2}$$

where $\Phi(\,\cdot\,; \mathcal{K}^1, \mathcal{K}^2)$ is the *density* function of the Gaussian distribution $\mathcal{N}(\mathcal{K}^1, \mathcal{K}^2)$ and

$$h^{i_1 \dots i_j}(\mathbf{t}) = \frac{1}{\Phi(\mathbf{t}; \mathcal{K}^1, \mathcal{K}^2)} \frac{\partial^j \Phi(\mathbf{t}; \mathcal{K}^1, \mathcal{K}^2)}{\partial x_{i_1} \cdots \partial x_{i_j}}$$

are (multivariate) Hermite polynomials.

Now, that we are interested in the first four moments, we truncate the series (2) to fourth order. Furthermore for simplicity of subsequent estimations, we consider the one-dimensional case, generalizing to arbitrary dimensions later on. The series therefore reduces to

$$f(t) = \Phi(t; \kappa^1, \kappa^2) \left\{ 1 + \frac{1}{3!} \kappa^3 h^3(t) + \frac{1}{4!} \kappa^4 h^4(t) \right\} \tag{3}$$

Note, that the corresponding first four moments are still exact, though f is not necessarily a valid probability density anymore.

3.1.4. Approximation of the Objective Function In contrast to SSA the present case exhibits no analytical form of the integral in the KL divergence. Instead we approximate a part of the objective function with a second order Taylor expansion:

$$\log(1+x) \approx x - \frac{x^2}{2}, \tag{4}$$

which then gives:

$$\begin{aligned}
L(B) &= \sum_{i=1}^{N} D_{\mathrm{KL}}\left[\, p_0 \,\middle\|\, p_i \,\right] \\
&\overset{(3)}{\simeq} \sum_{i=1}^{N} \int_{\mathbb{R}} \Phi(x;0,1) \left\{1 + \frac{1}{3!}\widetilde{\kappa}_0^3 h_0^3(x) + \frac{1}{4!}\widetilde{\kappa}_0^4 h_0^4(x)\right\} \\
&\quad \times \log\left(\frac{\Phi(x;0,1)\{1+\frac{1}{3!}\widetilde{\kappa}_0^3 h_0^3(x) + \frac{1}{4!}\widetilde{\kappa}_0^4 h_0^4(x)\}}{\Phi(x;\widetilde{\kappa}_i^1,\widetilde{\kappa}_i^2)\{1+\frac{1}{3!}\widetilde{\kappa}_i^3 h_i^3(x) + \frac{1}{4!}\widetilde{\kappa}_i^4 h_i^4(x)\}}\right) dx \\
&\overset{(4)}{\simeq} \sum_{i=1}^{N} \int_{\mathbb{R}} \Phi(x;0,1) \left\{1 + \frac{1}{3!}\widetilde{\kappa}_0^3 h_0^3(x) + \frac{1}{4!}\widetilde{\kappa}_0^4 h_0^4(x)\right\} \\
&\quad \times \left\{ \log\left(\frac{\Phi(x;0,1)}{\Phi(x;\widetilde{\kappa}_i^1,\widetilde{\kappa}_i^2)}\right) + \frac{1}{3!}\widetilde{\kappa}_0^3 h_0^3(x) + \frac{1}{4!}\widetilde{\kappa}_0^4 h_0^4(x) - \frac{1}{2}\left(\frac{1}{3!}\widetilde{\kappa}_0^3 h_0^3(x) + \frac{1}{4!}\widetilde{\kappa}_0^4 h_0^4(x)\right)^2 \right. \\
&\quad \left. - \frac{1}{3!}\widetilde{\kappa}_i^3 h_i^3(x) - \frac{1}{4!}\widetilde{\kappa}_i^4 h_i^4(x) + \frac{1}{2}\left(\frac{1}{3!}\widetilde{\kappa}_i^3 h_i^3(x) + \frac{1}{4!}\widetilde{\kappa}_i^4 h_i^4(x)\right)^2 \right\} dx
\end{aligned}$$

Solving integrals of the form above is trivial in the sense that the integrand is a product of a Gaussian density and a polynomial. Unfortunately, the polynomial is large and complicated, such that there is no compact solution which we could present in this paper due to limited space.

3.2. The Minimization

In order to make use of powerful standard analysis gradient descent methods for optimization an adjustment is necessary. Namely standard gradient methods perform additive updates. But the orthogonal matrices form no additive group as they are not closed under addition. So it would be hard to perform such updates ensuring to stay within the set $\mathsf{O}(D)$ of $D \times D$ orthogonal matrices. However, they form a multiplicative group such that multiplicative updates do not suffer from this problem. And since orthogonal matrices can be parameterized with additively closed anti-symmetric matrices we can circumvent the problems making use of the advantages of both sets, performing additive gradient descent in the parameter space of anti-symmetric matrices guaranteeing it to result into a multiplicative update within the orthogonal matrices (see [23] for a detailed version). Note that $\exp(A+B) = \exp(A)\exp(B) = \exp(B)\exp(A)$ is only guaranteed, if A and B commute. Therefore we perform gradient descent steps as a line search along the commutating subgroup generated by $\mathfrak{so}_H = \{tH | t \in \mathbb{R}\}$:

(i) Assume, we are at an initial point $B_0 \in \mathsf{O}(D)$

(ii) Denote by B the desired updated position after a line search step, such that B can be factorized as RB_0 with an orthogonal matrix $R = \exp(tH)$ with positive determinant (i.e. an element in $\mathsf{SO}(D)$) and an anti-symmetric matrix H (i.e. $H^\top = -H$) from the parameter space.

Journal of Physics: Conference Series **699** (2016) 012021 doi:10.1088/1742-6596/699/1/012021

(iii) We can consider the initial position B_0 as a constant and rewrite $L(B) = L_{B_0}(R) = (L_{B_0} \circ \exp)(tH)$. That is, $(L_{B_0} \circ \exp)(0) = L(B_0)$.

One can see that, in order to find H, we simply need to build the gradient of $L_{B_0} \circ \exp$ at position 0 for each step, since we can set $B_0 \leftarrow RB_0$ at the end of each step. The gradient is given by

$$\nabla(L_{B_0} \circ \exp)(0) = (\frac{\partial L_{B_0}}{\partial R}(I))I^\top - I(\frac{\partial L_{B_0}}{\partial R}(I))^\top,$$

where $\frac{\partial L_{B_0}}{\partial R}(I)$ can be derived by differentiating the solution from $L_{B_0}(R)$ along R making use of the fact that $R^\top R = I$. Note that $\nabla(L_{B_0} \circ \exp)(0)$ is antisymmetric. In summary we perform a conjugate gradient method along search directions as proposed by Polak and Ribière [26] as in Algorithm 1, where we denote by **vec** the vectorization of a matrix.

Algorithm 1 Conjugate gradient procedure for HOSSA

1: Choose $B_0 \in \mathsf{O}(D)$ at random and set $k = 0$.
2: **repeat**
3: Determine the next search direction $\mathbf{vec}(H_k) = \frac{g_k^\top(g_k - g_{k-1})}{g_{k-1}^\top g_{k-1}}$ for $k > 0$ and $\mathbf{vec}(H_0) = -g_0$, where $g_k = \mathbf{vec}(\nabla(L_{B_k} \circ \exp)(0))$.
4: Find $t \in \mathbb{R}$ that minimizes $L_{B_k}(\exp(tH_k))$ along $\exp(so_{H_k})$.
5: Update $B_{k+1} = \exp(tH_k)B_k$.
6: Set $k \leftarrow k + 1$.
7: **until** convergence

After minimization, the first component of the signal transformed by BW is one of the estimated stationary components. Further components can be extracted by successively applying the algorithm on the remaining estimated components, ending up with a set of orthogonal transformations $b_1, \ldots, b_{d_s}$ (of decreasing dimension). The final transformation is then given by

$$B := B_{d_s} \cdot B_{d_s - 1} \cdot \ldots \cdot B_1,$$

with the embedded transformations

$$B_i := \begin{cases} b_1 & , \quad i = 1 \\ \begin{bmatrix} I_{i-1} & 0 \\ 0 & b_i \end{bmatrix}, & \text{else} \end{cases},$$

where I_k is the k-dimensional identity matrix (further details can be found in [27]).

4. Experiments

The goal of the experiments is to show the validity of the proposed concept in comparison to second order SSA not only in general, but especially in the case, where non-stationarity is not measurable within the first two moments, where we expect second order SSA to fail. In order to analyze the latter case we need a controlled setting fulfilling the specification of the first two moments exhibiting no non-stationarity information. This means, that over all epochs the mean and covariance stay constant. For this purpose we generate a toy dataset with that property.

4.1. Toy Data Generation

Suppose we want to simulate N discriminable epochs of D dimensional data with $d_s < D$ stationary components, where the first two moments stay constant. Let $K = \lceil \log_2 N \rceil$, then we can generate the specified epochs using a Gaussian mixture model with $2K$ components pairing up each two of those. For each pair we generate two states of distribution, that exhibit the property of constant first two moments. Then by choosing one of the two states for all K pairs we can generate up to $2^K > N$ discriminable epochs with the required properties.

Let $\pi_1, \ldots, \pi_{2K} > 0$ be the mixture probabilities the $2K$ components with $\sum_{i=1}^{2K} \pi_i = 1$.

Furthermore let $X_1^0, \ldots, X_{2K}^0$ be independent random vectors with $X_i^0 \sim \mathcal{N}(a_i, \Sigma_i)$ with $a_{K+i} = -\frac{\pi_i}{\pi_{K+i}} a_i, i \in \{1, \ldots, K\}$ and $\Sigma_i = \widetilde{\Sigma}_i + \lambda_i a_i a_i^\top$ with $\widetilde{\Sigma}_i$ positive definite and $\lambda_i = \lambda_{K+i} > 0$.

For the other state of the distributions let $X_1^1, \ldots, X_{2K}^1$ be independent with

$$X_i^1 \sim \mathcal{N}(\sqrt{1+\lambda_i} a_i, \Sigma_i - \lambda_i a_i a_i^\top).$$

Now we choose for the l-th epoch one of the two states for each of the K pairs, i.e. $S^l \in \{0,1\}^K$. Then the Gaussian mixture model of the l-th epoch is given by

$$Y^l = \sum_{i=1}^{K} \left[\mathbb{1}_{\{P=i\}} X_i^{S_i^l} + \mathbb{1}_{\{P=K+i\}} X_{K+i}^{S_i^l} \right],$$

where P indicates the selected component based on the distribution $(\pi_i)_i$, $\mathbb{1}_{\{\cdot\}}$ denotes the indicator function and is independent of all X_i^j. For all $S^l, S^m \in \{0,1\}^K$ the following holds:

$$\mathbb{E}[Y^l] = \mathbb{E}[Y^m] = 0,$$

$$\mathrm{Cov}[Y^l] = \mathrm{Cov}[Y^m] = \sum_{i=1}^{2K} \pi_i \left[\Sigma_i + a_i a_i^\top \right]$$

Now, in order to enforce d_s of the D components to be unchanged in distribution, one can simply set the first d_s entries of each a_i to zero.

4.2. Identifiability

With the toy data from above, we test the subspace identifiability of second order SSA against higher order SSA for different numbers of epochs. As the performance measure we determine the angle between the unique true non-stationary subspace $A^{\mathfrak{n}}$ and its approximation $\hat{A}^{n}$, given by $\left[\hat{A}^{\mathfrak{s}} \;\; \hat{A}^{\mathfrak{n}}\right] = \hat{A} = \hat{B}^{-1}$.

The results in Figure 2 show that, not surprisingly, standard SSA fails to identify the true subspace, invariant of the epoch number. In contrast, higher order SSA converges, in the case of $D = 4$ dimensions, for more than ten epochs to a solution that lies close to the true subspace. Note, that especially the higher order part of the optimization suffers from the limits of approximation such that here the hope of finding and exact solution rather than an approximated one is unfeasible. This can be seen in Figure 2 for the case of $D = 6$ dimensions, where higher order SSA needs much more epochs to converge to a solution, that is close to the true subspace. However, if there is information contained in the first two moments, this information dominates the higher order SSA optimization, letting it act similar to standard SSA.

In fact, we can observe this behavior when we inject information within moments up to second order in a controlled manner. For this we modify the strict model from above, containing no

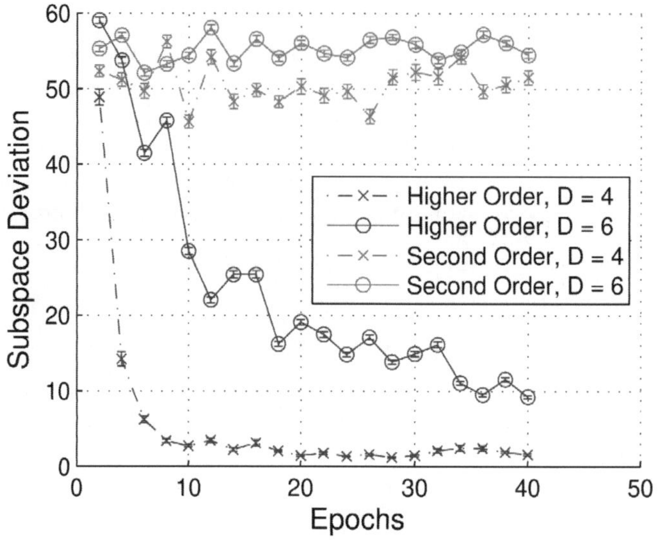

Figure 2: Average deviation of second and higher order SSA from the true solution, measured as the angle between the estimated and the true non-stationary subspace, when the first two moments exhibit no information. Setup: $d_s = 2$, exact (theoretical) Moments, 100 Repetitions.

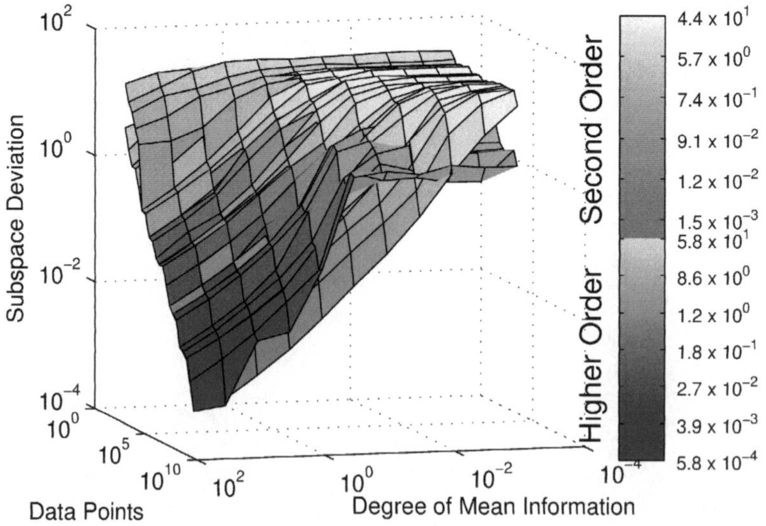

Figure 3: Average deviation of second and higher order SSA from the true solution, measured as the angle between the estimated and the true non-stationary subspace, for a varying amount of data per epoch and information content of first order moments. Setup: $d_s = 2$, $D = 4$, $N = 32$, 20 Repetitions.

information in moments up to second order, by allowing $\lambda_i \neq \lambda_{K+i}$ to a controlled extent. That is, set $\lambda_{K+i} \leftarrow \lambda_i + Cu_i$ for $u_i \sim \mathcal{U}[0,1]$ and a parameter $C \geq 0$, which controls the information content in the first moment. For $C = 0$ we have the strict model with no information in the mean and by increasing C accordingly more information accumulates in the first moment. When fixating a sufficiently large amount of epochs, we can observe in Figure 3 the influence of the parameter C (Degree of Mean Information) for a varying amount of data. Not surprisingly, second order SSA is more sensitive to the mean-information, starting to benefit for smaller values of C than higher order SSA. For a large C higher order SSA approaches second order SSA despite some approximation related inaccuracy. Since lower order moments are less prone to inexact estimation than higher order moments, for a growing C both methods become more robust with respect to noisy moments or small dataset size. In the area of smaller values of C - where second order SSA is not sensitive enough - we can see that higher order SSA gives usable results in a low noise setting, whereas second order SSA acts completely random. Figure 4 shows slices of Figure 3 for three dataset sizes, namely a too small, a medium and a sufficiently large size, resulting into higher order SSA performing worse, on par or significantly better, respectively, compared to second order SSA.

5. Discussion

Methods to decompose non-stationary data into their stationary and non-stationary subspaces have become important tools for modern data analysis. While the earlier SSA method has made use of the estimation of the second order moments only, the present conceptual extension HOSSA has contributed by employing in addition also higher order moments for decomposition.

Clearly, there is a well-known intrinsic trade-off between robustness and the ability to capture higher order information that we also encounter when comparing SSA and HOSSA. In cases where the information contained in the first two moments is insufficient, a higher order approach can better find the true solution compared to second order SSA – as expected. The proposed method was shown to yield improvement.

When second and higher order information is both hidden in the signals, we observe that the second order method generally provides a higher robustness. Note, however, that for lower dimensional systems, HOSSA has no significant disadvantage in performance. In higher dimensional examples we have demonstrated that the difficulty in accurately estimating higher order moments becomes the leading reason for a degradation of HOSSA; when the noise level is increased, then this effect becomes amplified. Then, clearly the second order method SSA becomes the more robust estimator for identifying the stationary subspace.

While we might initially think that the additional higher order moments in HOSSA provide additional information allowing an easier identifiability, the experiments show that this potential advantage does not materialize in practice.

Future work will aim to therefore improve the approximation part of HOSSA, in particular, we will study sampling based approaches. In the independent component analysis field, there have been used other non-Gaussian distributions (e.g. generalized Gaussian distribution in [28]) except for the Gram-Charlier expansion by [24]. More robust implementation of HOSSA might be possible with such a non-Gaussian model.

Acknowledgments

We would like to acknowledge support for this project by the German Ministry of Education and Research (BMBF) through the project ALICE II (Autonomous Learning in Complex Environments II) (01IB15001B) as well as the 'Adaptive BCI' Project, FKZ 01GQ1115, and by the German Research Foundation (DFG) through the project Co-BCI, 'Theoretical concepts for co-adaptive human machine interaction with application to BCI' (MU987/14-1). This work was also supported by the World Class University Program through the National Research Foundation of Korea funded by the Ministry of Education, Science, and Technology, under Grant R31-10008. MK was supported in part Japan Science and Technology Agency (German-Japanese cooperation program on computational neuroscience), the Ministry of Education, Culture, Sports, Science and Technology (Grant-in-Aid for Scientific Research B, 24300093) and by the Ministry of Internal Affairs and Communications.

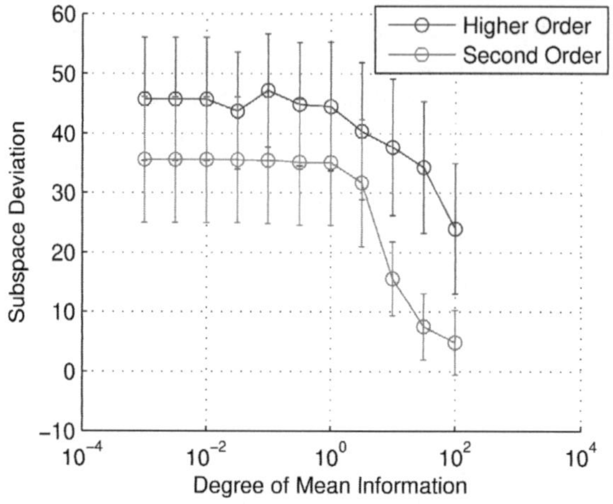

(a) Slice for 1000 data points per epoch

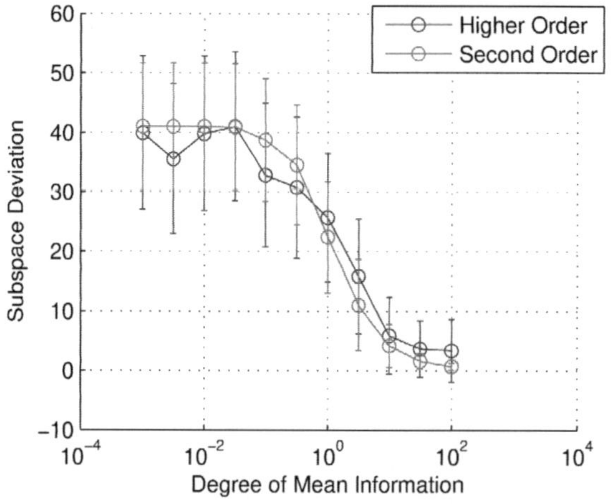

(b) Slice for $2.2 \cdot 10^4$ data points per epoch

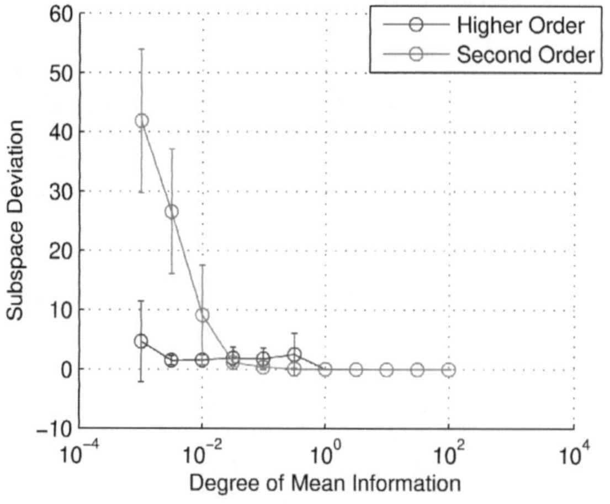

(c) Slice for $2.2 \cdot 10^9$ data points per epoch

Figure 4: Slices of Figure 3 along fixed data amounts, including approximated 95 % confidence intervals.

References

[1] Shenoy P, Krauledat M, Blankertz B, Rao R P N and Müller K R 2006 Towards adaptive classification for BCI *Journal of Neural Engineering* **3** *1* R13–23

[2] Samek W, Vidaurre C, Müller K R and Kawanabe M 2012 Stationary common spatial patterns for brain-computer interfacing *Journal of Neural Engineering* **9** *2* 026013

[3] Samek W, Meinecke F C and Müller K R 2013 Transferring subspaces between subjects in brain–computer interfacing *IEEE Transactions on Biomedical Engineering* **60** *8* 2289–98

[4] Samek W, Kawanabe M and Müller K R 2014 Divergence-based framework for common spatial patterns algorithms *IEEE Reviews in Biomedical Engineering* **7** 50–72

[5] Vidaurre C, Sannelli C, Müller K R and Blankertz B 2011 Machine-learning-based coadaptive calibration for brain-computer interfaces *Neural Computation* **23** *3* 791–816

[6] Sugiyama M, Krauledat M and Müller K R 2007 Covariate shift adaptation by importance weighted cross validation *Journal of Machine Learning Research* **8** 985–1005

[7] Granger C W J 1981 Some properties of time series data and their use in econometric model specification *Journal of Econometrics* **16** *1* 121-30

[8] Blankertz B, Kawanabe M, Tomioka R, Hohlefeld F, Nikulin V and Müller K R 2008 Advances in neural information processing systems 20 (MIT Press) ed J C Platt et al 113–20

[9] von Bünau P, Meinecke F C, Király F J and Müller K R 2009 Finding stationary subspaces in multivariate time series *Phys. Rev. Lett.* **103** *21* 213101

[10] Kawanabe M, Samek W, von Bünau P and Meinecke F C 2011 Artificial neural networks and machine learning – ICANN 2011 An information geometrical view of stationary subspace analysis *Lecture notes in computer science vol 6792* (Springer Berlin / Heidelberg) 397–404

[11] Hahne J, Dähne S, Hwang H J, Müller K R and Parra L 2015 Concurrent adaptation of human and machine improves simultaneous and proportional myoelectric control *IEEE Transactions on Neural Systems and Rehabilitation Engineering* **23** *4* 618–627

[12] Meinecke F C, von Bünau P, Kawanabe M and Müller K R 2009 Learning invariances with stationary subspace analysis *Lecture notes in computer science vol 6792 IEEE 12th International Conference on Computer Vision Workshops (ICCV Workshops)* 87–92

[13] Hara S, Kawahara Y, Washio T and von Bünau P 2010 Stationary subspace analysis as a generalized eigenvalue problem *Proceedings of the 17th international conference on neural information processing: theory and algorithms* **1** (Springer Berlin / Heidelberg) 422–9

[14] Hara S, Kawahara Y, Washio T, von Bünau P, Tokunaga T and Yumoto K 2012 Separation of stationary and non-stationary sources with a generalized eigenvalue problem *Neural Netw.* **33** (Elsevier Science Ltd.) 7–20

[15] Blythe D A J, von Bünau P, Meinecke F C and Müller K R 2012 Feature extraction for change-point detection using stationary subspace analysis *IEEE Transactions on Neural Networks and Learning Systems* **23** *4* 631–43

[16] Blythe D A J, Meinecke F C, von Bünau P and Müller K R 2013 Explorative data analysis for changes in neural activity *Journal of Neural Engineering* **10** *2* (IOP Publishing) 26018–33

[17] Müller J S, von Bünau P, Meinecke F C, Király F J and Müller K R 2011 The stationary subspace analysis toolbox *Journal of Machine Learning Research* **12** 3065–3069

[18] Jenssen R 2010 Kernel entropy component analysis *IEEE Transactions on Pattern Analysis and Machine Intelligence* **32** *5* 847–60

[19] Cover T M and Thomas J A 2006 Elements of Information Theory 2nd Edition (Wiley Series in Telecommunications and Signal Processing)

[20] Hyvärinen A, Karhunen J and Oja E 2001 Independent component analysis (New York: Wiley)

[21] Hyvärinen A, Hurri J and Hoyer P O 2009 Natural image statistics: a probabilistic approach to early computational vision (Springer publishing company, incorporated)

[22] Király F J, von Bünau P, Meinecke F C, Blythe D A J and Müller K R 2012 Algebraic geometric comparison of probability distributions *Journal of Machine Learning Research* **13** 855–903

[23] Plumbley M D 2005 Geometrical methods for non-negative ICA: manifolds, Lie groups and toral subalgebras *Neurocomputing* **67** (Elsevier Science Publishers B. V.) 161–97

[24] Amari S, Cichocki A and Yang H H 1996 A new learning algorithm for blind signal separation *Advances in Neural Information Processing Systems 8* ed D Touretzky et al (MIT Press) 757–63

[25] McCullagh P 1987 Tensor methods in statistics *Monographs on statistics and applied probability*

[26] Polak E 1971 Computational methods in optimization (Elsevier science & technology books)

[27] von Bünau P 2012 Stationary subspace analysis: towards understanding non-stationary data *Universitätsbibliothek*

[28] Lee T and Lewicki M S 2000 The generalized Gaussian mixture model using ICA *Proceedings of the 2nd International Workshop on Independent Component Analysis* 239–44

Journal of Physics: Conference Series **699** (2016) 012022 doi:10.1088/1742-6596/699/1/012022

Replica methods for loopy sparse random graphs†

ACC Coolen

Department of Mathematics, King's College London,
The Strand, London WC2R 2LS, United Kingdom

E-mail: ton.coolen@kcl.ac.uk

Abstract. I report on the development of a novel statistical mechanical formalism for the analysis of random graphs with many short loops, and processes on such graphs. The graphs are defined via maximum entropy ensembles, in which both the degrees (via hard constraints) and the adjacency matrix spectrum (via a soft constraint) are prescribed. The sum over graphs can be done analytically, using a replica formalism with complex replica dimensions. All known results for tree-like graphs are recovered in a suitable limit. For loopy graphs, the emerging theory has an appealing and intuitive structure, suggests how message passing algorithms should be adapted, and what is the structure of theories describing spin systems on loopy architectures. However, the formalism is still largely untested, and may require further adjustment and refinement.

1. Introduction

Networks and graphs are increasingly popular and effective tools for visualising and modelling large and complex processes and 'big' data sets. We know that many important graphical structures in the world (biological networks, computing and communication networks, resource grids, lattices in physics, etc) are not tree-like; they tend to have many short loops, and we know that processes on graphs are affected significantly by the presence of such loops. It is therefore problematic that most of our tools and algorithms for analysing (processes on) finitely connected graphs, such as cavity methods [1], belief propagation type algorithms [2, 3, 4], and conventional replica analyses [5, 6], require topologies that are locally tree-like. Some methods were extended with loop corrections [7, 8, 9, 10], but all tend to fail for graphs with many short loops. The exceptions, solvable spin models or spectrum calculations for loopy graphs, all rely on some special property of either the dynamics or the graph topology, and are thus non-generic: these include spherical models [11], one- and two-dimensional Ising systems [12, 13], loopy immune models that can be mapped to tree-like systems [14], and trees of loopy modules [15, 16].

In this paper I report on ongoing research aimed at the development of a new and more general statistical mechanical method that removes the restriction to tree-like graphs. It is designed to handle analytically ensembles of large and sparse random graphs with prescribed degree sequences and prescribed loop statistics (via their adjacency spectra), and stochastic processes on such graphs. It is based on an alternative flavour of the replica method, with imaginary replica dimensions, and produces explicit closed equations in the infinite size limit, leading to Shannon entropies and expressions for spectra of ensembles of sparse loopy graphs. The familiar equations describing tree-like graphs are recovered as a simple limiting case.

† This paper is dedicated to the memory of our colleague and friend Jun-Ichi Inoue, with whom the author has had the great pleasure and privilege of collaborating.

2. Definitions

We study simple nondirected N-node graphs characterised by adjacency matrices $\mathbf{c} = \{c_{ij}\}$, with $c_{ij} \in \{0,1\}$, $c_{ij} = c_{ji}$ and $c_{ii} = 0$ for all (i,j). They are drawn randomly from maximum entropy ensembles $p(\mathbf{c})$, in which we prescribe the values of all degrees $k_i(\mathbf{c}) = \sum_j c_{ij}$ and the eigenvalue spectrum $\varrho(\mu|\mathbf{c})$, subject to $\int \mathrm{d}\mu\ \mu\varrho(\mu|\mathbf{c}) = N^{-1}\sum_i c_{ii} = 0$ and $\int \mathrm{d}\mu\ \mu^2\varrho(\mu|\mathbf{c}) = N^{-1}\sum_{ij} c_{ij}c_{ji} = N^{-1}\sum_i k_i$. Since the spectrum controls the statistics of closed paths of all lengths via its moments, we are not limited to tree-like graphs. Modulo isomorphisms, many graphs are already determined by their spectra [17, 18, 19], and without external fields the free energy of spin systems on loopy graphs depends only on the loop statistics [20], so we expect that graphs from such ensembles can be tailored very effectively to model complex real-world systems. We impose the degrees as hard constraints, and the spectrum as a soft constraint, so

$$p(\mathbf{c}) \quad \propto \quad \mathrm{e}^{N\int \mathrm{d}\mu\ \hat{\varrho}(\mu)\varrho(\mu|\mathbf{c})} \prod_{i\leq N} \delta_{k_i, \sum_j c_{ij}} \tag{1}$$

We can write the relevant sums over graphs in terms of the generating function

$$\Phi[\hat{\varrho}] \quad = \quad \frac{1}{N}\log \sum_{\mathbf{c}} \mathrm{e}^{N\int \mathrm{d}\mu\ \hat{\varrho}(\mu)\varrho(\mu|\mathbf{c})} \prod_{i\leq N} \delta_{k_i, \sum_j c_{ij}} \tag{2}$$

The equation from which to solve the Lagrange parameter $\hat{\varrho}(\mu)$, i.e. $\varrho(\mu) = \sum_{\mathbf{c}} p(\mathbf{c})\varrho(\mu|\mathbf{c})$, and the ensemble entropy per node $S = -N^{-1}\sum_{\mathbf{c}} p(\mathbf{c})\log p(\mathbf{c})$, are both expressed in terms of (2):

$$\varrho(\mu) = \delta\Phi[\hat{\varrho}]/\delta\hat{\varrho}(\mu), \qquad S = \Phi[\hat{\varrho}] - \int \mathrm{d}\mu\ \hat{\varrho}(\mu)\rho(\mu) \tag{3}$$

We could alternatively start by choosing $\hat{\varrho}(\mu)$, and view the first equation of (3) as a tool for calculating the associated spectrum. Locally tree-like graphs correspond to $\hat{\varrho}(\mu) = 0$. For $\hat{\varrho}(\mu) = \alpha_3\mu^3$ with $\alpha_3 > 0$ we get loopy random graphs constrained by the degrees and the density of triangles. Adding higher order terms to $\hat{\varrho}(\mu)$ means controlling higher order closed path statistics. Imposing the full spectrum means constraining the numbers of closed paths of all lengths. The core of our problem is how to do analytically the sum over graphs in (2).

3. Calculation of the generating function

We can write (2) as an average over an Erdös-Rènyi graph ensemble [21] with average degree $\langle k\rangle = \frac{1}{N}\sum_i k_i$. Since the probabilities $p_{\mathrm{ER}}(\mathbf{c})$ of this ensemble depend on $\mathbf{c}$ only via $\sum_{i<j} c_{ij}$, we can use the short-hand $\langle f(\mathbf{c})\rangle_{\mathrm{ER}} = \sum_{\mathbf{c}} p_{\mathrm{ER}}(\mathbf{c})f(\mathbf{c})$ to write

$$\Phi[\hat{\varrho}] \quad = \quad \frac{1}{2}\langle k\rangle\big[\log\big(\frac{N}{\langle k\rangle}\big)+1\big] + \frac{1}{N}\log\Big\langle \mathrm{e}^{N\int \mathrm{d}\mu\ \hat{\varrho}(\mu)\varrho(\mu|\mathbf{c})} \prod_{i\leq N} \delta_{k_i, \sum_j c_{ij}} \Big\rangle_{\mathrm{ER}} + \mathcal{O}(\frac{1}{N}) \tag{4}$$

The factors induced by degree constraints are harmless. Our problem is the dependence on $\mathbf{c}$ via $\varrho(\mu|\mathbf{c})$. To handle this we use the following identity which can be derived in a few lines from the spectrum formula of [22], and was first presented in [23]:

$$\mathrm{e}^{N\int \mathrm{d}\mu\ \hat{\varrho}(\mu)\varrho(\mu|\mathbf{c})} \quad = \quad \lim_{\Delta,\varepsilon\downarrow 0}\ \lim_{n_\mu \to \frac{\mathrm{i}\Delta}{\pi}\frac{\mathrm{d}}{\mathrm{d}\mu}\hat{\varrho}(\mu)}\ \lim_{m_\mu\to -n_\mu} \prod_\mu \Big[Z(\mu{+}\mathrm{i}\varepsilon|\mathbf{c})^{n_\mu}\ \overline{Z(\mu{+}\mathrm{i}\varepsilon|\mathbf{c})}^{\,m_\mu}\Big] \tag{5}$$

where $\overline{z}$ indicates complex conjugation, and with the integrals $Z(\mu|\mathbf{c}) = \int_{\mathrm{IR}^N} \mathrm{d}\boldsymbol{\phi}\ \exp(-\frac{1}{2}\mathrm{i}\boldsymbol{\phi}\cdot[\mathbf{c}-\mu\mathbf{1}]\boldsymbol{\phi})$. We can now proceed via the replica method. We evaluate the graph average for integer

$\{n_\mu, m_\mu\}$, and take the limits to imaginary values via analytical continuation[1]. The tricky factor then becomes a product of integrals:

$$\prod_\mu \Big[Z(\mu+\mathrm{i}\varepsilon|\mathbf{c})^{n_\mu}\,\overline{Z(\mu+\mathrm{i}\varepsilon|\mathbf{c})}^{\,m_\mu}\Big]$$
$$= \prod_\mu \Big\{\Big[\prod_{\alpha_\mu=1}^{n_\mu}\int_{\mathbb{R}^N}\mathrm{d}\boldsymbol{\phi}_{\mu,\alpha_\mu}\mathrm{e}^{-\frac{1}{2}(\varepsilon-\mathrm{i}\mu)\boldsymbol{\phi}^2_{\mu,\alpha_\mu}}\Big]\Big[\prod_{\beta_\mu=1}^{m_\mu}\int_{\mathbb{R}^N}\mathrm{d}\boldsymbol{\psi}_{\mu,\beta_\mu}\mathrm{e}^{-\frac{1}{2}(\varepsilon+\mathrm{i}\mu)\boldsymbol{\psi}^2_{\mu,\beta_\mu}}\Big]\Big\}$$
$$\times \mathrm{e}^{\mathrm{i}\sum_{i<j}c_{ij}\sum_\mu\big[\sum_{\beta_\mu=1}^{m_\mu}\psi^i_{\mu,\beta_\mu}\psi^j_{\mu,\beta_\mu}-\sum_{\alpha_\mu=1}^{n_\mu}\phi^i_{\mu,\alpha_\mu}\phi^j_{\mu,\alpha_\mu}\big]} \qquad (6)$$

We can now do the graph summations. We abbreviate $\boldsymbol{\phi}^i = \{\phi^i_{\mu,\alpha_\mu}\}$ and $\boldsymbol{\psi}^i = \{\psi^i_{\mu,\beta_\mu}\}$, with $\boldsymbol{\phi}^i\!\cdot\!\boldsymbol{\phi}^j = \sum_\mu\sum_{\alpha_\mu\leq n_\mu}\phi^i_{\mu,\alpha_\mu}\phi^j_{\mu,\alpha_\mu}$ and $\boldsymbol{\psi}^i\!\cdot\!\boldsymbol{\psi}^j = \sum_\mu\sum_{\beta_\mu\leq m_\mu}\psi^i_{\mu,\beta_\mu}\psi^j_{\mu,\beta_\mu}$, and we introduce a matrix $\mathbf{M}$ with entries $M_{\mu,\alpha;\mu',\alpha'} = \mu\delta_{\mu\mu'}\delta_{\alpha\alpha'}$. Upon inserting (5) into (2), writing degree constraints in integral form via $\delta_{k_i,\sum_j c_{ij}} = (2\pi)^{-1}\int_{-\pi}^{\pi}\mathrm{d}\omega_i\;\mathrm{e}^{\mathrm{i}\omega_i(k_i-\sum_j c_{ij})}$, and using (6), we find that the order parameter of our problem is the distribution $\mathcal{P}(\boldsymbol{\phi},\boldsymbol{\psi},\omega) = N^{-1}\sum_i\delta(\boldsymbol{\phi}-\boldsymbol{\phi}^i)\delta(\boldsymbol{\psi}-\boldsymbol{\psi}^i)\delta(\omega-\omega_i)$, which we introduce into our formulae by inserting for each $(\boldsymbol{\phi},\boldsymbol{\psi},\omega)$, with $\omega\in[-\pi,\pi]$:

$$1 = \int\!\mathrm{d}\mathcal{P}(\boldsymbol{\phi},\boldsymbol{\psi},\omega)\;\delta\Big[\mathcal{P}(\boldsymbol{\phi},\boldsymbol{\psi},\omega) - \frac{1}{N}\sum_i\delta(\boldsymbol{\phi}-\boldsymbol{\phi}^i)\delta(\boldsymbol{\psi}-\boldsymbol{\psi}^i)\delta(\omega-\omega_i)\Big] \qquad (7)$$

Upon writing the δ-functions in integral form, we then obtain the following path integral representation of $\Phi[\hat{\varrho}]$ which is for large N evaluated via steepest descent, with $\lim_{N\to\infty}\epsilon_N = 0$:

$$\Phi[\hat{\varrho}] = \frac{1}{2}\langle k\rangle\log\big(\frac{N}{\langle k\rangle}\big) + \lim_{\Delta,\varepsilon\downarrow 0}\;\lim_{n_\mu\to\frac{\mathrm{i}\Delta}{\pi}\frac{\mathrm{d}}{\mathrm{d}\mu}\hat{\varrho}(\mu)}\;\lim_{m_\mu\to -n_\mu}\mathrm{extr}_{\{\mathcal{P},\hat{\mathcal{P}}\}}\Psi[\mathcal{P},\hat{\mathcal{P}}] + \epsilon_N \qquad (8)$$

$$\Psi[\mathcal{P},\hat{\mathcal{P}}] = \mathrm{i}\int\!\mathrm{d}\boldsymbol{\phi}\mathrm{d}\boldsymbol{\psi}\mathrm{d}\omega\;\hat{\mathcal{P}}(\boldsymbol{\phi},\boldsymbol{\psi},\omega)\mathcal{P}(\boldsymbol{\phi},\boldsymbol{\psi},\omega)$$
$$+\frac{1}{2}\langle k\rangle\int\!\mathrm{d}\boldsymbol{\phi}\mathrm{d}\boldsymbol{\psi}\mathrm{d}\omega\mathrm{d}\boldsymbol{\phi}'\mathrm{d}\boldsymbol{\psi}'\mathrm{d}\omega'\;\mathcal{P}(\boldsymbol{\phi},\boldsymbol{\psi},\omega)\mathcal{P}(\boldsymbol{\phi}',\boldsymbol{\psi}',\omega')\mathrm{e}^{-\mathrm{i}(\omega+\omega')+\mathrm{i}(\boldsymbol{\psi}\cdot\boldsymbol{\psi}'-\boldsymbol{\phi}\cdot\boldsymbol{\phi}')}$$
$$+\sum_k p(k)\log\int_{-\pi}^{\pi}\frac{\mathrm{d}\omega}{2\pi}\mathrm{e}^{\mathrm{i}k\omega}\int\!\mathrm{d}\boldsymbol{\phi}\mathrm{d}\boldsymbol{\psi}\;\mathrm{e}^{-\frac{1}{2}\boldsymbol{\phi}\cdot(\varepsilon\mathbf{I}-\mathrm{i}\mathbf{M})\boldsymbol{\phi}-\frac{1}{2}\boldsymbol{\psi}\cdot(\varepsilon\mathbf{I}+\mathrm{i}\mathbf{M})\boldsymbol{\psi}-\mathrm{i}\hat{\mathcal{P}}(\boldsymbol{\phi},\boldsymbol{\psi},\omega)} \qquad (9)$$

Working out the saddle-point equations of (9) shows that (8) can be written in the form

$$\Phi[\hat{\varrho}] = \frac{1}{2}\langle k\rangle\big[\log\big(\frac{N}{\langle k\rangle}\big)+1\big] + \sum_k p(k)\log\tilde{p}(k) + \epsilon_N$$
$$+ \lim_{\Delta,\varepsilon\downarrow 0}\;\lim_{n_\mu\to\frac{\mathrm{i}\Delta}{\pi}\frac{\mathrm{d}}{\mathrm{d}\mu}\hat{\varrho}(\mu)}\;\lim_{m_\mu\to -n_\mu}\sum_k p(k)\log\int\!\mathrm{d}\boldsymbol{\phi}\mathrm{d}\boldsymbol{\psi}\;\mathrm{e}^{-\frac{1}{2}\boldsymbol{\phi}\cdot(\varepsilon\mathbf{I}-\mathrm{i}\mathbf{M})\boldsymbol{\phi}-\frac{1}{2}\boldsymbol{\psi}\cdot(\varepsilon\mathbf{I}+\mathrm{i}\mathbf{M})\boldsymbol{\psi}}$$
$$\times\Big[\int\!\mathrm{d}\boldsymbol{\phi}'\mathrm{d}\boldsymbol{\psi}'\;\mathcal{W}(\boldsymbol{\phi}',\boldsymbol{\psi}')\mathrm{e}^{\mathrm{i}(\boldsymbol{\psi}\cdot\boldsymbol{\psi}'-\boldsymbol{\phi}\cdot\boldsymbol{\phi}')}\Big]^k \qquad (10)$$

Here $\tilde{p}(k) = \mathrm{e}^{-\langle k\rangle}\langle k\rangle^k/k!$, $\lim_{N\to\infty}\epsilon_N = 0$, and $\mathcal{W}(\boldsymbol{\phi},\boldsymbol{\psi})$ is to be solved from

$$\mathcal{W}(\boldsymbol{\phi},\boldsymbol{\psi}) = \sum_k\frac{k}{\langle k\rangle}p(k) \qquad (11)$$
$$\times\frac{\mathrm{e}^{-\frac{1}{2}\boldsymbol{\phi}\cdot(\varepsilon\mathbf{I}-\mathrm{i}\mathbf{M})\boldsymbol{\phi}-\frac{1}{2}\boldsymbol{\psi}\cdot(\varepsilon\mathbf{I}+\mathrm{i}\mathbf{M})\boldsymbol{\psi}}\Big[\int\mathrm{d}\boldsymbol{\phi}'\mathrm{d}\boldsymbol{\psi}'\;\mathcal{W}(\boldsymbol{\phi}',\boldsymbol{\psi}')\mathrm{e}^{\mathrm{i}(\boldsymbol{\psi}\cdot\boldsymbol{\psi}'-\boldsymbol{\phi}\cdot\boldsymbol{\phi}')}\Big]^{k-1}}{\int\mathrm{d}\boldsymbol{\phi}''\mathrm{d}\boldsymbol{\psi}''\mathrm{e}^{-\frac{1}{2}\boldsymbol{\phi}''\cdot(\varepsilon\mathbf{I}-\mathrm{i}\mathbf{M})\boldsymbol{\phi}''-\frac{1}{2}\boldsymbol{\psi}''\cdot(\varepsilon\mathbf{I}+\mathrm{i}\mathbf{M})\boldsymbol{\psi}''}\Big[\int\mathrm{d}\boldsymbol{\phi}'\mathrm{d}\boldsymbol{\psi}'\;\mathcal{W}(\boldsymbol{\phi}',\boldsymbol{\psi}')\mathrm{e}^{\mathrm{i}(\boldsymbol{\psi}''\cdot\boldsymbol{\psi}'-\boldsymbol{\phi}''\cdot\boldsymbol{\phi}')}\Big]^k}$$

[1] In standard replica analyses [24] the replica dimension n is taken to zero. Models with nonzero real-valued n describe systems with adiabatically separated timescales and multiple temperatures [25, 26, 27, 28, 29, 30]. Imaginary dimensions n have so far been used only in [31, 32, 33] and [34].

4. Replica symmetric theory

4.1. Replica symmetry ansatz

We assume that $\mathcal{W}(\boldsymbol{\phi},\boldsymbol{\psi})$ is symmetric under all permutations of $\{\phi_{\mu,1},\ldots,\phi_{\mu,n_\mu}\}$ and $\{\psi_{\mu,1},\ldots,\psi_{\mu,m_\mu}\}$. For integer $\{n_\mu,m_\mu\}$ equation (11) is also invariant under $\mathcal{W}(\boldsymbol{\phi},\boldsymbol{\psi})\to\overline{\mathcal{W}(\boldsymbol{\psi},\boldsymbol{\phi})}$, and we assume that the relevant solution of (11) is invariant under this transformation. It then follows via De Finetti's theorem [35] that

$$\mathcal{W}(\boldsymbol{\phi},\boldsymbol{\psi}) = \mathcal{C}\int\!\{\mathrm{d}\pi\}\mathcal{W}[\{\pi\}]\Big[\prod_\mu\prod_{\alpha_\mu=1}^{n_\mu}\pi(\phi_{\mu,\alpha_\mu}|\mu)\Big]\Big[\prod_\mu\prod_{\beta_\mu=1}^{m_\mu}\overline{\pi(\psi_{\mu,\beta_\mu}|\mu)}\Big] \tag{12}$$

$\mathcal{C}$ is a constant, and $\mathcal{W}[\{\pi\}]$ is a normalised measure on the space of conditioned distributions π with $\int\!\mathrm{d}x\;\pi(x|\mu)=1$ for all μ. We introduce the Fourier transforms $\hat{\pi}(\phi|\mu)=\int\!\mathrm{d}x\;\mathrm{e}^{-\mathrm{i}x\phi}\,\pi(x|\mu)$, the normalised distributions $P[\{\pi_1,\ldots,\pi_k\}]$, and the short-hand $\mathcal{A}[\{\pi_1,\ldots,\pi_k\}]$ as follows:

$$P(\phi|\mu,\pi_1,\ldots,\pi_k) = \frac{\mathrm{e}^{-\frac{1}{2}(\varepsilon-\mathrm{i}\mu)\phi^2}\prod_{\ell\le k}\hat{\pi}_\ell(\phi|\mu)}{\int\!\mathrm{d}x\;\mathrm{e}^{-\frac{1}{2}(\varepsilon-\mathrm{i}\mu)x^2}\prod_{\ell\le k}\hat{\pi}_\ell(x|\mu)} \tag{13}$$

$$\mathcal{A}[\{\pi_1,\ldots,\pi_k\}] = \prod_\mu\Big[\Big(\int\!\mathrm{d}x\;\mathrm{e}^{-\frac{1}{2}(\varepsilon-\mathrm{i}\mu)x^2}\prod_{\ell\le k}\hat{\pi}_\ell(x|\mu)\Big)^{n_\mu}\overline{\Big(\int\!\mathrm{d}x\;\mathrm{e}^{-\frac{1}{2}(\varepsilon-\mathrm{i}\mu)x^2}\prod_{\ell\le k}\hat{\pi}_\ell(x|\mu)\Big)^{m_\mu}}\Big] \tag{14}$$

This allows us to write equation (11) after some simple manipulations in a form which reveals that the ansatz (12) indeed solves (11), provided $\mathcal{W}$ and $\mathcal{C}$ obey

$$\mathcal{W}[\{\pi\}] = \frac{1}{\mathcal{C}^2}\sum_{k>0}p(k)\frac{k}{\langle k\rangle}\frac{\Big[\prod_{\ell<k}\int\{\mathrm{d}\pi_\ell\}\mathcal{W}[\{\pi_\ell\}]\Big]\mathcal{A}[\{\pi_1,\ldots,\pi_{k-1}\}]\delta_{\mathrm{F}}[\pi-P[\{\pi_1,\ldots,\pi_{k-1}\}]]}{\Big[\prod_{\ell\le k}\int\{\mathrm{d}\pi_\ell\}\mathcal{W}[\{\pi_\ell\}]\Big]\mathcal{A}[\{\pi_1,\ldots,\pi_k\}]} \tag{15}$$

$$\mathcal{C}^2 = \sum_{k>0}p(k)\frac{k}{\langle k\rangle}\frac{\Big[\prod_{\ell<k}\int\{\mathrm{d}\pi_\ell\}\mathcal{W}[\{\pi_\ell\}]\Big]\mathcal{A}[\{\pi_1,\ldots,\pi_{k-1}\}]}{\Big[\prod_{\ell\le k}\int\{\mathrm{d}\pi_\ell\}\mathcal{W}[\{\pi_\ell\}]\Big]\mathcal{A}[\{\pi_1,\ldots,\pi_k\}]} \tag{16}$$

with the functional delta-distribution $\delta_{\mathrm{F}}[g]\propto\prod_x\delta[g(x)]$. Our generating function becomes

$$\Phi_{\mathrm{RS}}[\hat{\varrho}] = \frac{1}{2}\langle k\rangle\big[\log\big(\frac{N}{\langle k\rangle}\big)+1\big]+\sum_k p(k)\log\tilde{p}(k)+\langle k\rangle\log\mathcal{C}+\epsilon_N + \sum_k p(k)\log\Big[\prod_{\ell\le k}\int\!\{\mathrm{d}\pi_\ell\}\mathcal{W}[\{\pi_\ell\}]\Big]\mathcal{A}[\{\pi_1,\ldots,\pi_k\}] \tag{17}$$

The functional (14) can be written alternatively in terms of (13) as

$$\mathcal{A}[\{\pi_1,\ldots,\pi_k\}]=\mathcal{A}[P[\{\pi_1,\ldots,\pi_k\}]] \tag{18}$$

in which we now define

$$\mathcal{A}[\{P\}]=\mathrm{e}^{-\sum_\mu\big[n_\mu\log P(0|\mu)+m_\mu\log\overline{P(0|\mu)}\big]} \tag{19}$$

This property is due to the fact that $\mathcal{A}[\{\pi_1,\ldots,\pi_k\}]$ can be expressed in terms of the denominators of the right-hand side of (13), which in turn can be written in terms of $P(0|\mu)$ as a consequence of the identity $\hat{\pi}_\ell(0|\mu)=\int\!\mathrm{d}x\;\pi(x|\mu)=1$ for all μ and all ℓ.

4.2. Formula for the spectrum

According to (3), working out $\varrho(\mu)$ requires differentiation of $\Phi_{\rm RS}[\hat{\varrho}]$ with respect to $\hat{\varrho}$. This was indeed the origin and purpose of the generating function. This differentiation is easier if we write $\Phi_{\rm RS}[\hat{\varrho}]$ as an extremum over $\mathcal{C}$ and $\mathcal{W}$, so that the dependences of $\mathcal{C}$ and $\mathcal{W}$ on $\hat{\varrho}$ will not affect these derivatives. If we also extremise over $\mathcal{C}$ for fixed $\mathcal{W}$, and eliminate $\mathcal{C}$, we obtain

$$\mathcal{C}^{-2} = \int\{{\rm d}\pi{\rm d}\pi'\}\mathcal{W}[\{\pi\}]\mathcal{W}[\{\pi'\}]\mathcal{B}[\{\pi,\pi'\}] \tag{20}$$

$$\begin{aligned}\Phi_{\rm RS}[\hat{\varrho}] =& \frac{1}{2}\langle k\rangle\big[\log\big(\frac{N}{\langle k\rangle}\big)+1\big] + \sum_k p(k)\log\tilde{p}(k) + \epsilon_N \\ &+ {\rm extr}_{\{\mathcal{W}\}}\Big\{\sum_k p(k)\log\Big[\Big[\prod_{\ell\leq k}\int\{{\rm d}\pi_\ell\}\mathcal{W}[\{\pi_\ell\}]\Big]\mathcal{A}[\{\pi_1,\ldots,\pi_k\}]\Big] \\ &-\frac{1}{2}\langle k\rangle\log\int\{{\rm d}\pi{\rm d}\pi'\}\mathcal{W}[\{\pi\}]\mathcal{W}[\{\pi'\}]\mathcal{B}[\{\pi,\pi'\}]\Big\}\end{aligned} \tag{21}$$

with

$$\mathcal{B}[\{\pi,\pi'\}] = \prod_\mu\Big\{\Big[\int{\rm d}\phi{\rm d}\phi'\pi(\phi|\mu)\pi'(\phi'|\mu){\rm e}^{-{\rm i}\phi\phi'}\Big]^{n_\mu}\overline{\Big[\int{\rm d}\phi{\rm d}\phi'\pi(\phi|\mu)\pi'(\phi'|\mu){\rm e}^{-{\rm i}\phi\phi'}\Big]}^{m_\mu} \tag{22}$$

Now $\hat{\varrho}$ appears only in $\mathcal{A}[\ldots]$ and $\mathcal{B}[\ldots]$, and we obtain from (3) the following formula:

$$\begin{aligned}\varrho_{\rm RS}(\mu) =& \sum_k p(k)\int\{{\rm d}\pi{\rm d}\pi'\}\mathcal{W}[\{\pi'\}]\mathcal{B}[\{\pi,\pi'\}]\Big\{\Big(\frac{\delta\log\mathcal{A}[\{\pi\}]}{\delta\hat{\varrho}(\mu)}+\frac{\delta\log\mathcal{B}\big[\{\pi,\pi'\}\big]}{\delta\hat{\varrho}(\mu)}\Big) \\ &\times\frac{\int\Big[\prod_{\ell<k}\{{\rm d}\pi_\ell\}\mathcal{W}[\{\pi_\ell\}]\Big]\delta_{\rm F}\big[\pi-P[\{\pi_1,\ldots,\pi_{k-1}\}]\big]\mathcal{A}[\{\pi\}]}{\int\Big[\prod_{\ell\leq k}\int\{{\rm d}\pi_\ell\}\mathcal{W}[\{\pi_\ell\}]\Big]\mathcal{A}[\{\pi_1,\ldots,\pi_k\}]}\Big\} \\ &-\frac{1}{2}\langle k\rangle\frac{\int\{{\rm d}\pi{\rm d}\pi'\}\mathcal{W}[\{\pi\}]\mathcal{W}[\{\pi'\}]\mathcal{B}[\{\pi,\pi'\}]\delta\log\mathcal{B}[\{\pi,\pi'\}]/\delta\hat{\varrho}(\mu)}{\int\{{\rm d}\pi{\rm d}\pi'\}\mathcal{W}[\{\pi\}]\mathcal{W}[\{\pi'\}]\mathcal{B}[\{\pi,\pi'\}]}\end{aligned} \tag{23}$$

Substituting the definitions of n_μ and m_μ, and taking $\Delta\downarrow 0$, the functional derivatives become

$$\frac{\delta\log\mathcal{A}[\{\pi\}]}{\delta\hat{\varrho}(\mu)} = -\frac{2}{\pi}\frac{\rm d}{{\rm d}\mu}{\rm Arg}\ \pi(0|\mu) \tag{24}$$

$$\frac{\delta\log\mathcal{B}[\{\pi,\pi'\}]}{\delta\hat{\varrho}(\mu)} = \frac{2}{\pi}\frac{\rm d}{{\rm d}\mu}{\rm Arg}\int{\rm d}\phi{\rm d}\phi'{\rm e}^{-{\rm i}\phi\phi'}\pi(\phi|\mu)\pi'(\phi'|\mu) \tag{25}$$

5. Exploiting the nature of the propagation

5.1. Gaussian propagated densities

Equation (15) can be seen as the fixed-point equation of a process in which densities $\pi(x|\mu)$ are mapped via (13). As in e.g. [36], this dynamics is seen to close for complex Gaussian functions:

$$\mathcal{W}[\{\pi\}] = \int\{{\rm d}x{\rm d}y\}\mathcal{W}[\{x,y\}]\ \delta_{\rm F}\big[\pi-\pi[\{x,y\}]\big] \tag{26}$$

Here $\{x\}$ and $\{y\}$ are complex functions of $\mu\in{\rm I\!R}$, with ${\rm Im}\ x(\mu)<0$ for all μ, and $\pi[\{x,y\}]$ is the μ-conditioned normalised density $\pi(\phi|x(\mu),y(\mu))$ defined via

$$\pi(\phi|x,y) = \frac{{\rm e}^{-\frac{1}{2}{\rm i}x\phi^2+{\rm i}y\phi}}{Z(x,y)}, \qquad Z(x,y) = \Big(\frac{2\pi}{|x|}\Big)^{\frac{1}{2}}{\rm e}^{\frac{1}{2}{\rm i}\ {\rm atan}[{\rm Re}(x)/{\rm Im}(x)]+\frac{1}{2}{\rm i}y^2/x} \tag{27}$$

The integrals leading to the normalisation factor $Z(x, y)$ are found in [37]. The Fourier transform of (27) is $\hat{\pi}(\phi|\mu) = \exp[\frac{1}{2}\mathrm{i}\phi^2/x(\mu) - \mathrm{i}\phi y(\mu)/x(\mu)]$, so (13) takes the form

$$P(\phi|\mu, \pi_1, \ldots, \pi_{k-1}) = \pi(\phi|F(\mu|x_1, \ldots, x_{k-1}), G(\mu|x_1, y_1, \ldots, x_{k-1}, y_{k-1})) \tag{28}$$

Here $F(\ldots)$ and $G(\ldots)$ are the following functions of μ:

$$F(\mu|x_1, \ldots, x_{k-1}) = -\mu - \mathrm{i}\varepsilon - \sum_{\ell<k} 1/x_\ell(\mu) \tag{29}$$

$$G(\mu|x_1, y_1, \ldots, x_{k-1}, y_{k-1}) = -\sum_{\ell<k} y_\ell(\mu)/x_\ell(\mu) \tag{30}$$

If all $x_\ell(\mu)$ have negative imaginary parts, so will $F(\mu|x_1, \ldots, x_{k-1})$. Hence all relevant integrals over ϕ exist and the propagation (13) indeed closes within the family (27).

5.2. Order parameter equations and spectrum formula

We work out the functionals $\mathcal{A}[\{\pi\}]$ and $\mathcal{B}[\{\pi, \pi'\}]$. Starting from (14,22) or (24,25) one derives

$$\mathcal{A}[\{x, y\}] = \mathrm{e}^{\frac{1}{\pi}\int \mathrm{d}\mu\ \hat{\varrho}(\mu)\frac{\mathrm{d}}{\mathrm{d}\mu}\left[\mathrm{atan}\left(\frac{\mathrm{Re}(x(\mu))}{\mathrm{Im}(x(\mu))}\right)+\mathrm{Re}\left(\frac{y^2(\mu)}{x(\mu)}\right)\right]} \tag{31}$$

$$\begin{aligned}\mathcal{B}[\{x, y; x', y'\}] &= \mathrm{e}^{\frac{1}{\pi}\int \mathrm{d}\mu\ \hat{\varrho}(\mu)\frac{\mathrm{d}}{\mathrm{d}\mu}\left[\mathrm{atan}\left(\frac{\mathrm{Re}(x(\mu)-1/x'(\mu))}{\mathrm{Im}(x(\mu)-1/x'(\mu))}\right)-\mathrm{atan}\left(\frac{\mathrm{Re}(x(\mu))}{\mathrm{Im}(x(\mu))}\right)\right]} \\ &\quad \times \mathrm{e}^{\frac{1}{\pi}\int \mathrm{d}\mu\ \hat{\varrho}(\mu)\frac{\mathrm{d}}{\mathrm{d}\mu}\mathrm{Re}\left(\frac{y^2(\mu)/x(\mu)+y'^2(\mu)/x'(\mu)-2y(\mu)y'(\mu)}{x(\mu)x'(\mu)-1}\right)}\end{aligned} \tag{32}$$

Similarly we can work out the order parameter equations (15) for $\mathcal{W}[\{x, y\}]$ and (20) for the normalisation factor $\mathcal{C}$. To compactify the result we introduce the short-hand

$$\mathcal{F}_k[\{x, y\}] = \left[\prod_{\ell\leq k}\int\{\mathrm{d}x_\ell \mathrm{d}y_\ell\}\mathcal{W}[\{x_\ell, y_\ell\}]\right]\delta_{\mathrm{F}}\left[\begin{matrix}x - F[x_1, \ldots, x_k] \\ y - G[x_1, y_1, \ldots, x_k, y_k]\end{matrix}\right] \tag{33}$$

Here $F[x_1, \ldots, x_k]$ and $G[x_1, y_1, \ldots, x_k, y_k]$ are the functions defined in (29,30). Now

$$\mathcal{W}[\{x, y\}] = \frac{1}{\mathcal{C}^2}\sum_{k>0} p(k)\frac{k}{\langle k\rangle}\frac{\mathcal{A}[\{x, y\}]\mathcal{F}_{k-1}[\{x, y\}]}{\int\{\mathrm{d}x'\mathrm{d}y'\}\mathcal{A}[\{x', y'\}]\mathcal{F}_k[\{x', y'\}]} \tag{34}$$

$$\mathcal{C}^{-2} = \int\{\mathrm{d}x\mathrm{d}y\mathrm{d}x'\mathrm{d}y'\}\mathcal{W}[\{x, y\}]\mathcal{W}[\{x', y'\}]\mathcal{B}[\{x, y; x', y'\}] \tag{35}$$

Finally we turn to our spectrum equation (23), which for the density (26) becomes

$$\begin{aligned}\varrho_{\mathrm{RS}}(\mu) &= \sum_k p(k)\int\{\mathrm{d}x\mathrm{d}y\mathrm{d}x'\mathrm{d}y'\}\mathcal{W}[\{x', y'\}]\mathcal{B}[\{x, y; x', y'\}] \\ &\quad \times \frac{\mathcal{A}[\{x, y\}]\mathcal{F}_{k-1}[\{x, y\}]}{\int\{\mathrm{d}x'\mathrm{d}y'\}\mathcal{A}[\{x', y'\}]\mathcal{F}_k[\{x', y'\}]}\left(\frac{\delta\log\mathcal{A}[\{x, y\}]}{\delta\hat{\varrho}(\mu)} + \frac{\delta\log\mathcal{B}[\{x, y; x', y'\}]}{\delta\hat{\varrho}(\mu)}\right) \\ &\quad -\frac{1}{2}\langle k\rangle\frac{\int\{\mathrm{d}x\mathrm{d}y\mathrm{d}x'\mathrm{d}y'\}\mathcal{W}[\{x, y\}]\mathcal{W}[\{x', y'\}]\ \delta\mathcal{B}[\{x, y; x', y'\}]/\delta\hat{\varrho}(\mu)}{\int\{\mathrm{d}x\mathrm{d}y\mathrm{d}x'\mathrm{d}y'\}\mathcal{W}[\{x, y\}]\mathcal{W}[\{x', y'\}]\mathcal{B}[\{x, y; x', y'\}]}\end{aligned} \tag{36}$$

with

$$\frac{\delta\log\mathcal{A}[\{x, y\}]}{\delta\hat{\varrho}(\mu)} = \frac{1}{\pi}\frac{\mathrm{d}}{\mathrm{d}\mu}\left[\mathrm{atan}\left(\frac{\mathrm{Re}(x)}{\mathrm{Im}(x)}\right) + \mathrm{Re}\left(\frac{y^2(\mu)}{x(\mu)}\right)\right] \tag{37}$$

$$\begin{aligned}\frac{\delta\log\mathcal{B}[\{x, y; x', y'\}]}{\delta\hat{\varrho}(\mu)} &= \frac{1}{\pi}\frac{\mathrm{d}}{\mathrm{d}\mu}\left[\mathrm{atan}\left(\frac{\mathrm{Re}(x(\mu) - 1/x'(\mu))}{\mathrm{Im}(x(\mu) - 1/x'(\mu))}\right) - \mathrm{atan}\left(\frac{\mathrm{Re}(x(\mu))}{\mathrm{Im}(x(\mu))}\right)\right. \\ &\quad \left. +\mathrm{Re}\left(\frac{y^2(\mu)/x(\mu) + y'^2(\mu)/x'(\mu) - 2y(\mu)y'(\mu)}{x(\mu)x'(\mu) - 1}\right)\right]\end{aligned} \tag{38}$$

5.3. Reflection symmetry in the origin

Our equations are invariant under $\mathcal{W}[\{x,y\}] \to \mathcal{W}[\{x,-y\}]$, which represents reflection in the origin of the complex plane of the centres of the $\pi(\phi|x,y)$. Assuming the true saddle-point to be of the invariant form $\mathcal{W}[\{x,y\}] = \mathcal{W}[\{x\}]\delta[\{y\}]$ implies that we consider the functions $\pi(\phi|x,y)$ to be centred at the origin, and we will find that this is a property that holds in the limit of treelike ensembles. It simplifies our equations to

$$\mathcal{W}[\{x\}] = \frac{1}{\mathcal{C}^2}\sum_{k>0} p(k)\frac{k}{\langle k\rangle}\frac{\mathcal{A}[\{x\}]\mathcal{F}_{k-1}[\{x\}]}{\int\{\mathrm{d}x'\}\mathcal{A}[\{x'\}]\mathcal{F}_k[\{x'\}]} \tag{39}$$

$$\mathcal{C}^{-2} = \int\{\mathrm{d}x\mathrm{d}x'\}\ \mathcal{W}[\{x\}]\mathcal{W}[\{x'\}]\mathcal{B}[\{x,x'\}] \tag{40}$$

with $\mathcal{F}_k[\{x,y\}] = \mathcal{F}_k[\{x\}]\delta[\{y\}]$, and

$$\mathcal{A}[\{x\}] = \mathrm{e}^{\frac{1}{\pi}\int\mathrm{d}\mu\ \hat{\varrho}(\mu)\frac{\mathrm{d}}{\mathrm{d}\mu}\mathrm{atan}\left(\frac{\mathrm{Re}(x(\mu))}{\mathrm{Im}(x(\mu))}\right)} \tag{41}$$

$$\mathcal{B}[\{x;x'\}] = \mathrm{e}^{\frac{1}{\pi}\int\mathrm{d}\mu\ \hat{\varrho}(\mu)\frac{\mathrm{d}}{\mathrm{d}\mu}\left[\mathrm{atan}\left(\frac{\mathrm{Re}(x(\mu)-1/x'(\mu))}{\mathrm{Im}(x(\mu)-1/x'(\mu))}\right)-\mathrm{atan}\left(\frac{\mathrm{Re}(x(\mu))}{\mathrm{Im}(x(\mu))}\right)\right]} \tag{42}$$

The spectrum becomes

$$\begin{aligned}\varrho_{\mathrm{RS}}(\mu) = &\sum_k p(k)\int\{\mathrm{d}x\mathrm{d}x'\}\mathcal{W}[\{x'\}]\mathcal{B}[\{x;x'\}] \\ &\times\frac{\mathcal{A}[\{x\}]\mathcal{F}_{k-1}[\{x\}]}{\int\{\mathrm{d}x'\}\mathcal{A}[\{x'\}]\mathcal{F}_k[\{x'\}]}\Big(\frac{\delta\log\mathcal{A}[\{x\}]}{\delta\hat{\varrho}(\mu)}+\frac{\delta\log\mathcal{B}\big[\{x;x'\}\big]}{\delta\hat{\varrho}(\mu)}\Big) \\ &-\frac{1}{2}\langle k\rangle\frac{\int\{\mathrm{d}x\mathrm{d}x'\}\mathcal{W}[\{x\}]\mathcal{W}[\{x'\}]\ \delta\mathcal{B}[\{x;x'\}]/\delta\hat{\varrho}(\mu)}{\int\{\mathrm{d}x\mathrm{d}x'\}\mathcal{W}[\{x\}]\mathcal{W}[\{x'\}]\mathcal{B}[\{x;x'\}]}\end{aligned} \tag{43}$$

Continuous bifurcations of states with $\mathcal{W}[\{x,y\}] \neq \mathcal{W}[\{x\}]\delta[\{y\}]$ can be located via a Guzai (i.e. functional moment) expansion [6]. One can show that such bifurcations do occur, but it is not yet clear whether they correspond to physically relevant transitions.

5.4. Regular graphs

For regular graphs our formulae simplify considerably. Here we find the order parameter equation (34) for $\mathcal{W}[\{x,y\}]$ and the spectrum reducing to

$$\mathcal{W}[\{x,y\}] = \frac{\mathcal{A}[\{x,y\}]\mathcal{F}_{k-1}[\{x,y\}]}{\int\{\mathrm{d}x'\mathrm{d}y'\}\mathcal{A}[\{x',y'\}]\mathcal{F}_{k-1}[\{x',y'\}]} \tag{44}$$

$$\varrho_{\mathrm{RS}}(\mu) = \frac{\int\{\mathrm{d}x\mathrm{d}y\mathrm{d}x'\mathrm{d}y'\}\mathcal{W}[\{x,y\}]\mathcal{W}[\{x',y'\}]\mathcal{B}[\{x,y;x',y'\}]\ \varrho(\mu|\{x,y;x',y'\})}{\int\{\mathrm{d}x\mathrm{d}y\mathrm{d}x'\mathrm{d}y'\}\mathcal{W}[\{x,y\}]\mathcal{W}[\{x',y'\}]\mathcal{B}[\{x,y;x',y'\}]} \tag{45}$$

with

$$\begin{aligned}\varrho(\mu|\{x,y;x',y'\}) = &\frac{1}{2\pi}\frac{\mathrm{d}}{\mathrm{d}\mu}\Big\{(1-\frac{1}{2}k)\Big[\mathrm{atan}(\frac{\mathrm{Re}(x(\mu)-1/x'(\mu))}{\mathrm{Im}(x(\mu)-1/x'(\mu))})+\mathrm{atan}(\frac{\mathrm{Re}(x'(\mu)-1/x(\mu))}{\mathrm{Im}(x'(\mu)-1/x(\mu))})\Big] \\ &+\frac{1}{2}k\Big[\mathrm{atan}(\frac{\mathrm{Re}(x(\mu))}{\mathrm{Im}(x(\mu))})+\mathrm{atan}(\frac{\mathrm{Re}(x'(\mu))}{\mathrm{Im}(x'(\mu))})\Big] \\ &+\mathrm{Re}\Big((2-k)\frac{y^2(\mu)/x(\mu)+y'^2(\mu)/x'(\mu)-2y(\mu)y'(\mu)}{x(\mu)x'(\mu)-1}+\frac{y^2(\mu)}{x(\mu)}+\frac{y'^2(\mu)}{x'(\mu)}\Big)\Big\}\end{aligned} \tag{46}$$

6. Interpretation and simple tests of the theory

6.1. Interpretation and link with message passing algorithms

In the limit $\hat{\varrho} \to 0$, where $\mathcal{C} = \mathcal{A}[\{x,y\}] = 1$ for all $\{x,y\}$, formula (34) acquires the standard form $\mathcal{W}[\{x,y\}] = \sum_{k>0} p(k)(k/\langle k\rangle)\mathcal{F}_{k-1}[\{x,y\}]$ of message passing algorithms on tree-like graphs. Also the more complex structure of (34) can be interpreted as describing the stationary state of a message passing process, but now extended with the nontrivial message acceptance probabilities

$$\mathcal{P}[\{x,y\}|k] \;\propto\; \frac{\mathcal{A}[\{x,y\}]}{\int\{\mathrm{d}x'\mathrm{d}y'\}\mathcal{A}[\{x',y'\}]\mathcal{F}_k[\{x',y'\}]} \tag{47}$$

(modulo a multiplicative constant). This is similar to the re-weighting of solutions in e.g. [38]. In tree-like graphs we accept *each* proposed message $\{x,y\}$. This interpretation shows us how to solve (34,35) numerically, and how we should adapt belief propagation and other message passing algorithms [2, 3, 4], to make these exact for stochastic processes on loopy graphs.

6.2. Locally tree-like graphs

Our theory should recover established results on locally tree-like graphs upon setting $\hat{\varrho} = 0$. This reduces our ensemble to a maximum entropy one in which only the degrees are prescribed. Now $\mathcal{A}[\{x,y\}] = \mathcal{B}[\{x,y;x',y'\}] = \mathcal{C} = 1$, and we indeed recover from (17) the correct ensemble entropy density of tree-like ensembles with prescribed degrees [39], with $\lim_{N\to\infty}\epsilon_N = 0$:

$$S \;=\; \frac{1}{2}\langle k\rangle\big[\log\big(\frac{N}{\langle k\rangle}\big)+1\big] + \sum_k p(k)\log\tilde{p}(k) + \epsilon_N \tag{48}$$

The spectra of locally tree-like graphs are also recovered correctly, with solutions of the simple form $\mathcal{W}[\{x,y\}] = \mathcal{W}[\{x\}]\delta[\{y\}]$ and real-valued $\{x\}$. Our present formalism predicts that

$$\varrho(\mu) \;=\; -\frac{\mathrm{d}}{\mathrm{d}\mu}\Big\{\frac{1}{2}\sum_k p(k)\int\{\mathrm{d}x\}\mathcal{F}_k[\{x\}]\mathrm{sgn}[x(\mu)]\Big] \tag{49}$$
$$+\frac{1}{2}\langle k\rangle\int\{\mathrm{d}x\mathrm{d}x'\}\mathcal{W}[\{x\}]\mathcal{W}[\{x'\}]\;\theta[x(\mu)x'(\mu)]\theta[1-x(\mu)x'(\mu)]\mathrm{sgn}[x(\mu)+x'(\mu)]\Big\}$$

in which

$$\mathcal{W}[\{x\}] = \sum_{k>0}p(k)\frac{k}{\langle k\rangle}\mathcal{F}_{k-1}[\{x\}], \qquad \mathcal{F}_k[\{x\}] = \Big[\prod_{\ell\leq k}\int\{\mathrm{d}x_\ell\}\mathcal{W}[\{x_\ell\}]\Big]\delta_{\mathrm{F}}\big[x-F[x_1,\ldots,x_k]\big] \tag{50}$$

For k-regular tree-like graphs our order parameter equation (50) and the spectrum (49) become

$$\mathcal{W}[\{x\}] \;=\; \Big[\prod_{\ell<k}\int\{\mathrm{d}x_\ell\}\mathcal{W}[\{x_\ell\}]\Big]\delta_{\mathrm{F}}\big[x-F[x_1,\ldots,x_{k-1}]\big] \tag{51}$$

$$\varrho(\mu) \;=\; -\frac{1}{2}\frac{\mathrm{d}}{\mathrm{d}\mu}\int\{\mathrm{d}x\mathrm{d}x'\}\mathcal{W}[\{x\}]\mathcal{W}[\{x'\}]$$
$$\times\theta[x(\mu)x'(\mu)]\mathrm{sgn}[x(\mu)+x'(\mu)]\Big[1+(k-2)\theta[1-x(\mu)x'(\mu)]\Big] \tag{52}$$

For $k=1$ equation (51) immediately leads to $\varrho(\mu) = \frac{1}{2}\delta(\mu-1) + \frac{1}{2}\delta(\mu+1)$, which is indeed the spectrum of a graph consisting of $N/2$ disconnected 2-node components (the only possible regular graph with $k=1$). Equation (51) can be solved analytically also for $k>1$. To find the spectrum we only need the distribution $\mathcal{W}(x|\mu)$ of each individual $x(\mu)$, for which we find:

$$|\mu| < 2\sqrt{k-1}: \qquad \mathcal{W}(x|\mu) = \frac{1}{\pi}\frac{\sqrt{k-1-\frac{1}{4}\mu^2}}{(x+\frac{1}{2}\mu)^2+k-1-\frac{1}{4}\mu^2} \tag{53}$$

$$|\mu| > 2\sqrt{k-1}: \qquad \mathcal{W}(x|\mu) = \delta\Big[x+\frac{1}{2}\mu+\frac{1}{2}\mu\sqrt{1-4(k-1)/\mu^2}\Big] \tag{54}$$

Inserting this result into (52) leads us to an integral that can again be evaluated analytically, and recovers the correct spectrum of [40] for k-regular treelike graphs:

$$\varrho(\mu) = \theta\big[2\sqrt{k-1}-|\mu|\big]\,\frac{k\sqrt{4(k-1)-\mu^2}}{2\pi(k^2-\mu^2)} \tag{55}$$

For graphs with arbitrary degree distributions $p(k)$ we can show that from (50) indeed follows the equation obtained in [41]. Thus our formalism passes the available tests in the tree-like limit.

7. Current and next stage of the research programme

7.1. Equations for order parameters and spectra of loopy ensembles

We now also have explicit equations for the analysis of graphs with extensively many short loops, where $\hat{\varrho}(\mu)\neq 0$. The order parameter equation and the spectrum are given by (44,45). In the tree-like limit we had $\mathcal{W}[\{x,y\}]=\mathcal{W}[\{x\}]\delta[\{y\}]$ with $\mathrm{Im}[x(\mu)]\uparrow 0$ for all μ. If we make the ansatz that also in the presence of loops we still have $\mathcal{W}[\{x,y\}]=\mathcal{W}[\{x\}]\delta[\{y\}]$ with $\mathrm{Im}[x(\mu)]\uparrow 0$ for all μ, and limit ourselves to regular graphs, we can simplify our order parameter equation to

$$\mathcal{W}[\{x\}] = \frac{\mathcal{A}[\{x\}]\Big[\prod_{\ell<k}\int\{\mathrm{d}x_\ell\}\mathcal{W}[\{x_\ell\}]\Big]\delta_{\mathrm{F}}\left[x-F[x_1,\ldots,x_{k-1}]\right]}{\int\{\mathrm{d}x'\}\mathcal{A}[\{x'\}]\Big[\prod_{\ell<k}\int\{\mathrm{d}x_\ell\}\mathcal{W}[\{x_\ell\}]\Big]\delta_{\mathrm{F}}\left[x'-F[x_1,\ldots,x_{k-1}]\right]} \tag{56}$$

with $\mathcal{A}[\{x\}]=\exp\left[-\frac{1}{2}\int\mathrm{d}\mu\;\hat{\varrho}(\mu)\frac{\mathrm{d}}{\mathrm{d}\mu}\mathrm{sgn}[x(\mu)]\right]$. The function $F[x_1,\ldots,x_{k-1}]$ is for $\varepsilon\downarrow 0$ defined as $F(\mu|x_1,\ldots,x_{k-1})=-\mu-\sum_{\ell<k}1/x_\ell(\mu)$. The spectrum formula simplifies to

$$\varrho_{\mathrm{RS}}(\mu) = \frac{\int\{\mathrm{d}x\mathrm{d}x'\}\mathcal{W}[\{x\}]\mathcal{W}[\{x'\}]\mathcal{B}[\{x;x'\}]\;\varrho(\mu|\{x;x'\})}{\int\{\mathrm{d}x\mathrm{d}x'\}\mathcal{W}[\{x\}]\mathcal{W}[\{x'\}]\mathcal{B}[\{x;x'\}]} \tag{57}$$

$$\varrho(\mu|\{x;x'\}) = -\frac{1}{4}\frac{\mathrm{d}}{\mathrm{d}\mu}\Big\{\Big[\mathrm{sgn}[x(\mu)]+\mathrm{sgn}[x'(\mu)]\Big]\Big[1+(k-2)\theta[1-x(\mu)x'(\mu)]\Big]\Big\} \tag{58}$$

in which

$$\mathcal{B}[\{x;x'\}] = \mathrm{e}^{\frac{1}{2}\int\mathrm{d}\mu\;\hat{\varrho}(\mu)\frac{\mathrm{d}}{\mathrm{d}\mu}\left(\left[\mathrm{sgn}[x(\mu)]+\mathrm{sgn}[x'(\mu)]\right]\theta[1-x(\mu)x'(\mu)]\right)} \tag{59}$$

It is not difficult to prove that (58) is normalised correctly. However, testing the validity of our equations in full is nontrivial for a number of reasons.

First, there appear to exist no benchmark solutions yet, in the form of exact and independent asymptotic spectrum derivations for ensembles of the form (1) with $\hat{\varrho}\neq 0$, against which to test our predictions. Second, solving the order parameter equations numerically for $\hat{\varrho}\neq 0$ is nontrivial, as it involves population dynamics algorithms in which we propagate functions rather than fields, and with nontrivial acceptance probabilities, that are quite hard to equilibrate accurately without sacrificing eigenvalue resolution. Thirdly, generating random graphs from (1) numerically, in order to then obtain their spectra by numerical diagonalisation, is itself nontrivial [42]; for relatively simple choices like $\hat{\varrho}(\mu)=\alpha_3\mu^3+\alpha_4\mu^4$ reliable algorithms and code do exist, but these ensembles are notorious for their complex phase transitions [43, 44, 45, 46]. For instance, for regular graphs with $k=3$ and $\hat{\varrho}(\mu)=\alpha_3\mu^3$ one observes in numerical simulations that, modulo finite size effects, the spectrum is of the form

$$\varrho(\mu) \approx (1-\gamma)\tilde{\varrho}(\mu)+\gamma\Big\{\frac{3}{4}\delta(\mu+1)+\frac{1}{4}\delta(\mu-3)\Big\} \tag{60}$$

in which $\gamma\in[0,1]$ and $\tilde{\varrho}(\mu)$ is McKay's $k=3$ formula (55) for tree-like regular random graphs. One confirms that $\langle\mu\rangle=0$ and $\langle\mu^2\rangle=3$ for any γ, as required. The second term in (60) is the spectrum of a fully connected simple graphlet of size four. Hence (60) describes graphical phase separation; the system increases the number of triangles by disconnecting a fraction γ of the nodes from the treelike bulk, to form $\gamma N/4$ disconnected 4-node cliques. Similarly, For regular graphs with $k=3$ and $\hat{\varrho}(\mu)=\alpha_4\mu^4$ one observes in numerical simulations spectra of the form

$$\varrho(\mu) \;\approx\; (1-\gamma)\tilde{\varrho}(\mu)+\gamma\Big\{\frac{2}{3}\delta(\mu)+\frac{1}{6}\delta(\mu-3)+\frac{1}{6}\delta(\mu+3)\Big\} \tag{61}$$

in which again $\tilde{\varrho}(\mu)$ is McKay's formula (55) for $k=3$. Also here $\langle\mu\rangle=0$ and $\langle\mu^2\rangle=3$ for any γ. The second term of (61) is the spectrum of the six node graphlet with adjacency matrix $a_{ij}=\delta_{i,j+1}+\delta_{i,j-1}+\delta_{i,j+3}+\delta_{i,j-3}$ $(i,j \bmod 6)$, which is a 6-node clique from which all triangles are removed. Here the system increases the number of squares by disconnecting a fraction γ of the nodes from the treelike bulk, to form $\gamma N/6$ suitable disconnected 6-node graphlets.

With the available numerical evidence at this moment, consisting of numerical solutions of order parameter equations and spectra for ensembles with $\hat{\varrho}(\mu)=\alpha_3\mu^3+\alpha_4\mu^4$ (obtained via population dynamics with nontrivial acceptance probabilities), together with spectra measured in numerical graph simulations, satisfactory agreement has not yet been obtained. Preliminary population dynamics computations suggest that the above phase separation phenomena are not yet captured. Yet the structure of the theory seems elegant and intuitive, with the richness to exhibit complex phase transition phenomenology, and it is correct in the limit $\hat{\varrho}\to 0$. The reason for the residual disagreement has not yet been identified. Apart from mundane explanations (e.g. imprecise or nonequilibrated numerical algorithms, either in population dynamics or graph generation, or saddle-points that are not of the assumed form), a possible candidate lies in the usage of complex logarithms. Some of the identities used in the present derivations, such as $\log\exp(Z)=Z$ or $\log(ZW)=\log(Z)+\log(W)$, hold for complex values only if we can steer away from the cut in the complex plane. A more careful re-derivation may well reveal extra terms that for some reason become irrelevant for $\hat{\varrho}\to 0$ (where we know the theory works) but may be important for $\hat{\rho}\neq 0$. To this author, the structure of the equations simply feels right as a description of loopy graph ensembles, and he is optimistic that the remaining discrepancy will soon be removed.

7.2. Processes on loopy graphs

Looking ahead towards the use of the presently proposed method for the modelling of spin systems on loopy graphs, one can see that the calculation of the disorder-averaged free energy density of such systems will involve further (non-complex) replicas, and requires the evaluation of the following more complicated generating function:

$$\Phi_K[\hat{\varrho},\{\boldsymbol{\sigma}\}] \;=\; \frac{1}{N}\log\sum_{\mathbf{c}} \mathrm{e}^{N\int\mathrm{d}\mu\;\hat{\varrho}(\mu)\varrho(\mu|\mathbf{c})+K\sum_{i<j}c_{ij}\boldsymbol{\sigma}_i\cdot\boldsymbol{\sigma}_j}\prod_{i\leq N}\delta_{k_i,\sum_j c_{ij}} \tag{62}$$

with $K=\beta J$ and the n-replicated spin vectors $\boldsymbol{\sigma}_i=(\sigma_i^1,\ldots,\sigma_i^n)$. Again also this quantity can be calculated, the limit $N\to\infty$ can be taken, and one finds a more complex closed set of equations in which now the graph order parameters and the spin order parameters are entangled. If we assume spin replica symmetry, in addition to graph replica symmetry, the final RS order parameter will take the form $\mathcal{W}[\{x,y\},v]$, in which v denotes an effective field. It is encouraging to confirm that upon assuming the physical saddle-point to be of the form $\mathcal{W}[\{x,y\},v]=\mathcal{W}[\{x,y\}]\mathcal{W}[v]$ (i.e. the statistical features of the process decouple from those of the graph) we indeed recover the previous equations of this paper for $\mathcal{W}[\{x,y\}]$, and the solution of [47] for the transition temperature on tree-like graphs (where indeed one would expect such decoupling to apply).

Acknowledgements
The author gratefully acknowledges valuable discussions with Alessia Annibale, Reimer Kühn, Federico Ricci-Tersenghi, and Pierpaolo Vivo.

References

[1] Mézard M and Parisi G 2001 *Eur. Phys. J.* **20** 217
[2] Pearl J 1988 *Probabilistic reasoning in intelligent systems* (San Mateo: Morgan Kaufman)
[3] Ihler A T, Fisher J W and Willsky A S 2005 *J. Mach. Learning Res.* **6** 905
[4] Mézard M and Montanari A 2009 *Information, physics, and computation* (Oxford: University Press)
[5] Pérez Vicente C J and Coolen A C C 2008 *J. Phys. A: Math. Theor.* 255003
[6] Coolen A C C, Skantzos N S, Pérez Castillo I, Perez Vicente C J, Hatchett J P L, Wemmenhove B and Nikoletopoulos T 2005 *J. Phys. A: Math. Gen.* **38** 8289
[7] Montanari A and Rizzo T 2005 *J. Stat. Mech.* P10011
[8] Parisi G and Slanina F 2006 *J. Stat. Mech.* L02003
[9] Chertkov M and Chernyak V Y 2006 *Phys. Rev.* E **73** 065102(R)
[10] Mooij J M and Kappen H J 2007 *J. Mach. Learning Res.* **8** 1113
[11] Berlin T H and Kac M 1952 *Phys. Rev.* **86** 821–835
[12] Onsager L 1944 *Phys. Rev.* **65** 117–149
[13] Baxter R J 1990 *Exactly solved models in statistical mechanics* (London: Academic Press)
[14] Agliari E, Annibale A, Barra A, Coolen A C C and Tantari D 2013 *J. Phys. A: Math. Theor.* **46** 415003
[15] Mozeika A and Coolen A C C 2015 *J. Phys. A: Math. Theor.* **48** 255001
[16] Metz F L, Neri I and Bollé D 2011 *Phys. Rev.* E **84** 055101(R)
[17] van Dam E R and Haemers W H 2003 *Linear Algebra Appl.* **373** 241–272
[18] van Dam E R and Haemers W H 2009 *Discrete Mathematics* **309** 576–586
[19] Brouwer A E and Haemers W H 2012 *Spectra of Graphs* (Berlin: Springer)
[20] Parisi G 1988 *Statistical field theory* (Redwood City: Addison-Wesley)
[21] Erdös P and Rényi A 1959 *Publ. Math.* **6** 290–297
[22] Edwards S F and Jones R C 1976 *J. Phys A: Math. Gen.* **9** 1595
[23] Roberts E S and Coolen A C C 2014 *ESAIM: Proceedings and Surveys* **47** 97–115
[24] Mézard M, Parisi G and Virasoro M A 1987 *Spin glass theory and beyond* (Singapore: World Scientific)
[25] Penney R W, Coolen A C C and Sherrington D 1993 *J. Phys. A: Math. Gen.* **26** 3681–3695
[26] Coolen A C C, Penney R W and Sherrington D 1993 *Phys. Rev.* B **48** 16116–16118
[27] Dotsenko V, Franz S and Mézard M 1994 *J. Phys. A: Math. Gen.* **27** 2351–2365
[28] van Mourik J and Coolen A C C 2001 *J. Phys. A: Math. Gen.* **34** L111–L117
[29] Uezu T and Coolen A C C 2002 *J. Phys. A: Math. Gen.* **35** 2761–2809
[30] Rabello S, Coolen A C C, Pérez-Vicente C J and Fraternali F 2008 *J. Phys. A: Math. Theor.* **41** 285004
[31] Ogure K and Kabashima Y 2004 *Prog. Theor. Phys.* **111** 661–688
[32] Obuchi T, Kabashima Y and Nishimori H 2009 *J. Phys. A: Math. Theor.* **42** 075004
[33] Obuchi T, Kabashima Y, Nishimori H and Ohzeki M 2011 *Physica E* **43** 786–789
[34] Derrida B and Mottishaw P 2014 arXiv:1410.1432
[35] Hewitt E and Savage L J 1955 *Trans. Am. Math. Soc.* **80** 470–501
[36] Kühn R 2008 *J. Phys A: Math. Theor.* **41** 295002
[37] Gradshteyn I S and Rhyzik I M 2007 *Table of integrals, series and products* (Burlington: Elsevier)
[38] Mézard M, Parisi G and Zecchina R 2002 *Science* **297** 812
[39] Annibale A, Coolen A C C, Fernandes L P, Fraternali F and Kleinjung J 2009 *J. Phys. A: Math. Theor.* **42** 485001
[40] McKay B D 1981 *Lin Algebra Appl.* **40** 203–216
[41] Dorogovtsev S N, Goltsev A V, Mendes J F F and Samukhin A N 2003 *Phys. Rev.* E **68** 046109
[42] Coolen A C C, De Martino A and Annibale A 2009 *J. Stat. Phys.* **136** 1035
[43] Strauss D A 1986 *SIAM Review* **28** 513–527
[44] Jonasson J 1999 *J. Appl. Prob.* **36** 852–867
[45] Burda Z, Jurkiewicz J and Krzywicki A 2004 *Phys. Rev.* E **69** 026106
[46] Burda Z, Jurkiewicz J and Krzywicki A 2004 *Phys. Rev.* E **70** 026106
[47] Bethe H A 1935 *Proc. Roy. Soc. London Ser* A **150** 552–575